Spectrum of Sex

Minoru Tanaka • Makoto Tachibana
Editors

Spectrum of Sex

The Molecular Bases that Induce Various Sexual Phenotypes

Springer

Editors
Minoru Tanaka
Graduate School of Science
Nagoya University
Nagoya, Aichi, Japan

Makoto Tachibana
Graduation School of Frontier Biosciences
Osaka University
Suita, Osaka, Japan

ISBN 978-981-19-5358-3 ISBN 978-981-19-5359-0 (eBook)
https://doi.org/10.1007/978-981-19-5359-0

This Springer imprint is published by the registered company Springer Nature Singapore Pte Ltd.
The registered company address is: 152 Beach Road, #21-01/04 Gateway East, Singapore 189721, Singapore

Preface

Many organisms employ a mode of sexual reproduction for the continuum of species by developing two types of sexes: female and male. Furthermore, sex development is a driving force in species evolution. Therefore, the determination and development of the two sexes have been a topic of great concern and a long-standing issue to be solved by biologists. Some populations of organisms, including mammals, use genetic systems to determine sex, called genetic sex determination (GSD). A long timescale, covering the time it takes for speciation, often creates different sex-determining genes during evolution.

However, if you look around, one can readily notice that many organisms do not adhere to the genetic system. Environmental factors, known as environmental sex determination (ESD), are critical for determining sex. One well-known factor is temperature. The optimal male or female inducing temperature, that embryos experience, also differs among species. The broad range of optimal temperatures, together with a variety of sex-determining genes, implies that many factors could cause sex determination. We are still understanding how different sex determinants (different sex determination genes and different environmental factors) have been acquired during evolution.

GSD and ESD are not the ends of the history of sex development. Recent studies have shown that sex must be maintained even after determination and claim that sex maintenance is key to understanding the mechanism of sex reversal. The mechanism underlying sex reversal provides insight into the understanding of sequential or simultaneous hermaphrodites and the involvement of sex hormones in the maintenance of sex. We now understand that the production of sex hormones modulates sex development as well as sex-specific perception and behavior.

We are beginning to draw a larger picture of how sex can be understood in view of both individuals and evolution. This picture is beyond the classical view that the sex determinant and sex hormone conserved among organisms initiate, establish, and fix females or males. Interwoven mechanisms allow for two types of sex exchange with graded phenotypes. This manifests as a sex spectrum.

This book overviews three elements that develop the sex spectrum: genetics, the endocrine system, and the environment. In Part I the genetic regulation during sex determination, which often results in a mixture of two sexes or sex reversal, is discussed. In addition, the evolutionary aspects of the genetic determinants were discussed. In Part II, the involvement of endocrine regulation in the sex spectrum, which covers a broad range of phenotypic events, including sexual behavior and metabolism, is presented. Interestingly, sex hormones can also act as sex determinants. At last, Part III shows that intrinsic factors such as sex-determining genes and sex hormones are not the only factors in sex development. The environment surrounding organisms, such as symbiosis and metabolism, acts on the sex as critical factors, generating the sex spectrum.

It would be our pleasure if the audience appreciates the examples of the sex spectrum seen in many different organisms and obtains recent images of sex that researchers have revealed.

Nagoya, Japan Minoru Tanaka
Osaka, Japan Makoto Tachibana

Contents

Part I
Genetic Regulation of the Sex Spectrum

Chapter 1
Spectrum of Sex in a Horn of the Japanese Rhinoceros Beetle

Shinichi Morita, Kazuki Sakura, and Teruyuki Niimi

Abstracts Sex has been traditionally considered to be classified into two categories, male and female. However, numerous reports have shown examples of insects unfamiliar to this traditional binary sex view. Recently, the view of the sex spectrum has been proposed as a revised version. In this view, sex is recognised as a continuum from male to female (or female to male), and maleness or femaleness of any sexual traits is quantitatively interpreted as 'a position on the continuum of sex'. This chapter discusses the molecular genetic mechanism defining a position on the continuum of sex based on the knowledge about the Japanese rhinoceros beetle *Trypoxylus dichotomus*.

Keywords Horned beetle · Trypoxylus dichotomus · RNAi · Sex differentiation

---

S. Morita
Division of Evolutionary Developmental Biology, National Institute for Basic Biology, Okazaki, Japan

Division of Biological Science, Graduate School of Science, Nagoya University, Nagoya, Japan

Department of Basic Biology, School of Life Science, The Graduate University for Advanced Studies, SOKENDAI, Okazaki, Japan
e-mail: shinichi@nibb.ac.jp

K. Sakura
Division of Evolutionary Developmental Biology, National Institute for Basic Biology, Okazaki, Japan
e-mail: kazuki@nibb.ac.jp

T. Niimi (✉)
Division of Evolutionary Developmental Biology, National Institute for Basic Biology, Okazaki, Japan

Department of Basic Biology, School of Life Science, The Graduate University for Advanced Studies, SOKENDAI, Okazaki, Japan
e-mail: niimi@nibb.ac.jp

M. Tanaka, M. Tachibana (eds.), *Spectrum of Sex*,
https://doi.org/10.1007/978-981-19-5359-0_1

1.1 Introduction

Most insect species show sexual dimorphism, and their appearances are distinctively different between males and females. For example, in some beetles, males have magnificently developed mandibles, while females do not develop well. Based on the facts like this, sex has been traditionally considered to be classified into two categories, male and female. However, to date, numerous individuals cannot be classified as male or female by their appearances. For example, in damselflies, some females show male-like coloration though they produce typical female traits except for the coloration (Gossum and Sherratt 2008). Other examples are the sexually mosaic phenotypes (gynandromorph) and sexually intermediate phenotypes (intersex) accidentally produced by developmental abnormalities (Narita et al. 2010). These examples are not included in the traditionally binary view of sex, which needs to be revised (Nong et al. 2020).

The sex spectrum is a view of recognising sex as a continuum from male to female (or from female to male) (Nong et al. 2020; Preface). In this view, the gynandromorph and the intersex are interpreted to be placed between male and female in the sex spectrum. Furthermore, male-like female coloration in damselflies may be placed between midpoint and female. Thus, almost all differences among sexual traits in insects can be explained as differences in the location on the sex spectrum.

How is the location of the sex spectrum defined in insects? In the fruit fly *Drosophila melanogaster*, a major sex-determining gene *doublesex* (*dsx*) mutation yields the gynandromorph and the intersex in both females and males (Hildreth 1965). It is well known that *dsx* is essential for sexual differentiation not only in *D. melanogaster* but also in other holometabolous insects. These facts indicate that genetic mechanisms for sex determination have a considerable effect on determining insects' location on the sex spectrum.

Here, we first overview the sex-determining mechanism in holometabolous insects. Then, based on an example of the Japanese rhinoceros beetle, we discuss a molecular mechanism to determine the location in the sex spectrum.

1.2 Sex-Determining Molecular Mechanism in Holometabolous Insects

This subsection first introduces a sex-determining molecular mechanism in the *D. melanogaster*, the most investigated model insect. Moreover, we also show the commonality of the sex-determining mechanism.

The primary signal for sex determination is the number of the X chromosome (Erickson and Quintero 2007). The sex of an individual with two X chromosomes, i.e., an XX individual, is finally determined as female; on the other hand, the sex of an individual with a single X chromosome (i.e., XY individual) is finally determined as male as a default state.

The initial signals from X chromosomes produce the sex-lethal (Sxl) protein only in XX early embryos (Cline 1978, 1986, 1988; Kramer et al. 1999; Sefton et al. 2000; Penalva and Sánchez 2003; Salz 2007). Sxl autoregulates and maintains the expression of another Sxl isoform (Cline 1984; Bell et al. 1988, 1991; Penalva and Sánchez 2003). The functional Sxl protein controls the RNA splicing of *tra*, and functional Tra protein is translated (Boggs et al. 1987; Bell et al. 1988; Penalva and Sánchez 2003). Then, Tra protein yields female-specific Dsx (DsxF) by regulating alternative splicing of *dsx* mRNA, which leads to female differentiation (Hoshijima et al. 1991).

On the other hand, in XY embryos, a non-functional Sxl protein is produced because of the lack of the initial signal from the X chromosome (Bell et al. 1988; Samuels et al. 1991; Keyes et al. 1992; Penalva and Sánchez 2003). Lack of functional Sxl leads to expression of non-functional Tra protein, and a male-specific Dsx (DsxM) is translated through lack of Tra-dependent splicing regulation, which leads to male differentiation (Bell et al. 1988).

Whether the sex-determining mechanism in *Drosophila* is conserved in other holometabolous insects has been recently studied by focusing on non-*Drosophila* insects. *Sxl* is a 'master switch' gene for sex determination in *D. melanogaster*. *Sxl* orthologues have been found in many holometabolous insects, including Diptera, Lepidoptera, Hymenoptera and Coleoptera (Traut et al. 2006). However, some dipteran species' research revealed that *Sxl* orthologues are not responsible for sex determination (Meise et al. 1998; Saccone et al. 1998; Sievert et al. 2000). Furthermore, the *Sxl* orthologue in the silkworm *Bombyx mori* does not contribute to sex determination but regulates spermatogenesis (Niimi et al. 2006; Sakai et al. 2019). Therefore, *Sxl* is not a broadly conserved sex determination gene in holometabolous insects.

tra is an intermediate factor in the sex determination cascade of *D. melanogaster*. *Tra* orthologues are identified in holometabolous insects such as some species in Diptera, Coleoptera and Hymenoptera (O'Neil and Belote 1992; Pane et al. 2002; Kulathinal et al. 2003; Lagos et al. 2007; Ruiz et al. 2007; Hasselmann et al. 2008; Concha and Scott 2009; Schmieder et al. 2012; Shukla and Palli 2012; Geuverink and Beukeboom 2014; Morita et al. 2019). Furthermore, in some of these species, *tra* orthologues regulate female determination (Pane et al. 2002; Hasselmann et al. 2008; Concha and Scott 2009; Hediger et al. 2010; Shukla and Palli 2012; Morita et al. 2019). On the other hand, *tra* orthologues seem to have been lost in Lepidoptera and in some species in Strepsiptera and Diptera (Salvemini et al. 2013; Geuverink and Beukeboom 2014). These findings indicated that although *tra* orthologues are not found in some species, the sex-determining function of *tra* is conserved in a much more comprehensive range of holometabolous insect species than that of *Sxl*.

dsx is a bottom factor in the sex determination cascade of *D. melanogaster*, which directly regulates the transcription of a battery of genes responsible for sex differentiation. *dsx* orthologues are conserved in all of the insects investigated so far (Price et al. 2015). Furthermore, the *dsx* has sex-specific transcripts and contributes to sex determination in the various holometabolous insects such as Diptera, Lepidoptera, Hymenoptera and Coleoptera (Schütt and Nöthiger 2000; Ohbayashi et al. 2001; Cho et al. 2007; Chen et al. 2008; Oliveira et al. 2009; Shukla and Palli 2012; Ito et al. 2013; Mine et al. 2017; Taracena et al. 2019). Therefore, *dsx* is the conserved regulatory factor in the sex determination cascade among holometabolous insects.

The above comparative analysis focusing on the conserved genes associated with the insect sex determination pathway suggests that downstream genes *tra* and *dsx* are the core regulatory genes conserved among holometabolous insects. Therefore, to understand the determination mechanism of position in the sex spectrum in holometabolous insects, we focus on the role of *tra* and *dsx* in sexual traits. In the next section, we discuss the role of *tra* and *dsx* in the sex spectrum using the Japanese rhinoceros beetle, in which we can manipulate the location in the sex spectrum by controlling the expression level of sex-determining genes using RNA interference (RNAi) methods.

1.3 Sex Spectrum in *T. dichotomus* Horn Visualised by Manipulating the Sex Determination Pathway

The horn of the Japanese rhinoceros beetle *Trypoxylus dichotomus* (Coleoptera, Scarabaeoidea, Scarabaeidae) exhibits sexual dimorphism (Fig. 1.1a, *EGFP*). A male has an exaggerated long horn on the head and a short horn on the first thoracic segment (pronotum). On the other hand, a female has no horn neither on the head nor on the pronotum but has three small projections in the rostral region of the head (clypeolabral region).

As described in the previous sections, sex-specific *tra* and *dsx* isoforms are essential genes to a sex-differentiating mechanism in holometabolous insects. In *T. dichotomus*, downregulation of *dsx* expression by RNAi results in short head horns in both males and females, while thoracic horns form in neither males nor females (Fig. 1.1a) (Ito et al. 2013). *tra* RNAi males showed no morphological changes, while females developed ectopic male-like horns on both the head and pronotum (Fig. 1.1a) (Morita et al. 2019). These facts indicated that in *T. dichotomus*, Tra and Dsx regulate sexual dimorphism in a horn (Fig. 1.1) (Morita et al. 2019).

Next, Morita et al. (2019) insufficiently suppressed *tra* expression levels in females. As a result, female-like and male-like traits coexist in a single *tra* RNAi female (Fig. 1.2) because *tra* regulates sex-specific splicing of *dsx* like other holometabolous insects (Fig. 1.1b) (Morita et al. 2019), and insufficient suppression of *tra* expression in *tra* RNAi females leads to incomplete switching from *dsxF* to *dsxM*.

Detailed observation of the *tra* RNAi phenotypes in Fig. 1.2 revealed that the degree of similarity to males is different between head and pronotum regions. This suggests that the location on the sex spectrum at the tissue level is different between head and pronotum regions. For example, in the head region, *tra* RNAi individuals show female-specific traits (three small projections) and male-specific traits (head

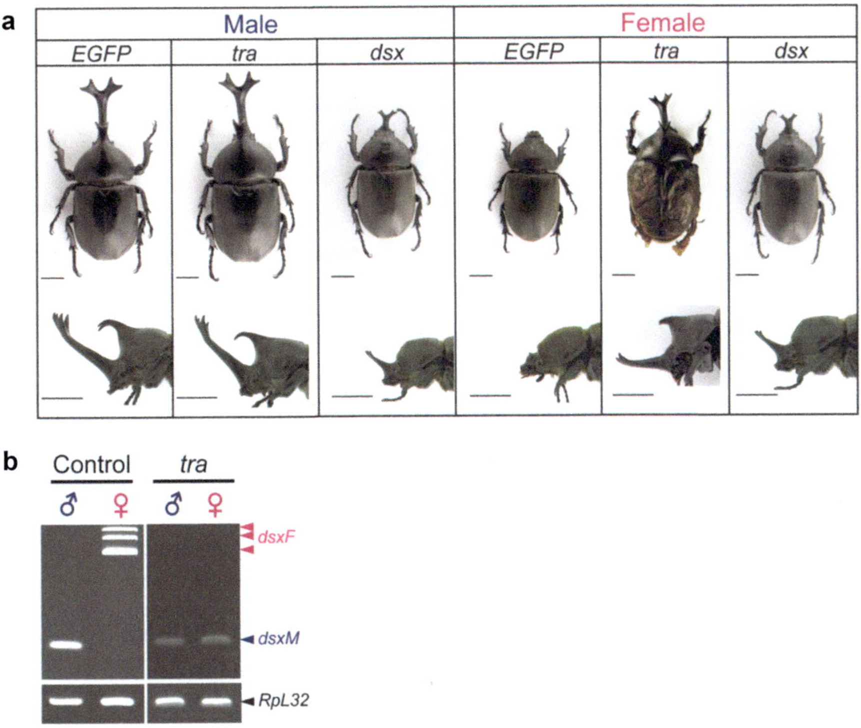

Fig. 1.1 RNAi-mediated loss of function of sex-determining genes and morphological change of horn primordia. (**a**) Representative individuals in *dsx* and *tra* RNAi treatment in males and females. *EGFP* RNAi treatment (negative control) showed no morphological defects. The upper row, the dorsal views of adults. The second row, the lateral views of adults. Scale bars are 1 cm. (**b**) Sex-specific splicing of *dsx* in RNAi treatments targeting *tra*. RpL32 was used as an internal control for RT-PCR. Blue arrowheads, male-specific splicing patterns (*dsxM*). Magenta arrowheads, female-specific splicing patterns (*dsxF*). Black arrowheads, RpL32. (Adapted from Morita et al. *PLOS Genet*., 15: e1008063, 2019)

horns) (Fig. 1.2b, c). However, in one individual (Fig. 1.2b), the ectopic head horn branched twice, forming a short stalk of head horns. On the other hand, the ectopic head horn branched only once in the other individual (Fig. 1.2c), and no stalk was formed. This suggests that the former phenotype was more similar to male-specific traits than the latter phenotype. Therefore, the locations on the sex spectrum of these *tra* RNAi phenotypes in the head region can be explained as shown in Fig. 1.3a.

Next, in the pronotum region, only the *tra* RNAi phenotype in Fig. 1.2b was that the ectopic thoracic horn was apparently like a male thoracic horn (Fig. 1.2d). The thoracic horn length is similar to that of a *tra* RNAi female injected with a sufficient amount of dsRNA (Fig. 1.2d). In contrast to the head horn in *tra* RNAi females, the incomplete formation of the thoracic horn has not been observed so far. This observation suggests that the thoracic horn phenotype is located closely at either endpoint on the sex spectrum (Fig. 1.3b).

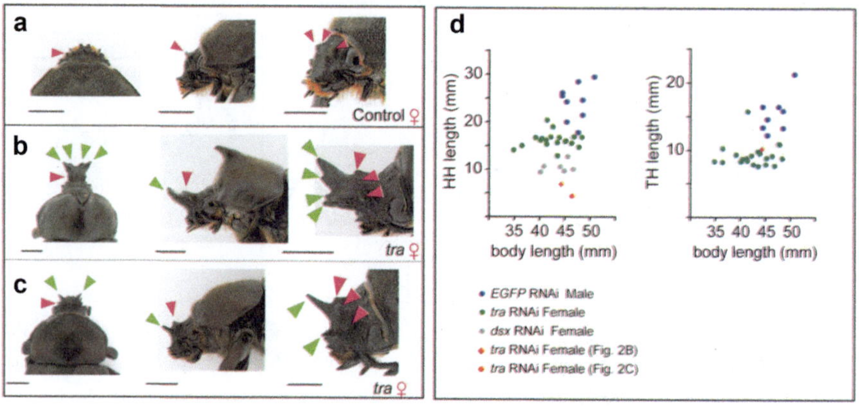

Fig. 1.2 *tra* RNAi phenotype induced by injection of small amounts of *tra* dsRNA. (**a–c**) Comparison of a wild-type female and ectopic intermediate sexual transformation of horns in females induced by *tra* RNAi treatments. (**a**) A wild-type female. (**b**, **c**) Small ectopic horn formation in *tra* RNAi females. Magenta arrowheads, three small protrusions formed in clypeolabrum. Green arrowheads, ectopic head horns formed in the region anterior to the three small protrusions in *tra* RNAi treatments. Scale bars are 5 mm. (**d**) Relationship between the head horn (left) and thoracic horn (right) length and body size in RNAi-treated individuals. The head and thoracic horn lengths and body sizes of control males (*EGFP*, blue dots), *dsx* RNAi-treated females (grey dots), and *tra* RNAi-treated females (green dots: **b** magenta diamond, **c** magenta hexagon) are plotted. HH, head horn; TH, thoracic horn. (Adapted from Morita et al. *PLOS Genet*., 15: e1008063, 2019)

These observations indicate that even within a single individual, the locations on the sex spectrums at the tissue level are tissue-dependent. Then, what mechanisms produce these differences? One possibility is a difference in the role of the sex-specific *dsx* isoforms among tissues. In the head region, male and female traits (the head horn and the three small projections, respectively) coexist in a single *tra* RNAi female (Fig. 1.2). In *T. dichotomus*, the *dsx* RNAi phenotype in both males and females showed the head horn was formed, albeit short (Fig. 1.1a). This head horn phenotype suggests that *dsxM* promotes the expression of the horn formation genes and *dsxF* represses (Ito et al. 2013). In Fig. 1.2, it is considered that *dsxF* and *dsxM* coexist in a single individual due to the insufficient suppression of *tra* expression. The antagonistic effects of *dsxM* and *dsxF* on the expression of horn formation genes might define the degree of similarity to males in the head region. In a region of the clypeolabrum where *dsxF* and *dsxM* coexist (Fig. 1.2b and c), *dsxM* functions to promote head horn formation, whereas *dsxF* functions to suppress head horn formation; therefore, this antagonistic effect would define the location on the sex spectrum at the tissue level in the head region. Conversely, male and female traits did not coexist in a *tra* RNAi female's thoracic horn (Fig. 1.2). In the *dsx* RNAi phenotype, the thoracic horn was not formed in either males or females. These phenotypes indicate that *dsxM* promotes thoracic horn formation, while *dsxF* does not contribute to thoracic horn formation. This fact means that thoracic horn formation is regulated depending on the only *dsxM*. In other words, in *tra* RNAi

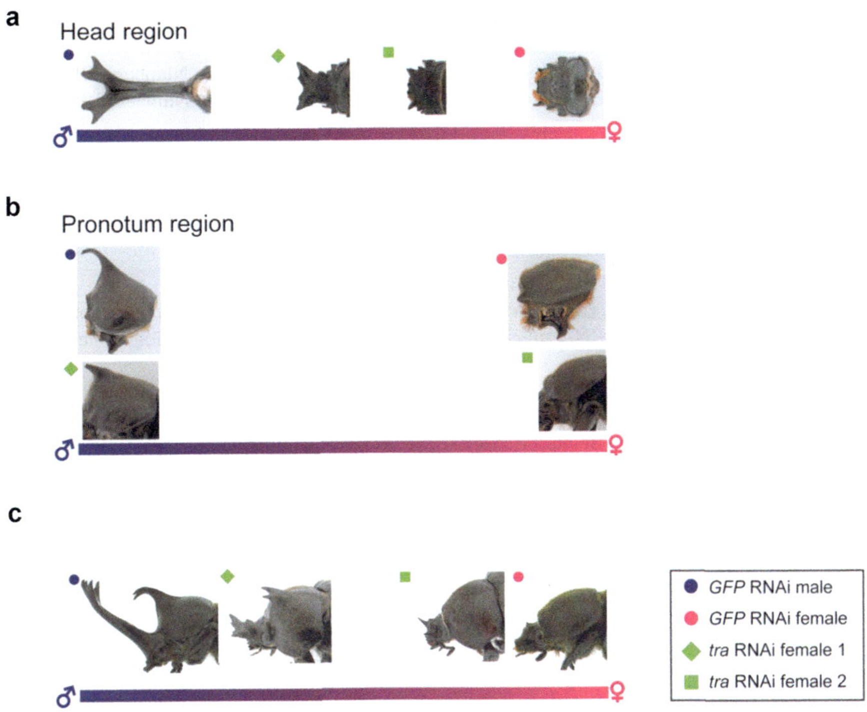

Fig. 1.3 Sex spectrum of *T. dichotomus* estimated from *tra* RNAi phenotypes. (**a**) The sex spectrum at the tissue level in the head region. (**b**) The sex spectrum at the tissue level in the pronotum region. The shorter head and thoracic horns in *tra* RNAi females than in control males may be due to differences in body size between males and females at the timing of injection. (**c**) The individual-level sex spectrum is understood from the sum of tissue-level sex spectrums. Blue dots, *GFP* RNAi male; magenta dots, *GFP* RNAi female; green diamond, *tra* RNAi female 1 (Fig. 1.2b); green square, *tra* RNAi female 2 (Fig. 1.2c). (Adapted from Morita et al. *PLOS Genet.*, 15: e1008063, 2019)

females with insufficiently suppressed *tra* expression, when the expression level of *dsxM* exceeds a certain threshold, thoracic horn formation is promoted; otherwise, the horn is not formed. Therefore, the phenotype of the thoracic horn by *tra* RNAi (Fig. 1.2) is considered to be located at either endpoint on the sex spectrum (Fig. 1.3b).

This section described how sex is interpreted as a spectrum of continuous phenotypes in *T. dichotomus*, using the horn as an indicator of sex by manipulating the expression level of a sex-determining gene *tra*. In addition, the sex spectrum at the tissue level showed different locations in the head and pronotum regions. In the concept of sex spectrum, the sum of the sex spectrum at the tissue level can be understood as the sex spectrum at the individual level (Preface). Therefore, the location of the sex spectrum at the individual level for *tra* RNAi in *T. dichotomus* could be shown as in Fig. 1.3c. From the view of the sex spectrum, rather than the traditional binary sex view, it can be uniformly explained for individuals that cannot be classified by their appearance, as in Fig. 1.2.

Acknowledgements The authors would like to thank Minoru Tanaka for the invitation to submit this monograph. In addition, we thank Toshiya Ando, Taro Nakamura and Yasuhiko Chikami for the helpful discussion. This work was supported by MEXT KAKENHI Grant Numbers 20H05944, 20H04933, 18H04766 and 16H01452 (to T. N.) and JSPS KAKENHI Grant Numbers 19 K16181 and 21K15135 (to S. M.).

References

Bell LR, Maine EM, Schedl P, Cline TW (1988) *Sex-lethal*, a Drosophila sex determination switch gene, exhibits sex-specific RNA splicing and sequence similarity to RNA binding proteins. Cell 55:1037–1046

Bell LR, Horabin JI, Schedl P, Cline TW (1991) Positive autoregulation of *sex-lethal* by alternative splicing maintains the female determined state in drosophila. Cell 65:229–239

Boggs RT, Gregor P, Idriss S, Belote JM, McKeown M (1987) Regulation of sexual differentiation in D. melanogaster via alternative splicing of RNA from the *transformer* gene. Cell 50:739–747

Chen SL, Dai SM, Lu KH, Chang C (2008) Female-specific *doublesex* dsRNA interrupts yolk protein gene expression and reproductive ability in oriental fruit fly, *Bactrocera dorsalis* (Hendel). Insect Biochem Mol Biol 38:155–165

Cho S, Huang ZY, Zhang J (2007) Sex-specific splicing of the honeybee *doublesex* gene reveals 300 million years of evolution at the bottom of the insect sex-determination pathway. Genetics 177:1733–1741

Cline TW (1978) Two closely linked mutations in *Drosophila melanogaster* that are lethal to opposite sexes and interact with daughterless. Genetics 90:683–697

Cline TW (1984) Autoregulatory functioning of a *drosophila* gene product that establishes and maintains the sexually determined state. Genetics 107:231–277

Cline TW (1986) A female-specific lethal lesion in an X-linked positive regulator of the Drosophila sex determination gene, *sex-lethal*. Genetics 113:641–663

Cline TW (1988) Evidence that *sisterless-a* and *sisterless-b* are two of several discrete "numerator elements" of the *X/a* sex determination signal in drosophila that switch *Sxl* between two alternative stable expression states. Genetics 119:829–862

Concha C, Scott MJ (2009) Sexual development in *Lucilia cuprina* (Diptera, Calliphoridae) is controlled by the transformer gene. Genetics 182:785–798

Erickson JW, Quintero JJ (2007) Indirect effects of ploidy suggest X chromosome dose, not the X: a ratio, signals sex in *drosophila*. PLoS Biol 5:e332

Geuverink E, Beukeboom LW (2014) Phylogenetic distribution and evolutionary dynamics of the sex determination genes *doublesex* and *transformer* in insects. Sex Dev 8:38–49

Gossum H, Sherratt TN (2008) A dynamic model of sexual harassment in damselflies and its implications for female-limited polymorphism. Ecol Model 210:212–220

Hasselmann M, Gempe T, Schiøtt M, Nunes-Silva CG, Otte M, Beye M (2008) Evidence for the evolutionary nascence of a novel sex determination pathway in honeybees. Nature 454:519–522

Hediger M, Henggeler C, Meier N, Perez R, Saccone G, Bopp D (2010) Molecular characterization of the key switch *F* provides a basis for understanding the rapid divergence of the sex-determining pathway in the housefly. Genetics 184:155–170

Hildreth PE (1965) Doublesex, recessive gene that transforms both males and females of drosophila into intersexes. Genetics 51:659–678

Hoshijima K, Inoue K, Higuchi I, Sakamoto H, Shimura Y (1991) Control of *doublesex* alternative splicing by *transformer* and *transformer-2* in *drosophila*. Science 252:833–836

Ito Y, Harigai A, Nakata M, Hosoya T, Araya K, Oba Y, Ito A, Ohde T, Yaginuma T, Niimi T (2013) The role of *doublesex* in the evolution of exaggerated horns in the Japanese rhinoceros beetle. EMBO Rep 14:561–567

Keyes LN, Cline TW, Schedl P (1992) The primary sex determination signal of drosophila acts at the level of transcription. Cell 68:933–943

Kramer SG, Jinks TM, Schedl P, Gergen JP (1999) Direct activation of *sex-lethal* transcription by the *drosophila* runt protein. Development 126:191–200

Kulathinal RJ, Skwarek L, Morton RA, Singh RS (2003) Rapid evolution of the sex-determining gene, *transformer*: structural diversity and rate heterogeneity among sibling species of drosophila. Mol Biol Evol 20:441–452

Lagos D, Koukidou M, Savakis C, Komitopoulou K (2007) The *transformer* gene in *Bactrocera oleae*: the genetic switch that determines its sex fate. Insect Mol Biol 16:221–230

Meise M, Hilfiker-Kleiner D, Dubendorfer A, Brunner C, Nothiger R, Bopp D (1998) *Sex-lethal*, the master sex-determining gene in *Drosophila*, is not sex-specifically regulated in *Musca domestica*. Development 125:1487–1494

Mine S, Sumitani M, Aoki F, Hatakeyama M, Suzuki MG (2017) Identification and functional characterization of the sex-determining gene *doublesex* in the sawfly, *Athalia rosae* (Hymenoptera: Tenthredinidae). Appl Entomol Zool 52:497–509

Morita S, Ando T, Maeno A, Mizutani T, Mase M, Shigenobu S, Niimi T (2019) Precise staging of beetle horn formation in *Trypoxylus dichotomus* reveals the pleiotropic roles of *doublesex* depending on the spatiotemporal developmental contexts. PLoS Genet 15:e1008063

Narita S, Pereira RAS, Kjellberg F, Kageyama D (2010) Gynandromorphs and intersexes: potential to understand the mechanism of sex determination in arthropods. Terr Arthropod Rev 3:63–96

Niimi T, Sahara K, Oshima H, Yasukochi Y, Ikeo K, Traut W (2006) Molecular cloning and chromosomal localization of the *bombyx sex-lethal* gene. Genome 49:263–268

Nong QD, Matsuura T, Kato Y, Watanabe H (2020) Two *Doublesex1* mutants revealed a tunable gene network underlying intersexuality in *Daphnia magna*. PLoS One 15:e0238256

Ohbayashi F, Suzuki MG, Mita K, Okano K, Shimada T (2001) A homologue of the *drosophila doublesex* gene is transcribed into sex-specific mRNA isoforms in the silkworm, *Bombyx mori*. Comp Biochem Physiol B Biochem Mol Biol 128:145–158

Oliveira DC, Werren JH, Verhulst EC, Giebel JD, Kamping A, Beukeboom LW, Van De Zande L (2009) Identification and characterization of the *doublesex* gene of *Nasonia*. Insect Mol Biol 18: 315–324

O'Neil MT, Belote JM (1992) Interspecific comparison of the *transformer* gene of drosophila reveals an unusually high degree of evolutionary divergence. Genetics 131:113–128

Pane A, Salvemini M, Bovi PD, Polito C, Saccone G (2002) The *transformer* gene in *Ceratitis capitata* provides a genetic basis for selecting and remembering the sexual fate. Development 129:3715–3725

Penalva LO, Sánchez L (2003) RNA binding protein sex-lethal (Sxl) and control of *Drosophila* sex determination and dosage compensation. Microbiol Mol Biol Rev 67:343–359

Price DC, Egizi A, Fonseca DM (2015) The ubiquity and ancestry of insect *doublesex*. Sci Rep 5:1–9

Ruiz MF, Milano A, Salvemini M, Eirín-López JM, Perondini AL, Selivon D, Polito C, Saccone G, Sanchez L (2007) The gene *transformer* of *Anastrepha* fruit flies (Diptera, Tephritidae) and its evolution in insects. PLoS One 2:e1239

Saccone G, Peluso I, Artiaco D, Giordano E, Bopp D, Polito LC (1998) The *Ceratitis capitata* homologue of the *Drosophila* sex-determining gene *sex-lethal* is structurally conserved, but not sex-specifically regulated. Development 125:1495–1500

Sakai H, Oshima H, Yuri K, Gotoh H, Daimon T, Yaginuma T, Sahara K, Niimi T (2019) Dimorphic sperm formation by *sex-lethal*. Proc Natl Acad Sci U S A 116:10412–10417

Salvemini M, D'Amato R, Petrella V, Aceto S, Nimmo D, Neira M, Alphey L, Polito LC, Saccone G (2013) The orthologue of the fruitfly sex behaviour gene *fruitless* in the mosquito *Aedes aegypti*: evolution of genomic organisation and alternative splicing. PLoS One 8:e48554

Salz HK (2007) Male or female? The answer depends on when you ask. PLoS Biol 5:e335

Samuels ME, Schedl P, Cline TW (1991) The complex set of late transcripts from the *Drosophila* sex determination gene sex-lethal encodes multiple related polypeptides. Mol Cell Biol 11: 3584–3602

Schmieder S, Colinet D, Poirié M (2012) Tracing back the nascence of a new sex-determination pathway to the ancestor of bees and ants. Nat Commun 3:1–7

Schütt C, Nöthiger R (2000) Structure, function and evolution of sex-determining systems in dipteran insects. Development 127:667–677

Sefton L, Timmer JR, Zhang Y, Beranger F, Cline TW (2000) An extracellular activator of the *drosophila* JAK/STAT pathway is a sex-determination signal element. Nature 405:970–973

Shukla JN, Palli SR (2012) Sex determination in beetles: production of all male progeny by parental RNAi knockdown of *transformer*. Sci Rep 2:1–9

Sievert V, Kuhn S, Paululat A, Traut W (2000) Sequence conservation and expression of the *sex-lethal* homologue in the fly *Megaselia scalaris*. Genome 43:382–390

Taracena ML, Hunt CM, Benedict MQ, Pennington PM, Dotson EM (2019) Downregulation of female *doublesex* expression by oral-mediated RNA interference reduces number and fitness of *Anopheles gambiae* adult females. Parasites Vectors 12:1–11

Traut W, Niimi T, Ikeo K, Sahara K (2006) Phylogeny of the sex-determining gene *sex-lethal* in insects. Genome 49:254–262

Chapter 2
Sexual Differentiation in Dragonflies and Damselflies

Ryo Futahashi

Abstract As represented by gynandromorphs (sexually mosaic individuals), sexual differentiation in insects proceeds primarily cell autonomously depending on sex chromosomes. Insect sex determination systems, although dominated by male heterogamety, are highly diverse. Dragonflies and damselflies (the order Odonata) are the most ancestral winged insects and have male heterogametic sex determination systems. Some species (e.g., *Crocothemis servilia*) have intraspecific polymorphisms in their karyotypes, such as switching from X0 to neo-XY sex chromosome system by chromosome fusion. In dragonflies and damselflies, adults of many species exhibit sexual color dimorphism, color transition upon adult maturation, and intraspecific color polymorphisms within the same sex. Molecular mechanisms underlying sex determination and sexual differentiation in insects have been investigated extensively in the fruit fly *Drosophila melanogaster*, but recent studies have revealed that the upstream genes of insect sex determination cascade are highly diverse. Most insects have sex-specific isoforms for *doublesex* (*dsx*) gene, which is important for sexual differentiation, and *dsx* gene plays important roles in masculinization not only for males but also for androchrome females in the damselfly *Ischnura senegalensis*. In this review, current knowledge on sex determination and sexual differentiation of insects is summarized, with particular focus on the most ancestral winged insects, dragonflies and damselflies.

Keywords Insects · Odonata · Dragonflies · Damselflies · Karyotype · Sex determination · Sexual differentiation · Polymorphism · Androchrome · Heterochrome

R. Futahashi (✉)
National Institute of Advanced Industrial Science and Technology (AIST), Tsukuba, Japan
e-mail: ryo-futahashi@aist.go.jp

M. Tanaka, M. Tachibana (eds.), *Spectrum of Sex*,
https://doi.org/10.1007/978-981-19-5359-0_2

2.1 Introduction

Gynandromorphs (sexually mosaic individuals) have been occasionally observed among various insects including dragonflies and damselflies (Fig. 2.1) (Narita et al. 2010; Martens and Wildermuth 2021). As represented by the presence of gynandromorphs, it has traditionally been considered that, in insects, sex chromosomes, rather than sex hormones, determine the sexual phenotype cell autonomously. It should be noted that vertebrate androgens and estrogens are detected from some insect species (Mechoulam et al. 1984; Denlinger et al. 1987), although the function of these hormones still remains largely unknown (Bear and Monteiro 2013; Das 2016). Meanwhile, juvenile hormone and ecdysone, which control molting and metamorphosis in insects, have been reported to participate in the maturation of adult reproductive organs and courtship behavior (De Loof 2006; Riddiford 2012), indicating that hormones are also partially associated with sexual maturation in insects. Most research on sex determination and sexual differentiation in insects has focused on differences between males and females at the onset of adulthood, such as the shape of the genitalia, and colors and patterns that do not change during adulthood. While molecular mechanisms underlying sex determination and sexual differentiation in insects have been studied primarily in the fruit fly *Drosophila melanogaster*, recent studies have revealed that the upstream genes of the sex determination cascade are highly diverse among insects. In this review, current knowledge on sex determination and sexual differentiation of insects is summarized, with particular focus on the order Odonata (dragonflies and damselflies), the most ancestral winged insects.

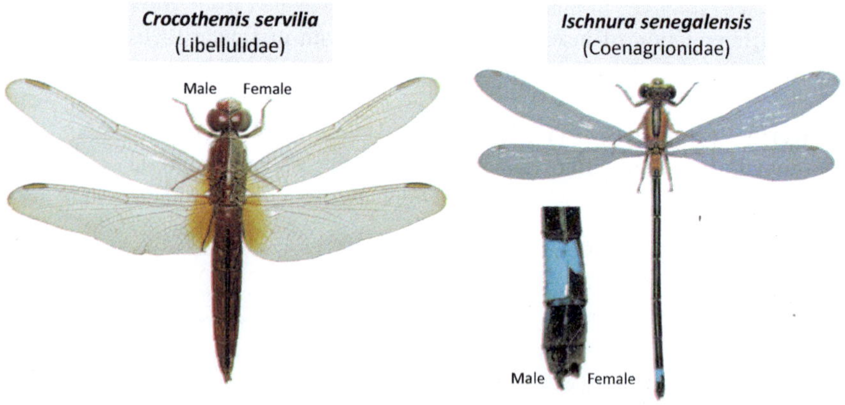

Fig. 2.1 Gynandromorphs of the dragonfly *Crocothemis servilia* and the damselfly *Ischnura senegalensis*. Photographs courtesy of Mitsutoshi Sugimura

2.2 Sex Determination System in Insects

In animals, females are defined as individuals that produce large gametes (with ovaries) and males as individuals that produce small gametes (with testes) (Bachtrog et al. 2014). In almost all insects, males and females are primarily defined by genotypic sex determination (Bachtrog et al. 2014; Blackmon et al. 2017), and hermaphrodites that possess both male and female sex organs in the same individual have been reported only from a few species of *Icerya* scale insects which belong to the order Hemiptera, although they differ from other hermaphrodite taxa by that the sperm-producing gonads are haploid (Ross et al. 2010; Gardner and Ross 2011). It should be noted that classic studies have shown that the larval growth temperature of the fly *Aedes stimulans* (feminize at high temperature) and the maternal nutritional status of the fly *Heteropeza pygmaea* (females develop under better nutritional conditions) are important for sex determination (Bergerard 1972; Went and Camenzind 1984).

In insects, sex determination is usually based on sex chromosomes, defined as male heterogamety (XX/XY) or female heterogamety (ZZ/ZW), and complete loss of Y or W results in X0 or Z0 system, respectively (Fig. 2.2) (Kiauta 1972; Kiauta and Kiauta 1982; Traut and Marec 1996; Papeschi and Bressa 2006; Poggio et al. 2007; Traut et al. 2007; Normark 2014; Tree of Sex Consortium 2014; Blackmon and Demuth 2015; Vershinina and Kuznetsova 2016; Blackmon et al. 2017; Kuznetsova and Golub 2020). Moreover, many insect orders have species with complex sex chromosomes, comprising multiple X, Y, Z, and W chromosomes (XX/XY_1Y_2 and $X_1X_1X_2X_2/X_1X_2Y$ as examples of complex XY (CXY), $X_1X_1X_2X_2/X_1X_2$ as an example of complex X0 (CX0), and $Z_1Z_1Z_2Z_2/Z_1Z_2W$ or ZZ/ZW_1W_2 as examples of complex ZW (CZW)) that can arise from the XY or ZW system by fusion of ancestral sex chromosomes with autosomes or by fission of ancestral sex chromosomes (Fig. 2.2) (Tree of Sex Consortium 2014; Blackmon et al. 2017). In some species, sex chromosomes are morphologically similar and are called homomorphic sex chromosomes (Fig. 2.2, Hom) (Tree of Sex Consortium 2014; Blackmon et al. 2017). Among insects, male heterogamety is a common mode of sex determination, whereas female heterogamety has been reported from all examined species of Trichoptera and Lepidoptera and a few species of Diptera (Fig. 2.2) (Tree of Sex Consortium 2014; Blackmon et al. 2017).

Instead of sex chromosomes, some insects (e.g., the orders Hymenoptera and Thysanoptera) determine sex on the basis of haplodiploidy (Fig. 2.2, HD), in which males develop from unfertilized eggs and thus haploid, and females develop from fertilized eggs and are diploid (Normark 2003; de la Filia et al. 2015). On the other hand, insects such as scale insects (the order Hemiptera) and booklice (the order Psocodea) determine sex on the basis of a pseudo-haplodiploidy, in which the male becomes functionally haploid through inactivation or loss of the paternal chromosomes (paternal genome elimination; PGE) after fertilization (de la Filia et al. 2015; Hodson et al. 2017). In addition, parthenogenesis, in which female embryos develop from unfertilized diploid eggs, has also been reported from various insects in small rates (Fig. 2.2, Parth) (Normark 2014; Blackmon et al. 2017).

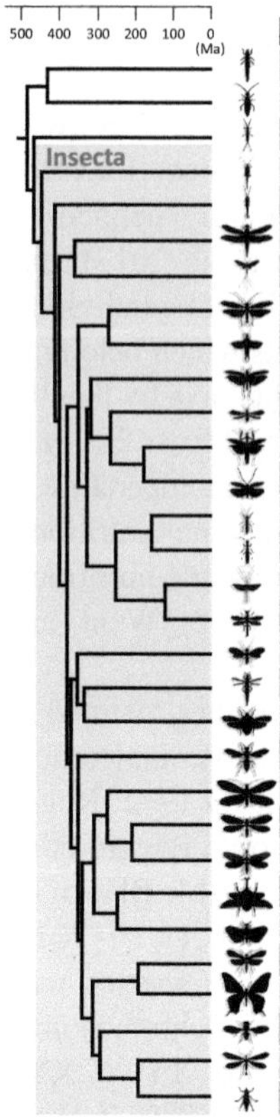

Insect order (example)	XY	X0	CXY	CX0	ZW	Z0	CZW	Hom	HD/PGE	Parth	Total
Protura (coneheads)	3							1		3	7
Collembola (springtails)		17							10	4	31
Diplura (two-pronged bristletails)											0
Archaeognatha (jumping bristletails)										5	5
Zygentoma (silverfish)		3								1	4
Odonata (dragonflies & damselflies)	30	662	1							1	693
Ephemeroptera (mayflies)	6	2								3	11
Plecoptera (stoneflies)	1	3		8						2	14
Dermaptera (earwigs)	22	2	27							1	52
Orthoptera (grasshoppers & crickets)	43	259	19							10	331
Zoraptera (ground lice)	1									1	2
Mantodea (praying mantids)	1	60	40							3	104
Blattodea (cockroaches & termites)	2	109	61					1		12	185
Mantophasmatodea (gladiators)		1									1
Grylloblattodea (ice crawlers)	2										2
Phasmatodea (stick & leaf insects)	14	70								17	101
Embioptera (webspinners)		8								3	11
Psocodea (parasitic lice & booklice)	3	91							15	37	147
Thysanoptera (thrips)									24	59	83
Hemiptera (true bugs & aphids)	890	299	66	6				3	247	420	1931
Hymenoptera (bees, wasps & ants)									1591	159	1750
Megaloptera (alderflies & dobsonflies)	4										4
Neuroptera (net-winged insects)	70	2		2							74
Raphidioptera (snakeflies)	6										6
Coleoptera (beetles)	3317	787	207	12					8	271	4602
Strepsiptera (twisted wing parasites)	1									1	2
Trichoptera (caddisflies)						15				6	21
Lepidoptera (butterflies & moths)					189	41	12			14	256
Diptera (true flies & mosquitoes)	1188	43	10		7			93	32	43	1416
Mecoptera (scorpionflies)		13	1								14
Siphonaptera (fleas)	2		4								6

Fig. 2.2 Sex determination systems in insects. Data are based on Kiauta (1972); Kiauta and Kiauta (1982); Traut and Marec (1996); Papeschi and Bressa (2006); Poggio et al. (2007); Traut et al. (2007); Normark (2014); Tree of Sex Consortium (2014); Blackmon and Demuth (2015); Vershinina and Kuznetsova (2016); Blackmon et al. (2017); Hodson et al. (2017); Kuznetsova and Golub (2020). Insect phylogeny is based on Misof et al. (2014) and Wang et al. (2016). *CXY* complex XY, *CX0* complex X0, *CZW* complex ZW, *Hom* homomorphic, *HD* haplodiploid, *PGE* paternal genome elimination, *Parth* parthenogenetic

2.3 Sex Chromosomes and Chromosome Number Variations in Dragonflies and Damselflies

Dragonflies and damselflies (the order Odonata) belong to the most ancestral winged insects (Fig. 2.2) (Misof et al. 2014; Wang et al. 2016; Bybee et al. 2021). Karyotypes have been examined in all 10 superfamilies of dragonflies and damselflies, and the basic type is assumed to be X0/XX, $n = 13$ (Fig. 2.3) (Kiauta 1972; Kiauta and Kiauta 1982; Kuznetsova and Golub 2020). In some families, species with the chromosome number $n = 12$ (Chlorocyphidae and Gomphidae) or $n = 14$ (Coenagrionidae and Aeshnidae) comprise the majority (Fig. 2.3) (Kiauta 1967; Kuznetsova and Golub 2020). Complex XY chromosome ($X_1X_1X_2X_2/X_1X_2Y$) has

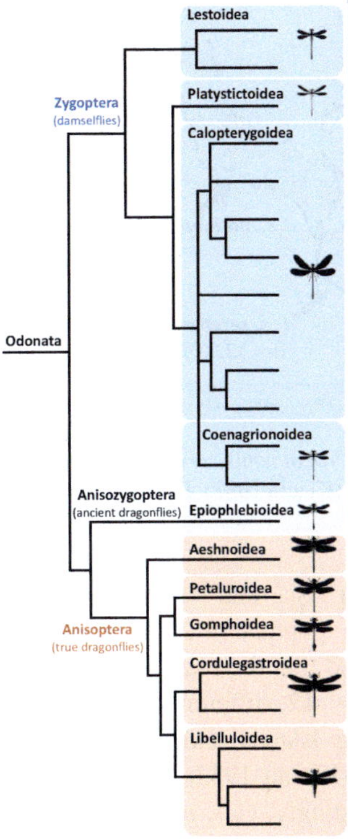

Family	Sex chromosomes			Chromosome numbers (n)												
	XY	XO	X_1X_2Y	3	6	7	8	9	10	11	12	13	14	15	19	21
Synlestidae	0	1	0	0	0	0	0	0	0	0	0	1	0	0	0	0
Lestidae	0	21	0	0	0	0	0	0	1	1	0	19	0	0	0	0
Platystictidae	0	4	0	0	0	0	0	0	0	0	0	4	0	0	0	0
Philogeniidae	0	1	0	0	0	0	0	0	0	0	0	1	0	0	0	0
Hypolestidae	0	1	0	0	0	0	0	1	0	0	0	0	0	0	0	0
Heteragrionidae	0	2	0	0	0	0	0	0	0	0	0	2	0	0	0	0
Polythoridae	0	2	0	0	0	0	0	0	0	0	2	0	0	0	0	0
Euphaeidae	0	4	0	0	0	0	0	0	0	0	0	4	0	0	0	0
Megapodagrionidae	0	3	0	0	0	0	0	0	0	0	0	3	0	0	0	0
Calopterygidae	0	26	0	0	0	0	0	0	0	0	1	23	2	0	0	0
Chlorocyphidae	0	10	0	0	0	0	0	0	0	0	8	2	0	0	0	0
Platycnemididae	0	27	0	0	0	0	0	0	0	1	1	25	0	0	0	0
Coenagrionidae	3	102	0	0	1	0	0	0	0	0	0	5	94	3	2	0
Epiophlebiidae	0	1	0	0	0	0	0	0	0	0	0	1	0	0	0	0
Aeshnidae	10	58	0	0	0	1	3	0	1	1	2	16	44	0	0	0
Petaluridae	0	4	0	0	0	0	0	2	1	0	0	1	0	0	0	0
Gomphidae	5	76	0	0	0	0	0	0	1	8	70	2	0	0	0	0
Chlorogomphidae	0	1	0	0	0	0	0	0	0	0	0	1	0	0	0	0
Cordulegastridae	0	9	0	0	0	0	0	0	0	0	1	8	0	0	0	0
Macromiidae	0	10	0	0	0	0	0	0	0	0	0	10	0	0	0	0
Corduliidae	2	28	0	0	1	2	0	0	2	2	0	21	2	0	0	0
Libellulidae	10	286	1	1	2	0	0	1	1	10	29	246	5	1	0	1
Total	30	677	1	1	4	3	3	4	7	23	114	395	147	4	2	1

Fig. 2.3 Sex chromosomes and chromosome numbers in dragonflies and damselflies. Data are based on Kiauta (1972); Kiauta and Kiauta (1982); Kuznetsova and Golub (2020). Odonata phylogeny is based on Bybee et al. (2021)

so far been reported in only one species, *Micrathyria ungulate* (Libellulidae) (Mola et al. 1999).

In several dragonflies and damselflies, variations in chromosome numbers have been reported within the same species and even within the same individual (Kiauta 1969a, 1969b, 1983; Mola 2007; Kuznetsova and Golub 2020). The intraspecific variation in chromosome number is partly due to the holocentricity (centromeres diffuse throughout the chromosome) of dragonflies and damselflies which allows normal chromosome segregation for both fragments after chromosome fission (Suzuki and Saitoh 1990; Nokkala et al. 2002). In many dragonflies, the presence of very small chromosomes namely microchromosomes (m-chromosomes) has been reported, presumably derived from fragments of autosomes (Kiauta 1969b; Mola 2007; Kuznetsova and Golub 2020).

A striking example of intraspecific changes in sex chromosomes and chromosome numbers has been reported in the scarlet skimmer dragonfly *Crocothemis servilia*

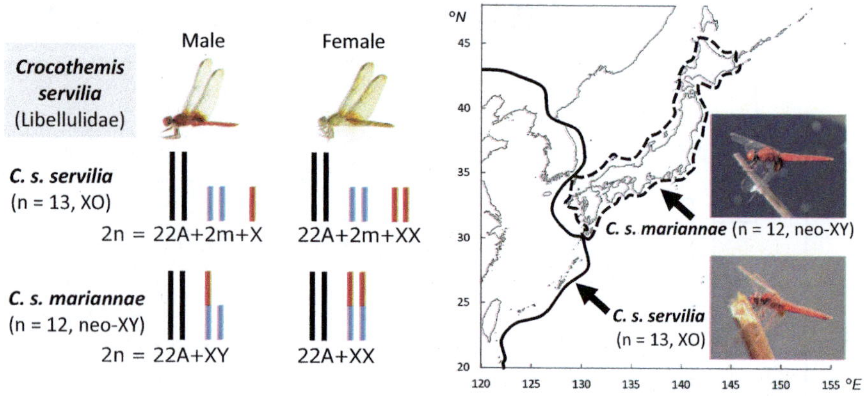

Fig. 2.4 Intraspecific variation of karyotypes in the dragonfly *Crocothemis servilia*. Blue and red lines indicate microchromosomes (m) and sex chromosomes (X) (Kiauta 1983). Distribution map is based on Higashi et al. (2001) and Ozono et al. (2021)

(Libellulidae) (Fig. 2.4), which is widely distributed in Asia, and has been recently introduced to North America and Hawaii. The Japanese population of this species is larger in body size, and karyotype observations revealed that the mainland Japanese population (including Tsushima Island) is XX/XY, $n = 12$, different from XX/X0, $n = 13$ in other regions (including Japanese Nansei Islands and Taiwan), thus designated as a different subspecies *C. s. mariannae* (Kiauta 1983; Katatani 1987; Higashi and Kayano 1993; Higashi et al. 2001). Interestingly, *C. s. mariannae* lacks microchromosomes, and Y chromosome in *C. s. mariannae* is approximately the same size as microchromosomes in *C. s. servilia*, suggesting that sex chromosomes of *C. s. servilia* fused with microchromosomes and shifted from X0 to neo-XY in *C. s. mariannae* (Fig. 2.4) (Kiauta 1983; Higashi and Kayano 1993). Subsequently, nuclear DNA analysis confirmed that populations of the Japanese mainland (including Tsushima Island) (i.e., *C. s. mariannae*) formed a clade independent of continental populations from South Korea and westward (including Japanese Nansei Islands and Taiwan) (i.e., *C. s. servilia*) (Fig. 2.4) (Futahashi 2011). The transition from X0 to neo-XY within the same species or closely related species has also been reported in several other dragonflies (Kiauta 1969a; Kuznetsova and Golub 2020).

2.4 Sexual Dimorphism, Adult Color Transition, and Female Color Polymorphism of Dragonflies and Damselflies

Dragonflies and damselflies, which primarily recognize their mates visually, often exhibit conspicuous differences in colors and patterns between males and females (Corbet 1999; Futahashi et al. 2015; Bybee et al. 2016; Futahashi 2016, 2020).

Compared with other insects, many species show the spectrum of sexual dimorphism such as remarkable color changes upon adult maturation and body-color polymorphism within the same sex (Tillyard 1917; Corbet 1999; Fincke et al. 2005; Bybee et al. 2016; Futahashi 2016). For example, in many dragonfly species that turn red or secrete whitish wax on their body surface upon adult maturation, immature adults look very similar in both sexes, but typically only males exhibit prominent color changes (Fig. 2.5). These color changes are attributed to male-specific ommochrome pigment reduction or wax production (Futahashi et al. 2012, 2019).

Notably, many species with sexual color dimorphism have female color polymorphisms, including the male-mimicking color morph called androchrome and the female-specific color morph called heterochrome or gynochrome (Figs. 2.5 and 2.6) (Corbet 1999; Fincke et al. 2005; Svensson et al. 2009; Svensson 2017; Blow et al. 2021). The evolution of female polymorphisms has been particularly studied in *Ischnura* damselflies (Coenagrionidae) (Willink et al. 2019; Sánchez-Guillén et al. 2020; Blow et al. 2021). *Ischnura* species have either one female morph (heterochrome or androchrome), two female morphs (heterochrome and androchrome), or even three female morphs (heterochrome-, androchrome-, and intermediate-type heterochrome) (Svensson et al. 2009; Willink et al. 2019; Blow et al. 2021). The third female morph has an androchrome-like thoracic pattern and is difficult to distinguish from androchrome during immature stages, but their body colors become heteromorphic with maturation (Fig. 2.6) (Svensson et al. 2009; Willink et al. 2019). The ancestral state of *Ischnura* species is assumed to be monomorphic with heterochrome female (Blow et al. 2021).

The female polymorphism of Japanese *Ischnura* species is shown in Fig. 2.6. *I. aurora* has two female morphs, but the frequency of androchrome is very rare (Ozono et al. 2021). *I. asiatica* and its closely related species *I. ezoin*, endemic to Ogasawara Islands, have one female morph (heterochrome). *I. senegalensis* and *I. elegans* are distributed allopatrically, with two and three female morphs,

Fig. 2.5 Sexual dimorphism, male color transition, and female color polymorphism in dragonflies *Sympetrum kunckeli* and *Orthetrum albistylum* (Libellulidae)

Fig. 2.6 Interspecific diversity of female color polymorphism among *Ischnura* damselflies (Coenagrionidae). *Ischnura* phylogeny is based on Blow et al. 2021

respectively. Although *I. senegalensis* and *I. elegans* diverged about five million years ago (Blow et al. 2021), it is possible to produce hybrids between these two species artificially in the laboratory (Okude et al. 2020). In contrast, some damselflies, such as the genus *Mnais* (Calopterygidae), have male color polymorphism with female-mimicking color morph (Corbet 1999; Córdoba-Aguilar and Cordero-Rivera 2005; Futahashi 2017). Comparative analysis of these damselflies is a promising approach for elucidating the molecular mechanisms underlying color polymorphism and sexual differentiation.

2.5 Sex Determination Cascades in Insects

Molecular mechanisms underlying sex determination and sexual differentiation of insects have been studied primarily in the fruit fly *Drosophila melanogaster* for male heterogamety XX/XY. In *D. melanogaster*, Y chromosome is required for male sexual maturation but is not involved in sex determination (Casper and Van Doren 2006), and the number of X chromosomes and the ploidy of autosomes are important in sex determination (Erickson and Quintero 2007). XXY karyotype is male in humans (Klinefelter's syndrome) but female in *D. melanogaster*, and X0 karyotype is female in humans (Turner's syndrome) but male in *D. melanogaster*.

In *D. melanogaster*, the master switch gene for sex determination is *Sex lethal* (*Sxl*), encoding an RNA binding protein that functions as a splicing regulator. In females with two X chromosomes, a high dose of X-linked signal elements produces functional Sxl proteins by skipping exon 3 containing in-frame stop codons. Sxl proteins control splicing of *Sxl* gene itself, and functional Sxl proteins are produced in females throughout development. Sxl proteins induce a female-specific isoform of the splicing regulator gene, *transformer* (*tra*) (Fig. 2.7a). In females, functional Tra protein is produced from a female-specific isoform of *tra* gene, skipping exon 2 containing in-frame stop codons. Conversely, male isoform of *tra* gene produces a nonfunctional truncated Tra protein. Tra protein forms a heterodimer with another splicing regulator protein encoded by *transformer-2* (*tra2*) gene expressed in both sexes and regulates sex-specific splicing of the transcription factor genes *doublesex* (*dsx*) and *fruitless* (*fru*). Dsx and Fru proteins control sex-specific morphology and behavior, respectively (Fig. 2.7a) (Sánchez 2008; Gempe and Beye 2010; Verhulst et al. 2010a; Hopkins and Kopp 2021). In addition, sex differentiation of female external morphology is mediated by the mediator complex component *intersex* (*ix*) gene (no differential expression between sexes) that functions in the downstream of *dsx* gene (Fig. 2.7a), and mutants of *ix* gene exhibit intermediate morphology

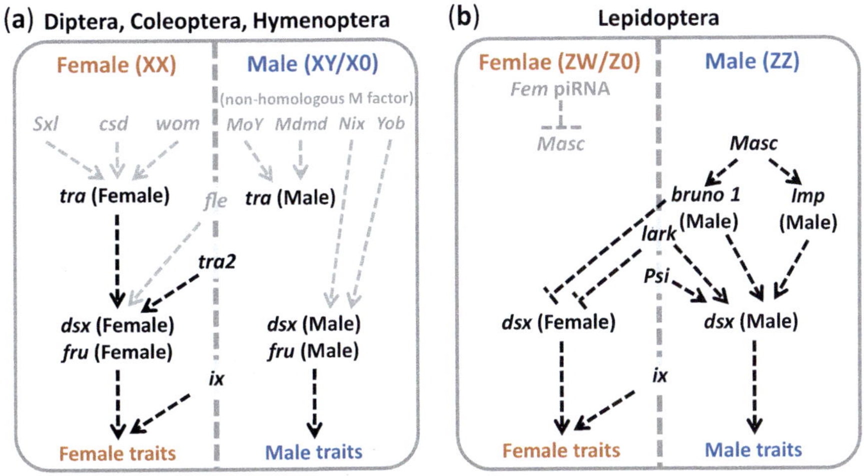

Fig. 2.7 Schematic diagram of insect sex determination cascade. (**a**) Sex determination cascade in insects with male heterogamety. *Sxl* Sex lethal, *csd* complementary sex determination, *wom* wasp overruler of masculinization, *MoY* Maleness on the Y, *Mdmd* Musca domestica male determiner, *fle* femaleless, *tra* transformer, *tra2* transformer-2, *dsx* doublesex, *fru* fruitless, *ix* intersex. (**b**) Sex determination cascade in insects with female heterogamety. *Fem* Feminizer, *Masc* Masculinizer, *Imp* Insulin-like growth factor II mRNA-binding protein, *Psi* P-element somatic inhibitor, *dsx* doublesex, *ix* intersex. The isoforms of *tra* and *dsx* are different between males and females, while there is a male-specific isoform in *bruno 1* and *Imp*. Pathways present only in specific species are shown in gray. (Figure modified from Siegal and Baker (2005), Gempe and Beye (2010), Kiuchi et al. (2014), Sakai et al. (2015), Zheng et al. (2019), Krzywinska et al. (2021), and Hopkins and Kopp (2021))

between sexes (male-like pigmentation appears in the abdomen of female *ix* mutants) (Siegal and Baker 2005).

Although the sex determination cascade in *D. melanogaster* has been described in many textbooks, it has become apparent that the upstream genes are not conserved even within the order Diptera. *Sxl* is not involved in sex determination other than in the family Drosophilidae (Sánchez 2008; Gempe and Beye 2010; Verhulst et al. 2010a; Hopkins and Kopp 2021). In *Drosophila*, Sxl protein is also involved in the translational repression of the *male-specific lethal 2* gene belonging to the male-specific lethal (MSL) complex that controls dosage compensation, although MSL complex-mediated dosage compensation is probably not conserved outside Drosophilidae (Ruiz et al. 2000; Keller Valsecchi et al. 2021). The medfly *Ceratitis capitata* has male heterogamety (XX/XY) like *D. melanogaster*, but analyses of partially Y-deficient individuals revealed that a male-determining factor (M factor) is present on Y chromosome (Willhoeft and Franz 1996). In females, functional *tra* gene is expressed since early embryonic developmental stage via maternal inheritance, and female-specific isoform of *tra* gene is maintained through autoregulation. In males, M factor modifies the splicing pattern of *tra* gene into nonfunctional types (Pane et al. 2002). The responsible gene for medfly M factor is *Maleness on the Y* (*MoY*), encoding a protein as small as 70 amino acids that are conserved only in the family Tephritidae (Meccariello et al. 2019). Importantly, the origin of M factor is highly diverse among the order Diptera (Fig. 2.7a). The housefly *Musca domestica* also typically has male heterogamety (XX/XY), and the sex determination cascade resembles *C. capitata*. However, there are multiple strains that translocate M factor from Y chromosome to autosomes (Dübendorfer et al. 2002), and the responsible gene for the housefly M factor is *Musca domestica male determiner* (*Mdmd*) originated from a duplication of the pre-mRNA splicing factor gene *nucampholin*, also known as CWC22 (Fig. 2.7a) (Sharma et al. 2017). Notably, there are the field populations where females emerge in spite of the presence of M factor, due to the constitutive activation by mutations in the *tra* gene itself (Hediger et al. 2010). In the yellow fever mosquito *Aedes aegypti*, the responsible gene for M factor is *Nix* derived from a duplication of *tra2* (Hall et al. 2015; Aryan et al. 2020). In the malaria mosquito *Anopheles gambiae*, the responsible gene for M factor is *Yob*, encoding a small protein of 56 amino acids (Krzywinska et al. 2016). Notably, *femaleless* (*fle*) gene, which belongs to the same clade as the *Nix* originated from a duplication of *tra2*, regulates sex-specific splicing of *dsx* and *fru* genes instead of *tra2* in *Anopheles* mosquitoes (Fig. 2.7a) (Krzywinska et al. 2021). In *A. gambiae*, *fle* is involved in the dosage compensation of females, and in *Anopheles stephensi*, the *Yob* ortholog *Guy1* gene is involved in the dosage compensation of males (Qi et al. 2019; Krzywinska et al. 2021). It should be noted that *A. aegypti* and *A. gambiae* genomes do not contain *tra* gene, and it is unclear whether *Nix* or *fle* requires a *tra*-like counterpart gene for the splicing control of *dsx* and *fru* genes (Fig. 2.7a). Thus, the most upstream gene for sex determination is highly diverse within the order Diptera.

In the order Coleoptera, genes involved in sexual differentiation have been functionally analyzed mainly in the red flour beetle *Tribolium castaneum* (Shukla and Palli 2012a, 2012b, 2013, 2014), the Japanese rhinoceros beetle *Trypoxylus dichotomus* (Ito et al. 2013; Morita et al. 2019), and the golden metallic stag beetle *Cyclommatus metallifer* (Gotoh et al. 2014, 2016). These studies indicate that genes downstream of *tra* are functionally conserved with Diptera (Fig. 2.7a), resulting in masculinization by *tra* RNAi in females, intermediate phenotype of both sexes by *dsx* RNAi, and intermediate phenotype by *ix* RNAi only in females.

Even in the order Hymenoptera, whose sex is determined by haplodiploidy, genes downstream of *tra* (also known as *feminizer*) are functionally conserved (Hasselmann et al. 2008; Gempe et al. 2009; Verhulst et al. 2010b; Nissen et al. 2012). In contrast, the most upstream gene for sex determination that control *tra* splicing is not conserved and has been identified as *complementary sex determination* (*csd*) originated from a duplication of *tra* gene in the honeybee *Apis mellifera* (Beye et al. 2003), or as *wasp overruler of masculinization* (*wom*), a recently evolved chimeric gene with a P53-like domain, in the parasitoid wasp *Nasonia vitripennis* (Zou et al. 2020). In *A. mellifera*, a female isoform of *tra* is expressed when alleles of *csd* gene are heterozygous (Beye et al. 2003), whereas in *N. vitripennis*, *wom* induces transcription of a female isoform of *tra* through genomic imprinting when there is a paternal genome (Fig. 2.7a) (Zou et al. 2020).

Compared with holometabolous insects described above, the molecular mechanisms underlying sexual differentiation have been less characterized in more ancestral hemimetabolous insects. Although *tra2*, *dsx*, *fru*, and *ix* genes are well conserved in many hemimetabolous insects, the amino acid sequence of Tra protein has low sequence similarity, making it difficult to identify (Laslo et al. 2022). In the brown planthopper *Nilaparvata lugens* (the order Hemiptera), *dsx* has sex-specific isoforms and plays important roles in sexual differentiation mainly in males (Zhuo et al. 2018). In *N. lugens*, two genes, *Female determinant factor* (*fmd*) and *Female determinant factor 2* (*fmd2*), control *dsx* splicing, similar to *tra* and *tra2* in holometabolous insects (Zhuo et al. 2021). It should be noted that *fmd* has a female-specific isoform with partial sequence similarity to Tra protein, whereas *fmd2*, an ortholog of the *Drosophila squid* gene, does not have a sex-specific isoform. In the German cockroach *Blattella germanica* (the order Blattodea), *dsx* has sex-specific isoforms and is involved in male differentiation (Wexler et al. 2019). Meanwhile, *tra* contributes to female differentiation, although there are no sex-specific isoforms (Wexler et al. 2019). It has also been reported that *fru* does not have sex-specific isoforms in the order Orthoptera (Salvemini et al. 2010; Watanabe 2019). Interestingly, in termites which are included in a monophyletic group of Blattodea (also referred to as the independent order Isoptera), *dsx* do not show sex-specific splicing but is expressed in a male-specific manner (Miyazaki et al. 2021). It should be noted that, in the branchiopod crustacean (belong to the sister group of insects) *Daphnia magna*, *dsx* is highly expressed in males with no sex-specific isoforms and is involved in male differentiation (Kato et al. 2011). In *D. magna*, *tra*-like gene is not differentially expressed between males and females and is not involved in sexual differentiation (Kato et al. 2010).

Among insects, a very unique sex determination cascade has developed in the order Lepidoptera with female heterogamety (Fig. 2.7b). In the silkworm *Bombyx mori*, a sex determinant is located on W chromosome (Hashimoto 1933). Although *dsx* gene has sex-specific isoforms similar to other insects (Ohbayashi et al. 2001; Suzuki et al. 2001, 2003, 2005), the splicing of *dsx* is regulated differently in other insects by splicing regulator protein genes *bruno 1* (also designated as *Aret*, or *RBP3*) and *lark* (also designated as *RBP1*) (Zheng et al. 2019), and by *P-element somatic inhibitor* (*Psi*) and *Insulin-like growth factor II mRNA binding protein* (*Imp*) genes (Suzuki et al. 2008, 2010, 2014). While *bruno1* and *Imp* have sex-specific isoforms, the expression of *lark* and *Psi* does not differ between sexes (Suzuki et al. 2010, 2014; Zheng et al. 2019). Moreover, male-specific isoforms of *Imp* and *bruno 1* are regulated by a lepidopteran-specific zinc finger protein gene *Masculinizer* (*Masc*), and *Masc* is inhibited by the master sex determiner *Feminizer* (*Fem*), a single female-specific piRNA present on the W chromosome (Kiuchi et al. 2014; Sakai et al. 2015; Zheng et al. 2019). *Masc* gene also controls *dsx* splicing and dosage compensation in several lepidopteran species (Lee et al. 2015; Fukui et al. 2018; Wang et al. 2019; Harvey-Samuel et al. 2020; Visser et al. 2021), but *Fem* piRNA target sequence in *Masc* gene is not conserved among Lepidoptera, suggesting that *Fem* controls sex determination only within the closely related species of *B. mori* (Lee et al. 2015). Notably, in *B. mori*, *Sxl* is required for the formation of two different types of sperm but is not involved in sex determination (Sakai et al. 2019), and *tra2* does not regulate splicing of *dsx* gene (Suzuki et al. 2012). Although *fru* is important for mating behavior, sex-specific isoforms have not been reported in *B. mori* (Xu et al. 2020). In *B. mori*, *ix* controls the development not only of female external genitalia but also of imaginal discs (Xu et al. 2019). Thus, sex determination in insects is significantly different between male heterogamety and female heterogamety (Fig. 2.7).

2.6 Molecular Mechanisms Underlying Sexual Differentiation in *Ischnura* Damselflies

In the order Odonata, genes involved in sexual differentiation and differentially expressed genes between sexes and female morphs have been investigated from genomics and transcriptomics (Chauhan et al. 2016, 2021; Futahashi et al. 2019; Takahashi et al. 2019, 2021; Willink et al. 2020). In the damselfly *I. elegans* with XX/X0 sex chromosome system, dosage compensation balances gene expression on X chromosome between males and females (Chauhan et al. 2021).

In the damselfly *I. senegalensis*, *dsx* gene shows differential expression of isoforms between sexes and female morphs (Fig. 2.8a) (Takahashi et al. 2019, 2021). Two isoforms (M1 and M2) are expressed mainly in males and androchrome females, and other two isoforms (F1 and F2, exon 3 is larger than M1 and M2 and contains stop codons) are expressed mainly in females (both heterochrome and androchrome) (Fig. 2.8a) (Takahashi et al. 2019). In some lepidopteran species, *dsx* is involved in female-limited wing pattern polymorphism, but unlike *I. senegalensis*, differences in expression level and/or amino acid sequence are important rather than isoform patterns (Kunte et al. 2014; Nishikawa et al. 2015).

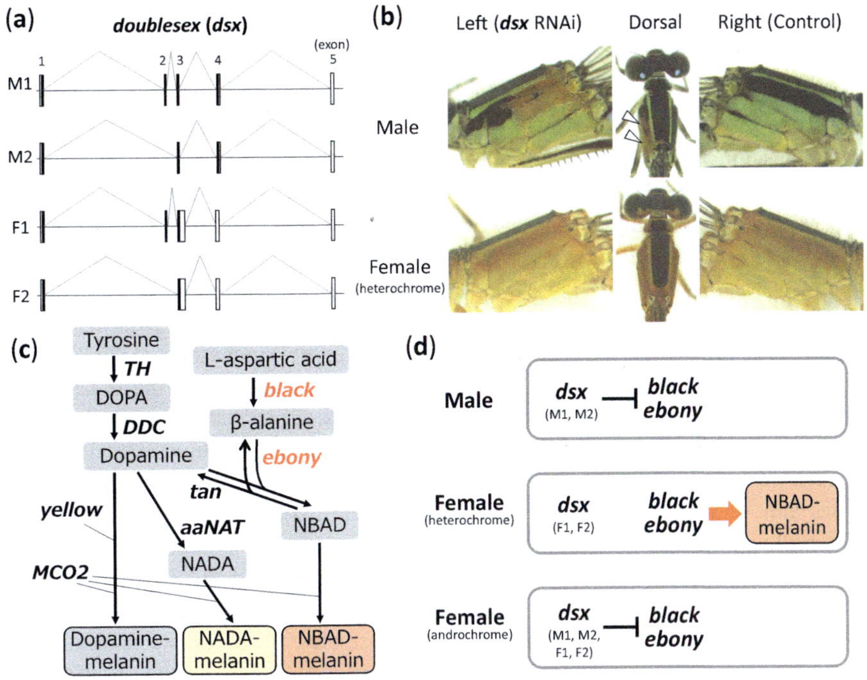

Fig. 2.8 Sex-specific isoform and inhibition of masculinization by *doublesex* (*dsx*) gene in the damselfly *Ischnura senegalensis*. (**a**) The structure of *dsx* gene estimated from the genome sequence of *I. elegans* (Takahashi et al. 2021; Chauhan et al. 2021). (**b**) Lateral and dorsal views of RNAi phenotypes on the thorax. Dotted lines and white arrowheads indicate the regions where color patterns have changed. siRNAs were designed at common regions for all isoforms (the first part of exon 3) (Takahashi et al. 2021). (**c**) Schematic diagram of melanin synthesis pathway in insects. (Figure modified from Okude and Futahashi (2021) and Futahashi and Osanai-Futahashi (2021)). (**d**) Summary of the regulation of melanin synthesis genes by *dsx* isoforms (Takahashi et al. 2021)

In dragonflies and damselflies, gene knockdown by RNAi is possible when electroporation is conducted to introduce small interfering RNA or double-stranded RNA into the cells, and the gene function is suppressed locally around the region where the positive electrode was placed for electroporation (Okude et al. 2021, 2022). RNAi of *dsx* (targeting all isoforms) induced feminization of body color in both males and androchrome females, but did not affect heterochrome females (Fig. 2.8b) (Takahashi et al. 2021). It has been also confirmed that *dsx* RNAi increased the expression of the light-colored NBAD melanin synthesis genes *ebony* and *black*, which are involved in orange thoracic coloration in heterochrome females (Fig. 2.8c) (Takahashi et al. 2019, 2021; Okude and Futahashi 2021; Futahashi and Osanai-Futahashi 2021). Thus, when M1 and/or M2 *dsx* isoforms are expressed, the expression of *black* and *ebony* gene is suppressed, resulting in masculinization of thoracic body color due to the absence of orange NBAD melanin (Fig. 2.8d). Meanwhile, female *dsx* isoforms (F1 and F2) are not involved in sexual differentiation, at least in color pattern. Together with the results of cockroaches and termites, *dsx* may play a major role in masculinization in ancestral insects.

2.7 Molecular Mechanisms Underlying Wax-Based Sexual Differentiation in Dragonflies

Many dragonfly and damselfly species exhibit remarkable sexual differences by secreting whitish wax on body surfaces of the abdomen, thorax, head, and/or wings as they mature (Fig. 2.5) (Futahashi 2020). In the wax-secreted regions, the surface fine structures reflect light at broad wavelengths, including ultraviolet (UV) light (Fig. 2.9a) (Futahashi et al. 2019; Futahashi 2020). For example, in the white-tailed skimmer dragonfly *Orthetrum albistylum*, the dorsal region of mature males is covered with scale- or plate-like structures, whereas the ventral region of mature females is covered with smaller fine structures (Fig. 2.9b) (Futahashi et al. 2019). This whitish color in wax-secreted regions disappeared when light scattering is disturbed by the application of acetone, suggesting that wax-based nanostructures are responsible for the whitish structural color with UV reflection (Futahashi et al. 2019).

Chemical analysis of the dragonfly wax has revealed that very long-chain aldehydes and very long-chain methyl ketones are the major components (Fig. 2.9c) (Futahashi et al. 2019), which totally differ from previously known organismal waxes. Notably, the chemical composition of the wax and the intensity of UV reflection differ between dragonfly species, sexes, and the dorsal/ventral sides of abdomen. In the ventral region of mature females of *S. darwinianum*, only tetracosanal, a very long-chain aldehyde, is detected, whereas, in mature males of *O. melania* and mature females of *O. albistylum*, several kinds of very long-chain aldehydes are identified (Fig. 2.9c). It should be noted that very long-chain methyl ketones are detected only in mature males of *O. albistylum* (Fig. 2.9c), which show strong UV reflection (Fig. 2.9a). The differences in wax composition may reflect their environmental and behavioral characteristics. Species that prefer sunny habitats tend to exhibit stronger UV reflections (e.g., mature males of *O. albistylum*), whereas females of *O. melania* that usually mate in shady places exhibit no UV reflection even on the ventral abdomen (Fig. 2.9a) (Futahashi et al. 2019). Since dragonflies that mate in sunny places expose the ventral abdomen of females to direct sunlight (Fig. 2.9a), the wax secreted on the ventral side of females may protect the abdomen, including the ovaries, against UV damage (Futahashi et al. 2019).

Notably, synthetic very long-chain methyl ketone, the main UV-reflective wax component of mature males of *O. albistylum*, spontaneously forms light-scattering fine structures with strong UV reflection and water repellency (Futahashi et al. 2019), suggesting that the secreted very long-chain methyl ketones should play a crucial role in UV reflection. In addition, the comparative transcriptome in different stages, sexes, and abdominal regions of *O. albistylum* has revealed that the *ELOVL17* gene, a member of the elongation of the very long-chain fatty acids (ELOVL) protein family, is specifically expressed in the region where very long-chain methyl ketones are present (Fig. 2.9d), suggesting that this gene is involved in the production of very long-chain methyl ketones (Futahashi et al. 2019). Expression levels of *ELOVL17* differed over 100-fold not only between sexes but also between heterochrome and androchrome females (Fig. 2.5) (Futahashi et al. 2019).

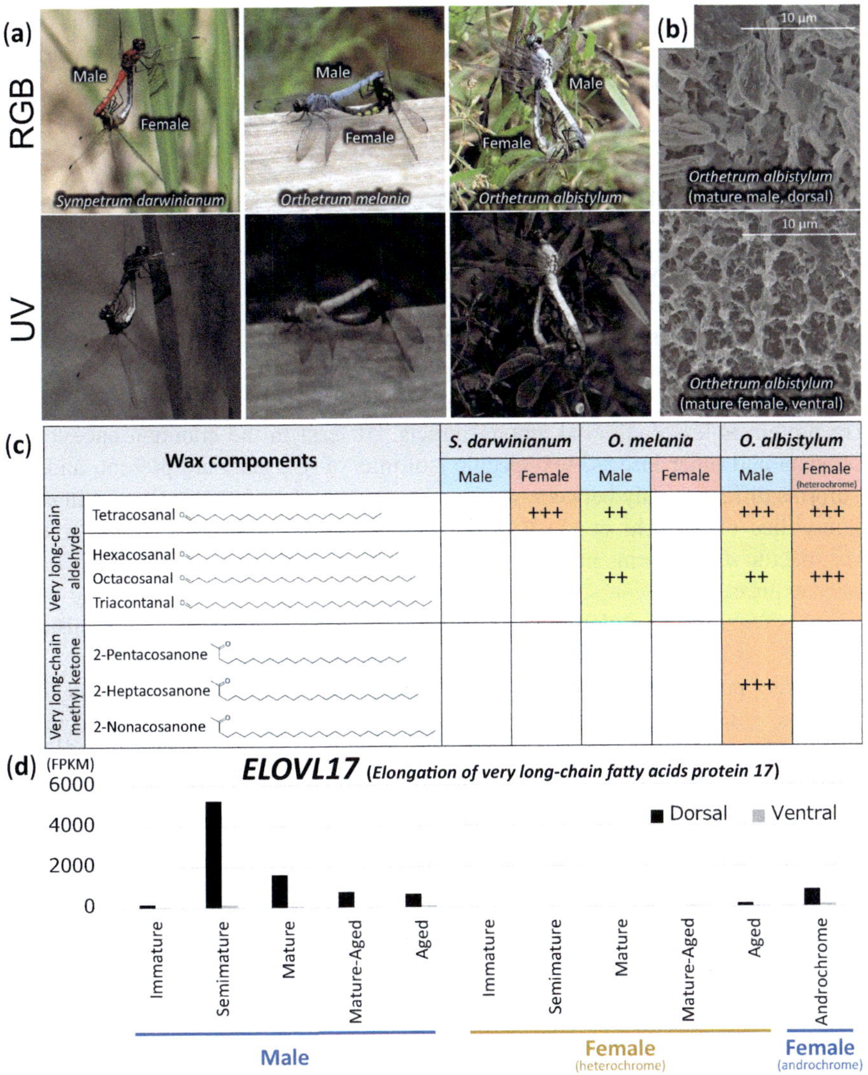

Fig. 2.9 UV reflection patterns, surface nanostructures, wax components, and expression pattern of wax-associated gene in dragonflies. (**a**) UV reflection pattern of three dragonfly species. Each image was photographed normally (above) or through a UV filter (below) in the field. Sex and species name are shown in each image. (**b**) Scanning electron microscope images of the abdominal surface of *Orthetrum albistylum*. (**c**) Summary of wax components. +++, high amount; ++, moderate amount. (**d**) Expression pattern of *elongation of very long-chain fatty acids protein 17* (*ELOVL17*) gene in abdominal epidermis of *O. albistylum*. (Figures modified from Futahashi et al. (2019))

Comparisons of gene expression between males and females as well as female color morphs are expected to identify genes contributing to sexual differences in various dragonflies and damselflies.

2.8 Conclusion and Perspective

Insects have a variety of sex determination system, and the upstream genes of the sex determination cascade are highly diverse. In general, splicing regulators play important roles in insect sexual differentiation (Sánchez 2008; Gempe and Beye 2010; Verhulst et al. 2010a; Hopkins and Kopp 2021; Laslo et al. 2022). In many insects, downstream of the splicing regulator gene *tra* is conserved, whereas the sequence of Tra protein is less conserved among insects. At least in the common ancestor of Odonata and other insects, sex-specific isoforms of *dsx* gene are present, and *dsx* isoforms differ not only between sex but also between female color morphs in some damselflies (Takahashi et al. 2019), indicating the conservation of the role of sex-specific *dsx* isoforms in sex determination. Since *tra* gene may not be differentially expressed between sexes in the ancestral hemimetabolous insects like cockroaches (Wexler et al. 2019; Hopkins and Kopp 2021), the molecular mechanism on how *dsx* isoforms are controlled in ancestral insects deserves future studies.

Female polymorphism in dragonflies is often genetically determined (e.g., *Ischnura* in the family Coenagrionidae), but in some species (e.g., *Aeshna* in the family Aeshnidae), blue (androchrome) and green (heterochrome) female polymorphism is observed continuously, and it has been reported that color polymorphism is regulated by the temperature immediately after adult emergence (Corbet 1999). In addition, the phenomenon of marked sexual dimorphism associated with adult maturation in several dragonflies and damselflies (e.g., *O. albistylum*) may involve certain hormonal factors, similar with sex hormones of vertebrates. Elucidation of the molecular mechanisms underlying the spectrum of sexual dimorphism in dragonflies and damselflies, the most ancestral winged insects, should provide understanding the origin of sexual differentiation in insects.

Acknowledgements The author would like to thank Mitsutoshi Sugimura for photograph of gynandromorphic dragonfly and damselfly, Genta Okude and Mizuko Osanai-Futahashi for helpful comments of the manuscript. The writing was partly supported by JSPS KAKENHI (JP18H02491 and JP20H04936).

References

Aryan A, Anderson MAE, Biedler JK, Qi Y, Overcash JM, Naumenko AN, Sharakhova MV, Mao C, Adelman ZN, Tu Z (2020) *Nix* alone is sufficient to convert female *Aedes aegypti* into fertile males and *myo-sex* is needed for male flight. Proc Natl Acad Sci U S A 117(30):17702–17709

Bachtrog D, Mank JE, Peichel CL, Kirkpatrick M, Otto SP, Ashman TL, Hahn MW, Kitano J, Mayrose I, Ming R, Perrin N, Ross L, Valenzuela N, Vamosi JC, Tree of Sex Consortium (2014) Sex determination: why so many ways of doing it? PLoS Biol 12(7):e1001899

Bear A, Monteiro A (2013) Both cell-autonomous mechanisms and hormones contribute to sexual development in vertebrates and insects. BioEssays 35(8):725–732

Bergerard J (1972) Environmental and physiological control of sex determination and differentiation. Annu Rev Entomol 17:57–74

Beye M, Hasselmann M, Fondrk MK, Page RE, Omholt SW (2003) The gene *csd* is the primary signal for sexual development in the honeybee and encodes an SR-type protein. Cell 114 (4):419–429

Blackmon H, Demuth JP (2015) Coleoptera karyotype database. Coleopts Bull 69:174–175

Blackmon H, Ross L, Bachtrog D (2017) Sex determination, sex chromosomes, and karyotype evolution in insects. J Hered 108(1):78–93

Blow R, Willink B, Svensson EI (2021) A molecular phylogeny of forktail damselflies (genus *Ischnura*) reveals a dynamic macroevolutionary history of female colour polymorphisms. Mol Phylogenet Evol 160:107134

Bybee S, Córdoba-Aguilar A, Duryea MC, Futahashi R, Hansson B, Lorenzo-Carballa MO, Schilder R, Stoks R, Suvorov A, Svensson EI, Swaegers J, Takahashi Y, Watts PC, Wellenreuther M (2016) Odonata (dragonflies and damselflies) as a bridge between ecology and evolutionary genomics. Front Zool 13:46

Bybee SM, Kalkman VJ, Erickson RJ, Frandsen PB, Breinholt JW, Suvorov A, Dijkstra KB, Cordero-Rivera A, Skevington JH, Abbott JC, Sanchez Herrera M, Lemmon AR, Moriarty Lemmon E, Ware JL (2021) Phylogeny and classification of Odonata using targeted genomics. Mol Phylogenet Evol 160:107115

Casper A, Van Doren M (2006) The control of sexual identity in the *Drosophila* germline. Development 133(15):2783–2791

Chauhan P, Swaegers J, Sánchez-Guillén RA, Svensson EI, Wellenreuther M, Hansson B (2021) Genome assembly, sex-biased gene expression and dosage compensation in the damselfly *Ischnura elegans*. Genomics 113(4):1828–1837

Chauhan P, Wellenreuther M, Hansson B (2016) Transcriptome profiling in the damselfly *Ischnura elegans* identifies genes with sex-biased expression. BMC Genomics 17(1):985

Corbet PS (1999) Dragonflies, behavior and ecology of odonata. Cornell University Press

Córdoba-Aguilar A, Cordero-Rivera A (2005) Evolution and ecology of Calopterygidae (Zygoptera: Odonata): status of knowledge and research perspectives. Neotro Entomol 34(6): 861–879

Das S (2016) Vertebrate hormones in insects: the role of estrogen in silkworm – a review. Turk J Zool 40:297–302

de la Filia AG, Bain SA, Ross L (2015) Haplodiploidy and the reproductive ecology of arthropods. Curr Opin Insect Sci 9:36–43

De Loof A (2006) Ecdysteroids: the overlooked sex steroids of insects? Males: the black box. Insect Sci 13:325–338

Denlinger DL, Brueggemeier RW, Mechoulam R, Katlic N, Yocum LB, Yocum GD (1987) Estrogens and androgens in insects. In: Law JH (ed) Molecular entomology. Alan R. Liss, New York, NY, USA, pp 189–199

Dübendorfer A, Hediger M, Burghardt G, Bopp D (2002) *Musca domestica*, a window on the evolution of sex-determining mechanisms in insects. Int J Dev Biol 46(1):75–79

Erickson JW, Quintero JJ (2007) Indirect effects of ploidy suggest X chromosome dose, not the X:A ratio, signals sex in *Drosophila*. PLoS Biol 5(12):e332

Fincke OM, Jödicke R, Paulson DR, Schultz TD (2005) The evolution and frequency of female color morphs in Holarctic Odonata: why are male-like females typically the minority? Int J Odonatol 8:183–212

Fukui T, Kiuchi T, Shoji K, Kawamoto M, Shimada T, Katsuma S (2018) *In vivo* masculinizing function of the *Ostrinia furnacalis Masculinizer* gene. Biochem Biophys Res Commun 503 (3):1768–1772

Futahashi R (2011) A revisional study of Japanese dragonflies based on DNA analysis. Tombo 53: 59–66

Futahashi R (2016) Color vision and color formation in dragonflies. Curr Opin Insect Sci 17:32–39

Futahashi R (2017) Molecular mechanisms underlying color vision and color formation in dragonflies. In: Sekimura T, Nijhout HF (eds) Diversity and evolution of butterfly wing patterns: an integrative approach. Springer, Singapore, pp 303–321

Futahashi R (2020) Diversity of UV reflection patterns in Odonata. Front Ecol Evol 8:201

Futahashi R, Kawahara-Miki R, Kinoshita M, Yoshitake K, Yajima S, Arikawa K, Fukatsu T (2015) Extraordinary diversity of visual opsin genes in dragonflies. Proc Natl Acad Sci U S A 112(11):E1247–E1256

Futahashi R, Kurita R, Mano H, Fukatsu T (2012) Redox alters yellow dragonflies into red. Proc Natl Acad Sci U S A 109(31):12626–12631

Futahashi R, Osanai-Futahashi M (2021) Pigments in insects. In: Hashimoto H, Goda M, Futahashi R, Kelsh R, Akiyama T (eds) Pigments, pigment cells, and pigment patterns. Springer, Singapore, pp 3–43

Futahashi R, Yamahama Y, Kawaguchi M, Mori N, Ishii D, Okude G, Hirai Y, Kawahara-Miki R, Yoshitake K, Yajima S, Hariyama T, Fukatsu T (2019) Molecular basis of wax-based color change and UV reflection in dragonflies. eLife 8:e43045

Gardner A, Ross L (2011) The evolution of hermaphroditism by an infectious male-derived cell lineage: an inclusive-fitness analysis. Am Nat 178:191–201

Gempe T, Beye M (2010) Function and evolution of sex determination mechanisms, genes and pathways in insects. BioEssays 33(1):52–60

Gempe T, Hasselmann M, Schiøtt M, Hause G, Otte M, Beye M (2009) Sex determination in honeybees: two separate mechanisms induce and maintain the female pathway. PLoS Biol 7(10):e1000222

Gotoh H, Miyakawa H, Ishikawa A, Ishikawa Y, Sugime Y, Emlen DJ, Lavine LC, Miura T (2014) Developmental link between sex and nutrition; *doublesex* regulates sex-specific mandible growth via juvenile hormone signaling in stag beetles. PLoS Genet 10(1):e1004098

Gotoh H, Zinna RA, Warren I, DeNieu M, Niimi T, Dworkin I, Emlen DJ, Miura T, Lavine LC (2016) Identification and functional analyses of sex determination genes in the sexually dimorphic stag beetle *Cyclommatus metallifer*. BMC Genomics 17:250

Hall AB, Basu S, Jiang X, Qi Y, Timoshevskiy VA, Biedler JK, Sharakhova MV, Elahi R, Anderson MA, Chen XG, Sharakhov IV, Adelman ZN, Tu Z (2015) A male-determining factor in the mosquito *Aedes aegypti*. Science 348(6240):1268–1270

Harvey-Samuel T, Norman VC, Carter R, Lovett E, Alphey L (2020) Identification and characterization of a *Masculinizer* homologue in the diamondback moth *Plutella xylostella*. Insect Mol Biol 29(2):231–240

Hashimoto H (1933) The role of the W-chromosome in the sex determination of *Bombyx mori*. Jpn J Genet 8:245–247

Hasselmann M, Gempe T, Schiøtt M, Nunes-Silva CG, Otte M, Beye M (2008) Evidence for the evolutionary nascence of a novel sex determination pathway in honeybees. Nature 454(7203): 519–522

Hediger M, Henggeler C, Meier N, Perez R, Saccone G, Bopp D (2010) Molecular characterization of the key switch *F* provides a basis for understanding the rapid divergence of the sex-determining pathway in the housefly. Genetics 184(1):155–170

Higashi K, Kayano H (1993) The distribution of distinct karyomorphs of *Crocothemis servilia* Drury (Anisoptera, Libellulidae) in Kyushu and the south-western islands of Japan. Japanese J Entomol 61:1–10

Higashi K, Lee CE, Kayano H, Kayano A (2001) Korea strait delimiting distribution of distinct karyomorphs of *Crocothemis servilia* (Drury) (Anisoptera: Libellulidae). Odonatologica 30 (3):265–270
Hodson CN, Hamilton PT, Dilworth D, Nelson CJ, Curtis CI, Perlman SJ (2017) Paternal genome elimination in *Liposcelis* booklice (Insecta: Psocodea). Genetics 206(2):1091–1100
Hopkins BR, Kopp A (2021) Evolution of sexual development and sexual dimorphism in insects. Curr Opin Genet Dev 69:129–139
Ito Y, Harigai A, Nakata M, Hosoya T, Araya K, Oba Y, Ito A, Ohde T, Yaginuma T, Niimi T (2013) The role of *doublesex* in the evolution of exaggerated horns in the Japanese rhinoceros beetle. EMBO Rep 14(6):561–567
Katatani N (1987) On the chromosomes of dragonflies, 1. Synopsis on the studies in some Japanese dragonflies. Aeschna 20:21–31
Kato Y, Kobayashi K, Oda S, Tatarazako N, Watanabe H, Iguchi T (2010) Sequence divergence and expression of a transformer gene in the branchiopod crustacean, *Daphnia magna*. Genomics 95(3):160–165
Kato Y, Kobayashi K, Watanabe H, Iguchi T (2011) Environmental sex determination in the branchiopod crustacean *Daphnia magna*: deep conservation of a *Doublesex* gene in the sex-determining pathway. PLoS Genet 7(3):e1001345
Keller Valsecchi CI, Marois E, Basilicata MF, Georgiev P, Akhtar A (2021) Distinct mechanisms mediate X chromosome dosage compensation in *Anopheles* and *Drosophila*. Life Sci Alliance 4 (9):e202000996
Kiauta B (1967) Considerations on the evolution of the chromosome complement in Odonata. Genetica 38(4):430–446
Kiauta B (1969a) Sex chromosomes and sex determining mechanisms in Odonata, with a review of the cytological conditions in the family Gomphidae, and reference to the karyotypic evolution in the order. Genetica 40(2):127–157
Kiauta B (1969b) Autosomal fragmentations and fusions in Odonata and their evolutionary implications. Genetica 40(2):158–180
Kiauta B (1972) Synopsis on the main cytotaxonomic data in the order Odonata. Odonatologica 1(2):73–102
Kiauta B (1983) The status of the Japanese *Crocothemis servilia* (Drury) as revealed by karyotypic morphology (Anisoptera: Libellulidae). Odonatologica 12:381–388
Kiauta B, Kiauta M (1982) List of species, with chromosome numbers and preliminary notes on the karyotypes of the Odonata, collected in May, 1979 and August, 1980 by the members of the Kansai Research Group of Odonatology, and examined by B. and M. Kiauta. Report for the Kansai Research Group of Odonatology, Osaka. Mimeographed. 8pp. Soc. Int. Odonatol., Utrecht
Kiuchi T, Koga H, Kawamoto M, Shoji K, Sakai H, Arai Y, Ishihara G, Kawaoka S, Sugano S, Shimada T, Suzuki Y, Suzuki MG, Katsuma S (2014) A single female-specific piRNA is the primary determiner of sex in the silkworm. Nature 509(7502):633–636
Krzywinska E, Dennison NJ, Lycett GJ, Krzywinski J (2016) A maleness gene in the malaria mosquito *Anopheles gambiae*. Science 353(6294):67–69
Krzywinska E, Ferretti L, Li J, Li JC, Chen CH, Krzywinski J (2021) *femaleless* controls sex determination and dosage compensation pathways in females of *Anopheles* mosquitoes. Curr Biol 31(5):1084–1091
Kunte K, Zhang W, Tenger-Trolander A, Palmer DH, Martin A, Reed RD, Mullen SP, Kronforst MR (2014) *doublesex* is a mimicry supergene. Nature 507(7491):229–232
Kuznetsova VG, Golub NV (2020) A checklist of chromosome numbers and a review of karyotype variation in Odonata of the world. Comp Cytogenet 14(4):501–540
Laslo M, Just J, Angelini DR (2022) Theme and variation in the evolution of insect sex determination. J Exp Zool B Mol Dev Evol. https://doi.org/10.1002/jez.b.23125

Lee J, Kiuchi T, Kawamoto M, Shimada T, Katsuma S (2015) Identification and functional analysis of a *Masculinizer* orthologue in *Trilocha varians* (Lepidoptera: Bombycidae). Insect Mol Biol 24(5):561–569

Martens A, Wildermuth H (2021) Gynandromorphism and intersexuality in Odonata: a review. Odonatologica 50(1/2):65–80

Meccariello A, Salvemini M, Primo P, Hall B, Koskinioti P, Dalíková M, Gravina A, Gucciardino MA, Forlenza F, Gregoriou ME, Ippolito D, Monti SM, Petrella V, Perrotta MM, Schmeing S, Ruggiero A, Scolari F, Giordano E, Tsoumani KT, Marec F, Windbichler N, Arunkumar KP, Bourtzis K, Mathiopoulos KD, Ragoussis J, Vitagliano L, Tu Z, Papathanos PA, Robinson MD, Saccone G (2019) Maleness-on-the-Y (MoY) orchestrates male sex determination in major agricultural fruit fly pests. Science 365(6460):1457–1460

Mechoulam R, Brueggemeier RW, Denlinger DL (1984) Estrogens in insects. Experientia 40:942–944

Misof B, Liu S, Meusemann K, Peters RS, Donath A, Mayer C, Frandsen PB, Ware J, Flouri T, Beutel RG, Niehuis O, Petersen M, Izquierdo-Carrasco F, Wappler T, Rust J, Aberer AJ, Aspöck U, Aspöck H, Bartel D, Blanke A, Berger S, Böhm A, Buckley TR, Calcott B, Chen J, Friedrich F, Fukui M, Fujita M, Greve C, Grobe P, Gu S, Huang Y, Jermiin LS, Kawahara AY, Krogmann L, Kubiak M, Lanfear R, Letsch H, Li Y, Li Z, Li J, Lu H, Machida R, Mashimo Y, Kapli P, McKenna DD, Meng G, Nakagaki Y, Navarrete-Heredia JL, Ott M, Ou Y, Pass G, Podsiadlowski L, Pohl H, von Reumont BM, Schütte K, Sekiya K, Shimizu S, Slipinski A, Stamatakis A, Song W, Su X, Szucsich NU, Tan M, Tan X, Tang M, Tang J, Timelthaler G, Tomizuka S, Trautwein M, Tong X, Uchifune T, Walzl MG, Wiegmann BM, Wilbrandt J, Wipfler B, Wong TK, Wu Q, Wu G, Xie Y, Yang S, Yang Q, Yeates DK, Yoshizawa K, Zhang Q, Zhang R, Zhang W, Zhang Y, Zhao J, Zhou C, Zhou L, Ziesmann T, Zou S, Li Y, Xu X, Zhang Y, Yang H, Wang J, Wang J, Kjer KM, Zhou X (2014) Phylogenomics resolves the timing and pattern of insect evolution. Science 346(6210):763–767

Miyazaki S, Fujiwara K, Kai K, Masuoka Y, Gotoh H, Niimi T, Hayashi Y, Shigenobu S, Maekawa K (2021) Evolutionary transition of *doublesex* regulation from sex-specific splicing to male-specific transcription in termites. Sci Rep 11(1):15992

Mola LM (2007) Cytogenetics of American Odonata. In: Tyagi BK (ed) Odonata: biology of dragonflies. Scientific Publishers, India, pp 153–173

Mola LM, Papeschi AG, Carrillo ET (1999) Cytogenetics of seven species of dragonflies. A novel sex chromosome determining system in *Micrathyria ungulate*. Hereditas 131:147–153

Morita S, Ando T, Maeno A, Mizutani T, Mase M, Shigenobu S, Niimi T (2019) Precise staging of beetle horn formation in *Trypoxylus dichotomus* reveals the pleiotropic roles of *doublesex* depending on the spatiotemporal developmental contexts. PLoS Genet 15(4):e1008063

Narita S, Pereira RAS, Kjellberg F, Kageyama D (2010) Gynandromorphs and intersexes: potential to understand the mechanism of sex determination in arthropods. Terr Arthropod Rev 3:63–96

Nishikawa H, Iijima T, Kajitani R, Yamaguchi J, Ando T, Suzuki Y, Sugano S, Fujiyama A, Kosugi S, Hirakawa H, Tabata S, Ozaki K, Morimoto H, Ihara K, Obara M, Hori H, Itoh T, Fujiwara H (2015) A genetic mechanism for female-limited Batesian mimicry in *Papilio* butterfly. Nat Genet 47(4):405–409

Nissen I, Müller M, Beye M (2012) The *Am-tra2* gene is an essential regulator of female splice regulation at two levels of the sex determination hierarchy of the honeybee. Genetics 192 (3):1015–1026

Nokkala S, Laukkanen A, Nokkala C (2002) Mitotic and meiotic chromosomes in *Somatochlora metallica* (Corduliidae, Odonata). The absence of localized centromeres and inverted meiosis. Hereditas 136:7–12

Normark BB (2003) The evolution of alternative genetic systems in insects. Annu Rev Entomol 48: 397–423

Normark BB (2014) Modes of reproduction. In: Shuker D, Simmons L (eds) The evolution of insect mating systems. University Press Oxford, Oxford (UK), pp 1–19

Ohbayashi F, Suzuki MG, Mita K, Okano K, Shimada T (2001) A homologue of the *Drosophila doublesex* gene is transcribed into sex-specific mRNA isoforms in the silkworm, *Bombyx mori*. Comp Biochem Physiol B Biochem Mol Biol 128(1):145–158

Okude G, Fukatsu T, Futahashi R (2020) Interspecific crossing between blue-tailed damselflies *Ischnura elegans* and *I. senegalensis* in the laboratory. Entomol Sci 23:165–172

Okude G, Fukatsu T, Futahashi R (2021) Electroporation-mediated RNA interference method in Odonata. J Vis Exp 168:e61952

Okude G, Futahashi R (2021) Pigmentation and color pattern diversity in Odonata. Curr Opin Genet Dev 69:14–20

Okude G, Moriyama M, Kawahara-Miki R, Yajima S, Fukatsu T, Futahashi R (2022) Molecular mechanisms underlying metamorphosis in the most-ancestral winged insect. Proc Natl Acad Sci U S A 119(9):e2114773119

Ozono A, Kawashima I, Futahashi R (2021) Dragonflies of Japan, Revised edition. Bunichi-Sogo Syuppan., Co. Ltd, Tokyo

Pane A, Salvemini M, Delli Bovi P, Polito C, Saccone G (2002) The *transformer* gene in *Ceratitis capitata* provides a genetic basis for selecting and remembering the sexual fate. Development 129(15):3715–3725

Papeschi AG, Bressa MJ (2006) Evolutionary cytogenetics in Heteroptera. J Biol Res 5:3–21

Poggio MG, Bressa MJ, Papeschi AG (2007) Karyotype evolution in Reduviidae (Insecta: Heteroptera) with special reference to Stenopodainae and Harpactorinae. Comp Cytogenet 1(2):159–168

Qi Y, Wu Y, Saunders R, Chen XG, Mao C, Biedler JK, Tu ZJ (2019) *Guy1*, a Y-linked embryonic signal, regulates dosage compensation in *Anopheles stephensi* by increasing X gene expression. elife 8:e43570

Riddiford LM (2012) How does juvenile hormone control insect metamorphosis and reproduction? Gen Comp Endocrinol 179(3):477–484

Ross L, Pen I, Shuker DM (2010) Genomic conflict in scale insects: the causes and consequences of bizarre genetic systems. Biol Rev 85:807–828

Ruiz MF, Esteban MR, Doñoro C, Goday C, Sánchez L (2000) Evolution of dosage compensation in Diptera: the gene *maleless* implements dosage compensation in Drosophila (Brachycera suborder) but its homolog in sciara (Nematocera suborder) appears to play no role in dosage compensation. Genetics 156(4):1853–1865

Sakai H, Oshima H, Yuri K, Gotoh H, Daimon T, Yaginuma T, Sahara K, Niimi T (2019) Dimorphic sperm formation by sex-lethal. Proc Natl Acad Sci U S A 116(21):10412–10417

Sakai H, Sakaguchi H, Aoki F, Suzuki MG (2015) Functional analysis of sex-determination genes by gene silencing with LNADNA gapmers in the silkworm, *Bombyx mori*. Mech Dev 137:45–52

Salvemini M, Polito C, Saccone G (2010) *fruitless* alternative splicing and sex behaviour in insects: an ancient and unforgettable love story? J Genet 89(3):287–289

Sánchez L (2008) Sex-determining mechanisms in insects. Int J Dev Biol 52(7):837–856

Sánchez-Guillén RA, Fadia-Ceccarelli S, Villalobos F, Neupane S, Rivas-Torres A, Sanmartín-Villar I, Wellenreuther M, Bybee SM, Velásquez-Vélez M, Realpe E, Chávez-Ríos JR, Dumont HJ, Cordero-Rivera A (2020) The evolutionary history of colour polymorphism in *Ischnura* damselflies (Odonata: Coenagrionidae). Odonatologica 49:333–370

Sharma A, Heinze SD, Wu Y, Kohlbrenner T, Morilla I, Brunner C, Wimmer EA, van de Zande L, Robinson MD, Beukeboom LW, Bopp D (2017) Male sex in houseflies is determined by *Mdmd*, a paralog of the generic splice factor gene *CWC22*. Science 356(6338):642–645

Shukla JN, Palli SR (2012a) Sex determination in beetles: production of all male progeny by parental RNAi knockdown of *transformer*. Sci Rep 2:602

Shukla JN, Palli SR (2012b) Doublesex target genes in the red flour beetle, *Tribolium castaneum*. Sci Rep 2:948

Shukla JN, Palli SR (2013) *Tribolium castaneum* Transformer-2 regulates sex determination and development in both males and females. Insect Biochem Mol Biol 43(12):1125–1132

Shukla JN, Palli SR (2014) Production of all female progeny: evidence for the presence of the male sex determination factor on the Y chromosome. J Exp Biol 217(Pt 10):1653–1655

Siegal ML, Baker BS (2005) Functional conservation and divergence of *intersex*, a gene required for female differentiation in *Drosophila melanogaster*. Dev Genes Evol 215(1):1–12

Suzuki KJ, Saitoh K (1990) A revised chromosome study of Japanese Odonates (II). Chromosomes of 13 species belonging to two families. Sci Rep Hirosaki Univ 37(2):111–123

Suzuki MG, Ohbayashi F, Mita K, Shimada T (2001) The mechanism of sex-specific splicing at the *doublesex* gene is different between *Drosophila melanogaster* and *Bombyx mori*. Insect Biochem Mol Biol 31(12):1201–1211

Suzuki MG, Funaguma S, Kanda T, Tamura T, Shimada T (2003) Analysis of the biological functions of a *doublesex* homologue in *Bombyx mori*. Dev Genes Evol 213(7):345–354

Suzuki MG, Funaguma S, Kanda T, Tamura T, Shimada T (2005) Role of the male BmDSX protein in the sexual differentiation of *Bombyx mori*. Evol Dev 7(1):58–68

Suzuki MG, Imanishi S, Dohmae N, Nishimura T, Shimada T, Matsumoto S (2008) Establishment of a novel in vivo sex-specific splicing assay system to identify a *trans*-acting factor that negatively regulates splicing of *Bombyx mori dsx* female exons. Mol Cell Biol 28(1):333–343

Suzuki MG, Imanishi S, Dohmae N, Asanuma M, Matsumoto S (2010) Identification of a male-specific RNA binding protein that regulates sex-specific splicing of *Bmdsx* by increasing RNA binding activity of BmPSI. Mol Cell Biol 30(24):5776–5786

Suzuki MG, Kobayashi S, Aoki F (2014) Male-specific splicing of the silkworm *Imp* gene is maintained by an autoregulatory mechanism. Mech Dev 131:47–56

Suzuki MG, Suzuki K, Aoki F, Ajimura M (2012) Effect of RNAi-mediated knockdown of the *Bombyx mori transformer-2* gene on the sex-specific splicing of *Bmdsx* pre-mRNA. Int J Dev Biol 56(9):693–699

Svensson EI (2017) Back to basics: using colour polymorphisms to study evolutionary processes. Mol Ecol 26:2204–2211

Svensson EI, Abbott JK, Gosden TP, Coreau A (2009) Female polymorphisms, sexual conflict and limits to speciation processes in animals. Evol Ecol 23:93–108

Takahashi M, Okude G, Futahashi R, Takahashi Y, Kawata M (2021) The effect of the *doublesex* gene in body colour masculinization of the damselfly *Ischnura senegalensis*. Biol Lett 17 (6):20200761

Takahashi M, Takahashi Y, Kawata M (2019) Candidate genes associated with color morphs of female-limited polymorphisms of the damselfly *Ischnura senegalensis*. Heredity 122(1):81–92

Tillyard RJ (1917) The biology of dragonflies. Cambridge University Press, Cambridge

Traut W, Marec F (1996) Sex chromatin in Lepidoptera. Q Rev Biol 71(2):239–256

Traut W, Sahara K, Marec F (2007) Sex chromosomes and sex determination in Lepidoptera. Sex Dev 1(6):332–346

Tree of Sex Consortium (2014) Tree of sex: a database of sexual systems. Sci Data 1:140015

Verhulst EC, van de Zande L, Beukeboom LW (2010a) Insect sex determination: it all evolves around *transformer*. Curr Opin Genet Dev 20(4):376–383

Verhulst EC, Beukeboom LW, van de Zande L (2010b) Maternal control of haplodiploid sex determination in the wasp *Nasonia*. Science 328(5978):620–623

Vershinina AO, Kuznetsova VG (2016) Parthenogenesis in Hexapoda: Entognatha and non-holometabolous insects. J Zool Syst Evol Res 54:257–268

Visser S, Voleníková A, Nguyen P, Verhulst EC, Marec F (2021) A conserved role of the duplicated *Masculinizer* gene in sex determination of the Mediterranean flour moth, *Ephestia kuehniella*. PLoS Genet 17(8):e1009420

Wang YH, Chen XE, Yang Y, Xu J, Fang GQ, Niu CY, Huang YP, Zhan S (2019) The *Masc* gene product controls masculinization in the black cutworm. *Agrotis ipsilon* Insect Sci 26(6):1037–1044

Wang YH, Engel MS, Rafael JA, Wu HY, Rédei D, Xie Q, Wang G, Liu XG, Bu WJ (2016) Fossil record of stem groups employed in evaluating the chronogram of insects (Arthropoda: Hexapoda). Sci Rep 6:38939

Watanabe T (2019) Evolution of the neural sex-determination system in insects: does *fruitless* homologue regulate neural sexual dimorphism in basal insects? Insect Mol Biol 28(6):807–827

Went DF, Camenzind R (1984) Sex determination in the dipteran insect *Heteropeza pygmaea*. Genetica 52(1):373–377

Wexler J, Delaney EK, Belles X, Schal C, Wada-Katsumata A, Amicucci MJ, Kopp A (2019) Hemimetabolous insects elucidate the origin of sexual development via alternative splicing. Elife 8:e47490

Willhoeft U, Franz G (1996) Identification of the sex-determining region of the *Ceratitis capitata* Y chromosome by deletion mapping. Genetics 144:737–745

Willink B, Duryea MC, Svensson EI (2019) Macroevolutionary origin and adaptive function of a polymorphic female signal involved in sexual conflict. Am Nat 194:707–724

Willink B, Duryea MC, Wheat C, Svensson EI (2020) Changes in gene expression during female reproductive development in a color polymorphic insect. Evolution 74(6):1063–1081

Xu J, Liu W, Yang D, Chen S, Chen K, Liu Z, Yang X, Meng J, Zhu G, Dong S, Zhang Y, Zhan S, Wang G, Huang Y (2020) Regulation of olfactory-based sex behaviors in the silkworm by genes in the sex-determination cascade. PLoS Genet 16(6):e1008622

Xu J, Yu Y, Chen K, Huang Y (2019) *Intersex* regulates female external genital and imaginal disc development in the silkworm. Insect Biochem Mol Biol 108:1–8

Zheng ZZ, Sun X, Zhang B, Pu J, Jiang ZY, Li M, Fan YJ, Xu YZ (2019) Alternative splicing regulation of *doublesex* gene by RNA-binding proteins in the silkworm *Bombyx mori*. RNA Biol 16(6):809–820

Zhuo JC, Hu QL, Zhang HH, Zhang MQ, Jo SB, Zhang CX (2018) Identification and functional analysis of the *doublesex* gene in the sexual development of a hemimetabolous insect, the brown planthopper. Insect Biochem Mol Biol 102:31–42

Zhuo JC, Zhang HH, Hu QL, Zhang JL, Lu JB, Li HJ, Xie YC, Wang WW, Zhang Y, Wang HQ, Huang HJ, Lu G, Chen JP, Li JM, Tu ZJ, Zhang CX (2021) A feminizing switch in a hemimetabolous insect. Sci Adv 7(48):eabf9237

Zou Y, Geuverink E, Beukeboom LW, Verhulst EC, van de Zande L (2020) A chimeric gene paternally instructs female sex determination in the haplodiploid wasp *Nasonia*. Science 370 (6520):1115–1118

Chapter 3
Dimorphic Female-Limited Batesian Mimicry in Two *Papilio* Butterflies

Haruhiko Fujiwara and Shinya Komata

Abstract To deceive predators, palatable species often resemble colour patterns and morphology of unpalatable species, which is called Batesian mimicry. In two closely related swallowtail butterflies, *Papilio polytes* and *P. memnon*, only females mimic model poisonous butterflies, and females have mimetic and non-mimetic types. Responsible loci (*H* for *P. polytes* and *A* for *P. memnon*) for mimicry were identified in the homologous chromosomal region as supergene containing *doublesex* (*dsx*) and a few genes. The supergene sequences between *H* and *h* (*A* and *a*) alleles are highly diversified due to recombination suppression, which is caused by chromosomal inversion in *P. polytes*, but no inversion in *P. memnon*. In both species, higher expression of mimetic *dsx* (*dsx-H* and *dsx-A*) in female pupal wings induced mimetic colouration, and the expression of genes inside the supergene was similar to that of mimetic *dsx*. In contrast, colouration patterns and supergene structure were considerably different between the two species. Furthermore, the chemical features of pale-yellow in hindwings, which are important both for sexual and mimicry strategies in *P. polytes*, seemed different in *P. memnon*. Here, we summarise the similarities and differences in mimetic traits between two butterflies and discuss how two mimicry supergenes have evolved.

Keywords Female-limited Batesian mimicry · Supergene · In vivo electroporation-mediated RNAi · Cost–benefit balance of mimicry · *Papilio* butterflies · *doublesex*

3.1 General Introduction

Various body colours and patterns can be observed in many animals, including insects. These patterns and colours play an important role in their life history, such as sexual strategies between males and females, as well as escaping predators as protective and warning colours (Ruxton et al. 2004). As many insects are the target

H. Fujiwara (✉) · S. Komata
Department of Integrated Biosciences, Graduate School of Frontier Sciences, The University of Tokyo, Kashiwa, Chiba, Japan
e-mail: haruh@edu.k.u-tokyo.ac.jp; komata@edu.k.u-tokyo.ac.jp

M. Tanaka, M. Tachibana (eds.), *Spectrum of Sex*,
https://doi.org/10.1007/978-981-19-5359-0_3

of predation by animals, they have developed the strategy "mimicry" to avoid predation by imitating the colours, shapes, patterns, or behaviours of other organisms (Ruxton et al. 2004; Vane-Wright 1980). Palatable species mimic distantly related poisonous species, which is known as Batesian mimicry and is seen in a wide range of organisms (Bates 1862). Mullerian mimicry, in which poisonous species resemble each other, is also well-known and well-studied, especially in bees and butterflies (Müller 1879). Darwin and Wallace, pioneers of evolutionary biology more than 150 years ago, were interested in the fact that in large butterflies in Papilionidae or Pieridae families, only females but not males show Batesian mimicry (Wallace 1865). Among these butterflies, males have high flight ability, whereas females with eggs have lower flight ability, and females have been considered to have a high merit of mimicry to reduce predation pressure (Ohsaki 2005; Kunte 2009a).

Clarke et al. investigated the distribution of various types of Batesian mimicry among the *Papilio* butterflies and found that although some of them do not mimic, some exhibit several types of Batesian mimicry: a mimetic pattern in both sexes, different patterns between males and females, and female-limited mimicry (Clarke and Sheppard 1960; Clarke and Sheppard 1972; Clarke et al. 1968). It is noteworthy, however, that male-limited mimicry was not found in the Darwin era. Among female-limited Batesian mimicry in *Papilio* butterflies, there are several types; only mimetic females exist in many species, both mimetic and non-mimetic females coexist in wild populations, such as the common mormon *Papilio polytes* and the great mormon *P. memnon* (Fig. 3.1). There are polymorphic females, each of which resembles different poisonous model species, such as *P. dardanus* (Clarke and Sheppard 1960). It is of interest that the variety of Batesian mimicry types in *Papilio* butterflies described above is not necessarily monophyletic from the common ancestors and often appears discontinuously during evolution (Kunte 2009b). Both female-limited Batesian mimicry and polymorphic Batesian mimicry have been fascinating topics in the field of ecology and evolution, but the molecular mechanisms underlying these phenomena still remain to be identified after a long time. In this review, we primarily focus on two dimorphic female-limited Batesian mimic butterflies: *P. polytes* and *P. memnon*, which are closely related species (Fig. 3.1). Here, we introduce the molecular mechanisms underlying mimicry, which have been and are currently being studied.

The genes that regulate butterfly mimicry have long been genetically studied (Clark et al. 2008; Joron et al. 2011). In *P. memnon* and *P. polytes*, both of which fascinated Darwin and others, genetic analyses showed that the mimetic phenotype is dominant and that the causative locus exists in a specific autosomal region, which is mentioned as *H* in *P. polytes* and *A* in *P. memnon* (in chromosome 25) in this study, and is transmitted to the next generation, according to Mendel's inheritance (Clarke and Sheppard 1972; Clarke et al. 1968). It is noteworthy that in females of the two species, *HH* (or *AA*) and *Hh* (*Aa*) are mimetic but *hh* (*aa*) is non-mimetic, but all males are non-mimetic in any genotype (Fig. 3.1).

The phenotypes that appear in Batesian mimicry include multiple traits, such as wing colour patterns, presence or absence of hindwing tails and flying behaviour

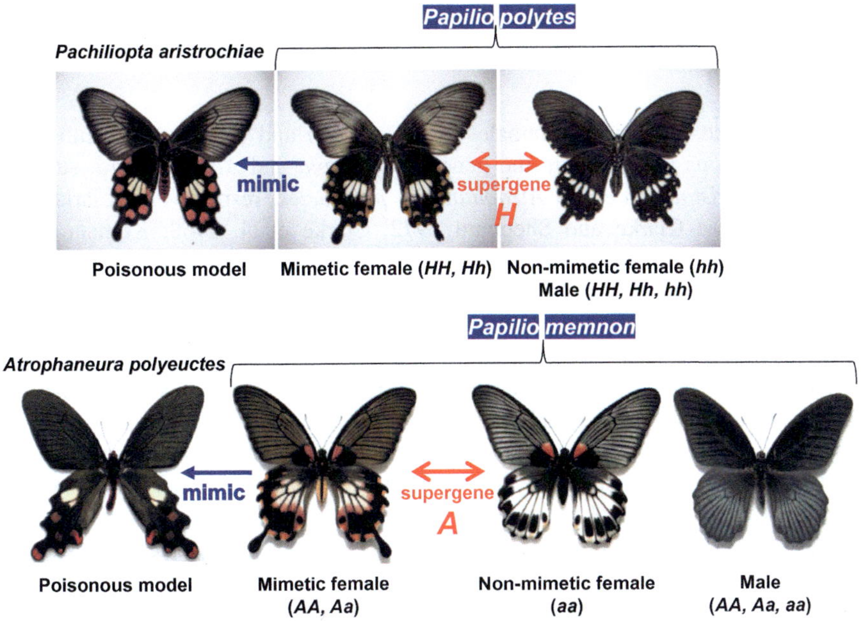

Fig. 3.1 Adult wing colour patterns of *Papilio polytes*, *P. memnon*, and their model butterflies. The top photographs show the model butterfly *Pachliopta aristolochiae*, mimetic female, and non-mimetic female/male of *P. polytes*. The bottom photographs show the model butterfly *Atrophaneura polyeuctes*, mimetic females, non-mimetic females, and males of *P. memnon*. Genotypes are shown in parentheses

(Clarke and Sheppard 1960; Clarke and Sheppard 1972; Clarke et al. 1968; Kitamura and Imafuku 2015). As these complex traits in *P. memnon* are controlled by a single locus, Fisher proposed the hypothesis that multiple adjacent genes as a cluster on the chromosome are integrated to control mimicry traits, and this gene cluster is named "supergene" (Fisher 1930; Charlesworth and Charlesworth 1975). The existence of supergenes has been reported in a wide range of species, connected with various adaptive complex traits, such as the pistil length of plants, self-incompatibility in plants, protective colours in fishes, social behaviour in fire ants, and sexual behaviour of birds (Joron et al. 2011; Wang et al. 2013; Kupper et al. 2016; Lamichhaney et al. 2016; Li et al. 2016; Tuttle et al. 2016; Pearse et al. 2019; Shore et al. 2019; Yan et al. 2020). As described later, recent advances in genome sequencing and analyses have contributed to the identification of the supergene structure on a chromosome locus, which is composed of multiple adjacent genes (Gutiérrez-Valencia et al. 2021).

3.2 Characteristic Traits of Mimetic/Non-mimetic *P. polytes* and *P. memnon* Females and Males

It is considered that mimetic females of *P. memnon* and *P. polytes* resemble the model poisonous butterflies in the genera *Pachliopta* and *Atrophaneura*, such as *Pachliopta aristolochiae* and *Atrophaneura polyeuctes*, respectively (Clarke and Sheppard 1960; Clarke and Sheppard 1972; Clarke et al. 1968; Komata et al. 2017). In both mimetic females, there are red spots on the edges and whitish (mentioned as pale-yellow thereafter) spots in the centre of the hindwings and white streaky stripes on the forewings (Fig. 3.1). In addition, hindwing tails are observed only in mimetic females in *P. memnon* and some subspecies of *P. polytes*, whereas the structure is observed even in non-mimetic females and males in *P. polytes* that inhabit the Okinawa islands in Japan. The abdomen of *Papilio* butterflies is usually blackish, but only mimetic females of *P. memnon* show an orange abdominal colour, which is similar to the warning colour in the model poisonous butterfly (Fig. 3.1).

In contrast to the mimetic type, the non-mimetic females and males of *P. polytes* show almost the same wing pattern and shape; loss of (or reduced) peripheral red spots; and centred pale-yellow stripe in the hindwings. It is noteworthy that the centred pale-yellow regions of *P. polytes* hindwings have different responses to ultraviolet (UV) rays between mimetic and non-mimetic females, with the former reflecting UV rays, whereas the latter absorb UV rays and emit blue fluorescence (Nishikawa et al. 2013; Yoda et al. 2021). When observed using this property, there are slight differences between non-mimetic females and males. In contrast to *P. polytes*, there are large differences in wing patterns between mimetic and non-mimetic females and non-mimetic females and males in *P. memnon* (Fig. 3.1) (Komata et al. 2016). In addition, the UV responses in *P. memnon* are not very different between mimetic and non-mimetic females and males. Thus, although closely related *P. polytes* and *P. memnon* show similar dimorphic female-limited Batesian mimicry, the phenotypic characteristics and their controls seem quite different between both species, suggesting that the differences may be correlated with their lifestyles.

3.3 Chemical and Physical Properties of Pigments in Wing Scales in Two *Papilio* Butterflies

Colours on butterfly wings are roughly divided into those based on pigments and structural colours (Stavenga et al. 2015). Previous studies have shown that some blue spots on the wings of *Papilio* butterflies are dependent on the structural colours, but most colours involved in wing patterns and body colours of *P. polytes* and P. *memnon* and model poisonous butterflies are considered to be dependent on pigments.

Black and brown, which make up most of the body colour in the above butterflies, are primarily due to melanin pigments. Conversely, in the *Papilio* butterflies, pigments derived from kynurenine and N-beta alanyl dopamine (NBAD) are often used for body colours, such as pale-yellow and red (Rembold and Umebachi 1985). We identified that the band-shaped pale-yellow in the hindwings of non-mimetic females and males in *P. polytes* is composed primarily of papiliochrome II (Nishikawa et al. 2013). This pigment is a combination of kynurenine and NBAD and absorbs UV light and emits blue fluorescence (Umebachi 1985). In contrast, the centred pale-yellow region of mimetic female hindwings is considered to consist of a melanin-related pigment, whereas UV reflectance originates from the basic scale structure, which contributes to the UV reflection rather than the pigment substance (Yoda et al. 2021). The centred pale-yellow region of the model butterfly *P. aristolochiae* shows chemical and physical properties similar to those of mimetic females of *P. polytes* (Yoda et al. 2021).

The peripheral red spots on mimetic females of *P. polytes* were reported to contain both kynurenine and NBAD, named papiliochrome R (Umebachi 1985), but it seems highly cross-linked and insoluble; thus, its detailed chemical structure has not been clarified yet. Conversely, it has been reported that the red spots in the model butterfly *P. aristolochiae* contain "A-type pigment", which contains NBAD but not kynurenine (Umebachi 1978), but its detailed structure is yet to be clarified.

The above results indicate that pale-yellow pigments in hindwings are synthesised differently between *P. polytes* mimetic and non-mimetic females, but similarly between mimetic females and the model *P. aristolochiae.* However, the peripheral red pigments in hindwings are synthesised differently between mimetic females and the model *P. aristolochiae.* These features imply that the Batesian mimicry and convergent evolution have been established. However, the chemical and physical properties of pigments in *P. memnon* have not yet been studied.

3.4 Identification of Responsible Genes and Supergene Structure for the *H* and *A* Locus

From classical genetic studies in *P. polytes* and *P. memnon*, it was reported that the locus responsible for *H* and *A* is located in a single region in an autosome (Clarke and Sheppard 1972; Clarke et al. 1968). Recently, more accurate genetic techniques, linkage analyses using DNA polymorphisms (such as AFLP), and RAD sequencing have revealed that the *H* locus in *P. polytes* is located near the *doublesex* (*dsx*) gene on chromosome 25 (Fig. 3.2) (Kunte et al. 2014; Nishikawa et al. 2015). Furthermore, whole genome sequencing of a heterozygous *Hh* female of *P. polytes* clarified that the sequences around *dsx* were highly diversified between *h* and *H* alleles, which are named highly diversified regions (HDRs), as well as the sex chromosomes (Z and W) (Nishikawa et al. 2015). The HDR around *dsx* extends to 130 kb, which is presumed to be a supergene. It was unexpected that the same region was identified

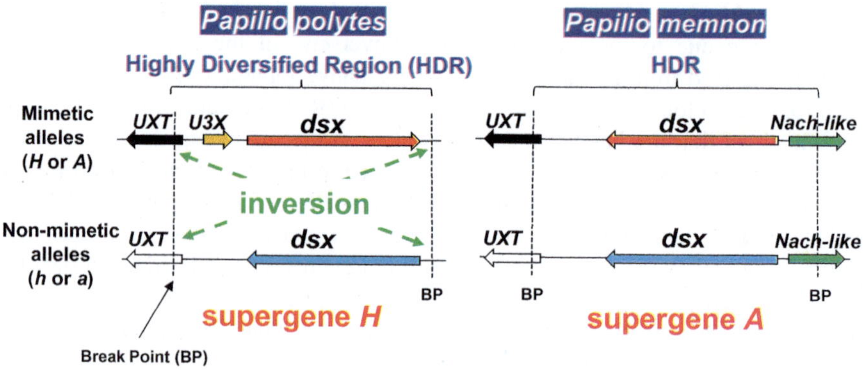

Fig. 3.2 Structure of the mimicry supergene in two *Papilio* species. The structure of the highly diversified region (HDR) on chromosome 25 was compared between *H* and *h* (left, *P. polytes*) and between *A* and *a* (right, *P. memnon*). The breakpoint (BP) between HDR and flanking conserved regions is shown as a vertical dotted line in the two species. *dsx, doublesex*; *UXT, ubiquitously expressed transcript*; *U3X, unknown 3 exons*; *Nach-like, natrium channel-like*

by two different approaches: DNA polymorphism search and whole genome sequencing. As the latter method is quicker and easier at present, whole genome sequencing of intraspecies polymorphic alleles is efficient for identifying supergene candidates. Using this idea, we sequenced the whole genome of *AA* and *aa* individuals of *P. memnon* and found three HDRs in different autosomes (Iijima et al. 2018). Among the three HDRs, we identified the *dsx*, including HDR, in chromosome 25 as the mimicry-associated one using a genome-wide association study.

Detailed analysis of whole genome sequences revealed approximately 130 kb HDR in *P. polytes* and approximately 150 kb HDR in *P. memnon* as a region corresponding to the mimicry supergene (Fig. 3.2). Both HDRs contain full-length *dsx*. It is intriguing that the direction of *dsx* is reversed between *H* and *h* alleles in *P. polytes*, whereas the direction is the same in *P. memnon* (Iijima et al. 2018). This clarifies that the former has a chromosomal inversion, but the latter has no inversion. In addition to *dsx*, both HDRs contain the 5′-UTR of *ubiquitously expressed transcript* (*UXT*) (Nishikawa et al. 2015; Iijima et al. 2018); the left breakpoint (or left junction site) of inversion (or HDR) is inside the *UXT* at almost the same position in both species. Only in *P. polytes*, *untranslated 3 exon* (*U3X*), which is a long non-coding RNA, was shown in *H* but not *in h* alleles, but no homologous sequence was found in other chromosomal regions or in other species' genomes, suggesting that this new sequence appeared during supergene evolution. Conversely, the HDR of *P. memnon* contained a *natrium channel-like* (*nach-like)* gene (Iijima et al. 2018), whose sequence is diversified between the *A* and *a* allele; thus, the right junction site of HDR is inside the *nach-like*, whereas the right breakpoint of inversion of *P. polytes* was found between *dsx* and *nach-like* (Fig. 3.2).

3.5 Evolutionary Process of the Mimicry Supergene in Two *Papilio* Butterflies

As the gene synteny and each gene orientation of the *h* allele in *P. polytes* are homologous to those of other Lepidoptera, the *H* allele has been newly generated from the *h* allele and evolved due to the suppression of chromosomal recombination by inversion (Fig. 3.3). It is presumed that suppression of recombination accelerates the accumulation of mutations and diversification of sequences of *H* and *h* alleles and functions to fix their sequence heterogeneity (Gutiérrez-Valencia et al. 2021). In fact, we observed that the sequence homology was considerably low within the HDR between the *h* and *H* alleles, but nearly 100% outside the HDR (Nishikawa et al. 2015; Iijima et al. 2018). In *P. memnon*, in which no chromosomal inversion is observed in the mimicry supergene, the linkage disequilibrium in the HDR is at the repressed level compared to that of the surrounding regions; thus, the recombination is also suppressed without inversion, which leads to diversification of sequences between the *A* and *a* allele (Fig. 3.3). Although the mechanism of recombination suppression in the *A/a* alleles has not been clarified yet, many repetitive sequences and various transposable elements are accumulated in the HDR, and recombination suppression is initiated by the insertion/deletion of some transposable elements (Iijima et al. 2018). However, the sequence diversification between *A* and *a* seems lower than that between *H* and *h*, and the ability to suppress the *A/a* alleles may be weaker by inversion in the *H/h* alleles.

Comparing the amino acid sequences of Dsx between *H* and *h* in *P. polytes*, we observed 14 amino acid substitutions; the amino acid changes occurred primarily in Dsx-H but rarely in Dsx-h based on the Dsx conserved sequence among Lepidoptera, suggesting that the function of Dsx-H has evolved to cause Batesian mimicry.

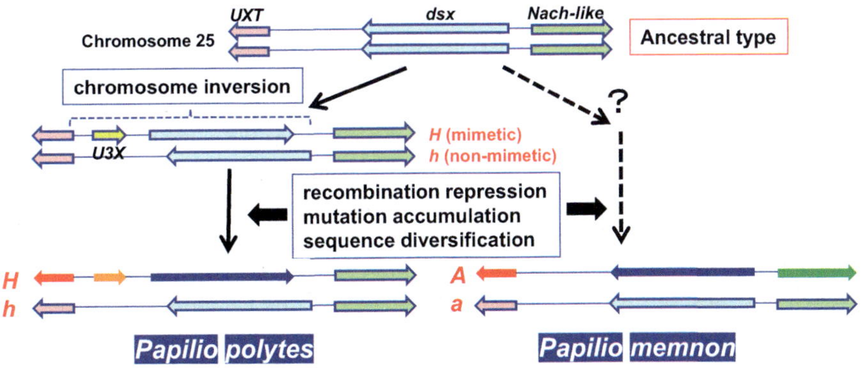

Fig. 3.3 Hypothetical evolutionary process of the mimicry supergene in two *Papilio* species. In *P. polytes*, chromosomal inversion occurred in the mimetic locus (H), and recombination between H and h alleles was repressed, which caused mutation accumulation and sequence diversification in both alleles. In *P. memnon*, some transposable elements or repetitive sequence integration in A/a alleles might have caused sequence diversification between them

Conversely, the Dsx amino acid changes between *A* and *a* in *P. memnon* are found only at four sites, which occur both in Dsx-A and Dsx-a at a similar rate to the conserved Lepidopteran sequence, suggesting that mutations have occurred randomly in both Dsxs (Iijima et al. 2018). Most of the mutations in the above Dsx sequences are present outside the DNA-binding and ligand regions, indicating that the basic function of Dsx is not lost. It is noteworthy that no mutation sites were conserved in the four Dsx sequences in either species and were found independently between Dsx-H and Dsx-A (Iijima et al. 2018). This indicates that there seems to be no specific amino acid involved in mimicry traits. Considering the difference in the mode of existence of chromosomal inversion between *H* and *A* alleles, independent changes in the Dsx amino acid sequences described above suggest that mimicry supergenes in the two butterflies have evolved independently. However, another group reported that Dsx-H and Dsx-A seemed to be different from the ancestral sequences based on the comparison of the Dsx sequences among closely related species of *P. memnon* (Palmer and Kronforst 2020). Therefore, further studies are necessary to verify which hypothesis is appropriate.

3.6 Expression Profiles of Genes Inside the Mimicry Supergene

The expression of *dsx* genes was similar in both Batesian mimic *Papilio* butterflies. In the hindwings of mimetic females (*Aa* or *Hh*), mimetic *dsx* (*dsx-A* or *dsx-H*) is strongly expressed during the early pupal stage which is an important time for the commitment of mimicry traits (Fig. 3.4a) (Nishikawa et al. 2015; Iijima et al. 2018; Komata et al. 2022a, b, c). In contrast, non-mimetic *dsx* (*dsx-a* or *dsx-h*) is expressed at a lower level in the early pupa but strongly in the late pupa. It is noteworthy that the abdomen of *P. memnon* shows a yellow colour that resembles the warning colour in the abdomen of the model butterfly, and *dsx-A* is expressed at a higher level in the abdomen of *P. memnon* (Komata et al. 2022b). These facts indicate that the expression of mimetic *dsx*-A and *dsx-H* is correlated with mimetic traits in both butterflies. Kunte et al. reported that a Dsx antibody stains the corresponding region of white stripes in the forewings of *P. polytes*, a pattern similar to the forewings of the model butterfly and is thought to be one of the mimetic traits, which supports the above idea (Kunte et al. 2014). The important fact is that the mimetic *dsx-H* and *dsx-A* are rarely expressed in males, even *H* and *A* alleles are included in the genome. It is not currently known why only females, but not males, with the same genotype *Hh* or *Aa* show mimetic traits. As the *dsx* gene produces female- and male-specific isoforms, which usually control sexual differentiation, the female-specific isoforms of mimetic Dsx may cause mimetic traits in females. However, the repression of the *dsx* gene in the male tissues involved in mimetic traits seems to be a major reason for female-limited mimicry.

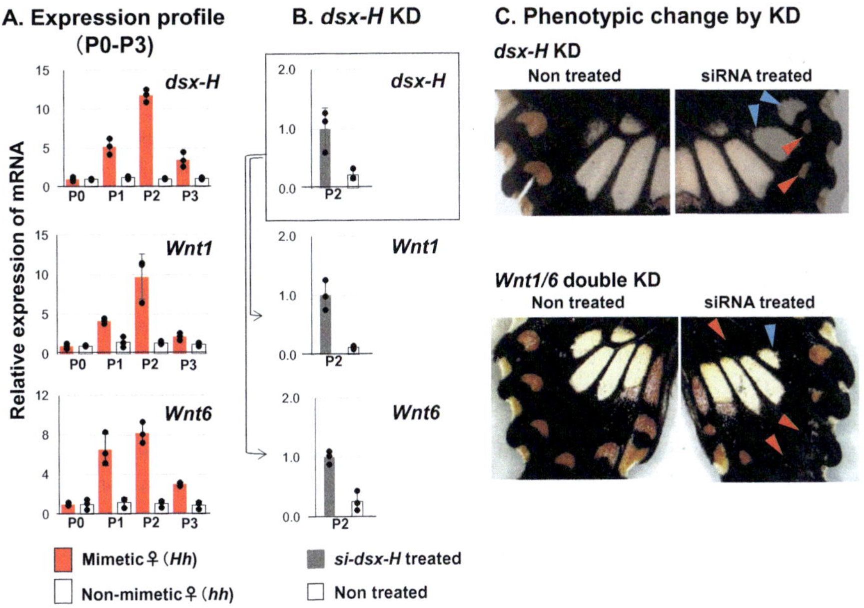

Fig. 3.4 Functional analyses of *dsx-H* and its downstream *Wnt1/Wnt6* on *P. polytes* mimetic female wings. (**a**) The expression levels of *dsx-H*, *Wnt1*, and *Wnt6* in hindwings from day 0 (P0) to day 3 (P3) after pupal ecdysis in mimetic (red) and non-mimetic (blank) females of *P. polytes*. (**b**) The expression levels of *Wnt1* and *Wnt6* after knockdown (KD) of *dsx-H* in hindwings of mimetic females by the electroporation-mediated functional analysis (EMFA) method: grey box, the wings of si-dsx-H-treated butterflies; blank, non-treated wings. (**c**) Phenotypic changes in mimetic female hindwings following gene knockdown. siRNA-treated (right) and non-treated wings (left) in the same individual were compared after *dsx-H* knockdown (top) and *Wnt1/Wnt6* double knockdown (bottom). When compared with non-treated wings, a pale-yellow band appeared (blue arrowhead) and red spots or centred pale-yellow diminished (red arrowhead) in siRNA-treated wings. (This figure was prepared and arranged based on the figures of Nishikawa et al. (2015) and Iijima et al. (2019))

Conversely, when examining internal genes within the HDR of the mimicry supergene, *UXT* and *U3X* (only in *P. polytes*) showed similar expression trends as mimetic *dsx-A* and *dsx-H* genes (Komata et al. 2022a). The *UXT* commonly included in both *Papilio* supergenes is highly expressed in the hindwings of mimetic females during early pupa. Similarly, *U3X* was highly expressed in mimetic females of *P. polytes*. The expression of *nach-like*, which is inside the *P. memnon* HDR but outside the *P. polytes* HDR, is very low in wings and most organs except for gonads. As the hindwings show the most visible phenotypic difference between mimetic and non-mimetic individuals, we analysed the expression of the above genes primarily in the wings or abdomen. However, mimetic traits involve not only body colour but also flight behaviour. Thus, the expression and function of internal genes of HDR should be clarified in tissues other than the wings or abdomen.

3.7 Functional Analysis of the *dsx* Gene and Downstream Genes Controlled under Mimetic Dsx

Recently, we developed an RNAi knockdown system, named electroporation-mediated functional analysis (EMFA), which is a rapid method for observing phenotypic changes, especially in body colour patterns (Ando and Fujiwara 2013). To determine the functional involvement of each gene inside the mimicry supergene, we first analysed the function of *dsx* (Yoda et al. 2021; Nishikawa et al. 2015; Iijima et al. 2019). After injection and electroporation of siRNA specific for *dsx-H* into a mimetic female hindwing of *P. polytes* immediately after pupal ecdysis, we found that the mimetic colour pattern changed to a non-mimetic one in the emerged butterfly (Fig. 3.4c). This result suggests that *dsx-H* not only induces the mimetic pattern but also represses the non-mimetic pattern and that both prepatterns are ready for use at the pupal ecdysis. We did not observe any effects of *dsx-H* knockdown in non-mimetic females. Recently, we found that knockdown of *dsx-h* in non-mimetic female hindwings causes the female colour pattern to resemble that of males. This suggests that the default status in hindwings is a male pattern, which is converted to a female pattern by the *dsx-h* function (Komata et al. 2022a). Similarly, *dsx-A* and *dsx-a* in *P. memnon* are involved in mimetic pattern and non-mimetic pattern formation in female wings from the male pattern as default (Komata et al. 2022b). In addition, the abdominal orange warning colour of *P. memnon* mimetic females also changed to black by RNAi knockdown of *dsx* (Komata et al. 2022b). From these observations, the mimetic-type *dsx* gene in the mimicry supergene *dsx-H* and *dsx-A* is considered to be the main molecule responsible for mimetic traits.

Since Dsx functions as a transcription factor, mimetic traits are considered to be caused by the gene network controlled by mimetic *dsx*. To understand this gene network, we knocked down *dsx-H* by EMFA in one hindwing just after pupal ecdysis (day 0) in the mimetic female (*Hh*) and prepared total RNAs from the wings from day 1 to 3 after injection (Iijima et al. 2019). After comparing the transcriptome by RNA-seq between siRNA injected and non-injected wings, we found that over 100 genes, including genes for transcription factors and signal transduction, were upregulated and downregulated. *Wnt1* and *Wnt6*, which form colour patterns in other organisms, are induced by *dsx-H* at early pupal stages and are involved in central white and peripheral red colouration in the mimetic female hindwings (Fig. 3.4a). The fact that the expression levels of *Wnt1* and *Wnt6* decreased in the mimetic hindwings after knockdown of *dsx-H* strongly suggests that *Wnt1/6* is under the control of *dsx-H* (Fig. 3.4b). Furthermore, when *Wnt1/6* are double-knocked down in mimetic female wings, the centred pale-yellow region changes its shape as a non-mimetic pattern but not elongated, as shown by the knockdown of *dsx-H* (Fig. 3.4c), suggesting that genes other than *Wnt1/6* are necessary for reproducing the non-mimetic wing pattern. In contrast, *Abd-A* is repressed by *dsx-H* at the early pupal stages, which is involved in the non-mimetic hindwing patterns. Although most of the candidate genes listed by the comparative transcriptome approach described above, some homeotic genes, such as *al (alistaless), cut,* and *Ems (empty spiracle),* were also identified as members of the gene network regulated by *dsx-H* (unpublished).

On the other hand the chemical and physical features of pale-yellow regions in hindwings are critically different between mimetic and non-mimetic *P. polytes*, as described above. The knockdown of *dsx-H* in mimetic females shows the transition of mimetic type pale-yellow to non-mimetic type pale-yellow (papiliochrome II pigment) (Yoda et al. 2021). The knockdown of some genes for enzymes involved in papiliochrome II synthesis, such as tyrosine hydroxylase, dopa decarboxylase, and laccase 2, results in the loss of non-mimetic pale-yellow colouration (Yoda et al. 2021). In addition, in the mimetic type of the pale-yellow region, which was treated with siRNA for *dsx-H*, gene expression of enzymes involved in papiliochrome II synthesis, TH, DDC, ebony, vermilion, and kynurenine formamidase was all reduced. These results indicate that *dsx-H* represses multiple genes involved in papiliochrome II synthesis in the mimetic pale-yellow regions and the release of this repression in non-mimetic females or males makes the non-mimetic pale-yellow region, whereas the gene (or genes) that directly regulates the repression under the *dsx-H* function has not yet been identified.

3.8 Cost–Benefit Balance in Batesian Mimicry of *P. polytes*: Sexual Strategy Versus Mimicry

In *P. polytes* and *P. memnon*, there are two types of mimetic and non-mimetic females; however, the molecular mechanisms underlying the formation and maintenance of this polymorphism are unclear. From an ecological point of view, polymorphisms are thought to be maintained by negative frequency-dependent selection via predation pressure (Kunte 2009a; Huheey 1988). In addition, the costs of mimetic females could also affect the maintenance of polymorphisms. There are two hypotheses regarding the costs of mimetic females: one is that males prefer non-mimetic females more than mimetic females in *P. polytes* (Ohsaki 2009; Westerman et al. 2019). Although some researchers argue against this observation by Dr. Osaki, if this is true, both females have benefits and costs. The mimetic females that escape predation are disadvantageous for mating, and the non-mimetic females that are advantageous for mating are easy to predate. The populations of the two types of females may be balanced in the wild; however, the balance may be shifted under different circumstances. The pale-yellow region of the hindwings is considered an important signal for mate choice in *P. polytes*. We have recently found that the reflection spectrum of this region differs significantly between mimetic and non-mimetic colours (Yoda et al. 2021). Based on the spectrum, we compared the relative excitation patterns of photoreceptors (UV, S, M, L) in humans, birds, and *Papilio* butterflies and simulated how to perceive the corresponding pale-yellow colours in males and mimetic and non-mimetic females of *P. polytes*, and in model butterflies. We found a possibility that although four colours look similar to human

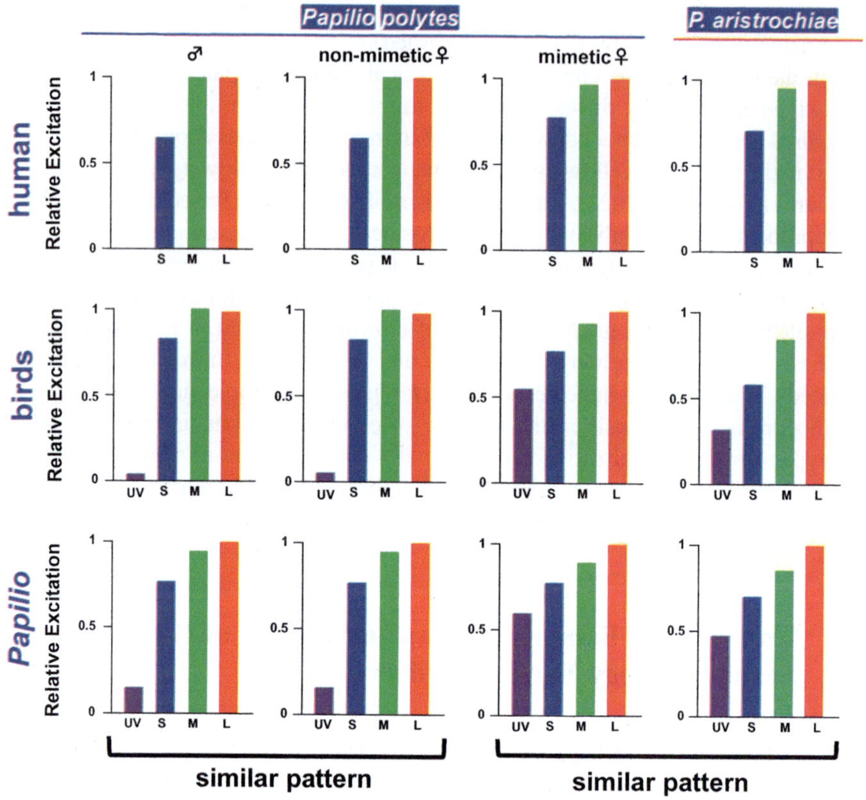

Fig. 3.5 Comparison of photoreceptor sensitivity in three animals for pale-yellow colours of hindwings. The relative excitation of the set of photoreceptors (S, M, L, or UV) for pale-yellow colouration in males, non-mimetic females, and mimetic females of *P. polytes* and model *P. aristolochiae* are shown. Based on the reflectance spectra of each pale-yellow colour, the relative excitation for each photoreceptor was calculated in humans, birds (blue tits), and *Papilio* butterflies (*P. xuthus*). (This figure is a modified version of Fig. 7 of Yoda et al. (2021))

eyes, they look significantly different to the eyes of birds and *Papilio* butterflies, which can see UV (Fig. 3.5). In birds and butterflies, the relative excitation patterns for four pale-yellow colours were similar between non-mimetic females and males and between mimetic females and model butterflies. This suggests that the former similarity may contribute to the sexual preference of males in *Papilio* butterflies and that the latter similarity may contribute to avoiding predation by birds. However, in *P. memnon*, we could not find a critical difference in the mating preference of males between mimetic and non-mimetic females (Komata et al. 2017).

Another hypothesis is that the formation and maintenance of mimetic traits come at some costs. Osaki suggested that mimetic females of *P. polytes* have a short lifespan, although detailed results have not yet been published (Ohsaki 2005). Recently, we compared the lifespan of male and female *P. polytes* bred in the laboratory and found that the lifespan of non-mimetic females (*hh*) (approximately

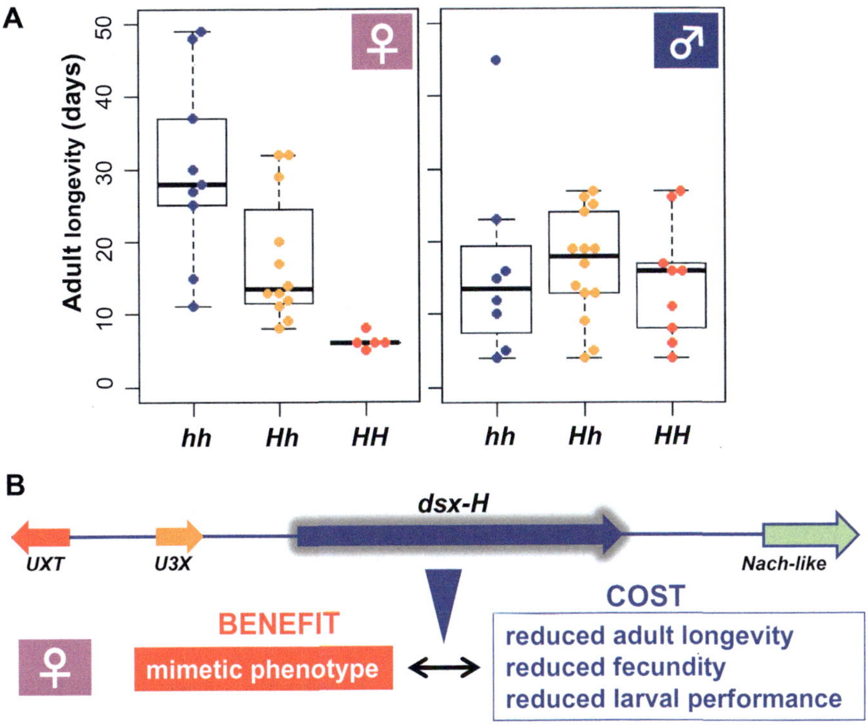

Fig. 3.6 Comparison of adult longevity in each *dsx* genotype of *P. polytes* (**a**) and cost–benefit balance model (**b**). (**a**) The female-specific effect of *dsx-H* on adult longevity. (**b**) In mimetic females of *P. polytes*, mimetic *dsx* (*dsx-H*) yields a benefit of reducing predation pressure by Batesian mimicry, accompanied by physiological costs, such as reduced adult longevity. ((**a**) is a modified version of Fig. 3 of Komata et al. (2020))

3 weeks) was longer than that of mimetic females *Hh* (approximately 2 weeks) and *HH* (approximately 1 week) (Komata et al. 2020). However, the lifespan of male butterflies was approximately the same (approximately 2 weeks) among different genotypes (*hh, Hh,* and *HH*) (Fig. 3.6a). It is noteworthy that *dsx-H* is rarely expressed in wings and other organs of males (data not shown). In addition, larval performance and fecundity are reduced in individuals with the *H* allele, suggesting that the expression of *dsx-H* affects the mimetic females physiologically. These results indicate that *dsx-H* expressed in mimetic females brings a benefit to cause mimetic traits but at some physiological costs simultaneously (Fig. 3.6b). We do not yet know what type of mechanisms underlie the costs or how *dsx-H* affects physiology, and further studies are necessary to answer these questions. However, we did not observe any physiological costs in mimetic females (or other morphs) in *P. memnon* (Komata et al. 2018). Both the cost–benefit balance between mimicry and physiological cost and mating preference of males are suggested in *P. polytes* but not in *P. memnon*. Thus, it is of interest that non-mimetic females of *P. memnon* inhabit the Honshu islands of Japan, but we could not find mimetic females. This

suggests some physiological costs, even in *P. memnon*. Further physiological and ecological studies will clarify the environmental or lifestyle factors that affect the differences in cost–benefit balancing between the two species.

References

Ando T, Fujiwara H (2013) Electroporation-mediated somatic transgenesis for rapid functional analysis in insects. Development 140:454–458

Bates HW (1862) Contributions to an insect fauna of the Amazon Valley (lepidoptera: Heliconidae). Trans Linn Soc Lond 23:495–556

Charlesworth D, Charlesworth B (1975) Theoretical genetics of Batesian mimicry II. Evolution of supergenes. J Theoretic Biol 55:305–324

Clark R, Brown SM, Collins SC, Jiggins CD, Heckel DG, Vogler AP (2008) Colour pattern specification in the mocker swallowtail Papilio dardanus: the transcription factor invected is a candidate for the mimicry locus H. Proc R Soc B 275:1181–1188

Clarke CA, Sheppard PM (1960) The genetics of Papilio dardanus Brown. II. Races dardanus, polytrophus, meseres, and tibullus. Genetics 45:439–456

Clarke CA, Sheppard PM (1972) The genetics of the mimetic butterfly *Papilio polytes* L. Philos Trans R Soc Lond Ser B Biol Sci 26:431–458

Clarke CA, Sheppard PM, Thornton IWB (1968) The genetics of the mimetic butterfly *Papilio memnon* L. Philos Trans R Soc Lond Ser B Biol Sci 254:37–89

Fisher RA (1930) The genetical theory of natural selection. Clarendon Press, Oxford

Gutiérrez-Valencia J, Hughes PW, Berdan EL, Slotte T (2021) The genomic architecture and evolutionary fates of supergenes. Genom Biol Evol 13:evab057

Huheey JE (1988) Mathematical models of mimicry. Am Nat 131:S22–S41

Iijima T, Kajitani R, Komata S, Lin CP, Sota T, Itoh T, Fujiwara H (2018) Parallel evolution of Batesian mimicry supergene in two *Papilio* butterflies, *P. polytes* and *P. memnon*. Sci Adv 4: eaao5416

Iijima T, Yoda S, Fujiwara H (2019) The mimetic wing pattern of *Papilio polytes* butterflies is regulated by a *doublesex*orchestrated gene network. Commun Biol 2:1–10

Joron M et al (2011) Chromosomal rearrangements maintain a polymorphic supergene controlling butterfly mimicry. Nature 477:203–206

Kitamura T, Imafuku M (2015) Behavioural mimicry in flight path of Batesian intraspecific polymorphic butterfly *Papilio polytes*. Proc Biol Sci 282:20150483

Komata S, Kitamura T, Fujiwara H (2020) Batesian mimicry has evolved with deleterious effects of the pleiotropic gene *doublesex*. Sci Rep 10:1–8

Komata S, Lin CP, Iijima T, Fujiwara H, Sota T (2016) Identification of *doublesex* alleles associated with the female-limited Batesian mimicry polymorphism in *Papilio memnon*. Sci Rep 6:1–7

Komata S, Lin CP, Sota T (2017) Temporal dynamics of the mimetic allele frequency at the doublesex locus, which controls polymorphic Batesian mimicry in *Papilio memnon* butterflies. Sci Rep 7:1–6

Komata S, Lin CP, Sota T (2018) Do juvenile developmental and adult body characteristics differ among genotypes at the *doublesex* locus that controls female-limited Batesian mimicry polymorphism in *Papilio memnon*?: a test for the "cost of mimicry" hypothesis. J Insect Physiol 107:1–6

Komata S, Yoda S, KonDo Y, Shinozaki S, Tamai K, Fujiwara H (2022a) Functional involvement of multiple genes as members of the supergene unit in the female-limited Batesian mimicry of *Papilio polytes*. bioRxiv. https://doi.org/10.1101/2022.02.21.480812

Komata S, Lin CP, Fujiwara H (2022b) *doublesex* controls both hindwing and abdominal mimicry traits in the female-limited batesian mimicry of *Papilio memnon*. Front Insect Sci 2:2929518

Komata S, Kajitani R, Itoh T, Fujiwara H (2022c) Genomic architecture and functional unit of mimicry supergene in female limited Batesian mimic *Papilio* butterflies. Phil Trans Royal Soc B 377:1856

Kunte K (2009a) Female-limited mimetic polymorphism: a review of theories and a critique of sexual selection as balancing selection. Anim Behav 78:1029–1036

Kunte K (2009b) The diversity and evolution of Batesian mimicry in *Papilio* swallowtail butterflies. Evolution 63:2707–2716

Kunte K, Zhang W, Tenger-Trolander A, Palmer DH, Martin A, Reed RD, Mullen SP, Kronforst MR (2014) *doublesex* is a mimicry supergene. Nature 507:229–232

Kupper C et al (2016) A supergene determines highly divergent male reproductive morphs in the ruff. Nat Genet 48:79–83

Lamichhaney S et al (2016) Structural genomic changes underlie alternative reproductive strategies in the ruff (Philomachus pugnax). Nat Genet 48:84–88

Li J et al (2016) Genetic architecture and evolution of the S locus supergene in *Primula vulgaris*. Nat Plants 2:16188

Müller F (1879) Ituna and Thyridia: a remarkable case of mimicry in butterflies. Trans Entomol Soc Lond 1879:20–29

Nishikawa H, Iga M, Yamaguchi J, Saito K, Kataoka H, Suzuki Y, Sugano S, Fujiwara H (2013) Molecular basis of wing coloration in a Batesian mimic butterfly, *Papilio polytes*. Sci Rep 3:1–9

Nishikawa H et al (2015) A genetic mechanism for female-limited Batesian mimicry in *Papilio* butterfly. Nat Genet 47:405–409

Ohsaki N (2005) A common mechanism explaining the evolution of female-limited and both-sex Batesian mimicry in butterflies. J Anim Ecol 74:728–734

Ohsaki N (2009) Gitai no Shinka: Darwin mo Gokai shita 150 Nen no Nazo wo Toku (the evolution of mimicry: solve the 150-year-old mystery that has puzzled biologists ever since Charles Darwin) Kaiyusha, Tokyo

Palmer DH, Kronforst MR (2020) A shared genetic basis of mimicry across swallowtail butterflies points to ancestral co-option of *doublesex*. Nat Commun 11:6

Pearse DE et al (2019) Sex-dependent dominance maintains migration supergene in rainbow trout. Nat Ecol Evol 3:1731–1742

Rembold H, Umebachi Y (1985) In: Schlossberger HG, Kochen W, Linzen B, Steinhart H (eds) The structure of papiliochrome II, the yellow wing pigment of the papilionid butterflies. In Progress in tryptophan and serotonin research. Walter de Gruyter, Berlin, pp 743–746

Ruxton GD, Sherratt TN, Speed M (2004) Avoiding attack: the evolutionary ecology of crypsis, Warning Signals and Mimicry. Oxford University Press, Oxford

Shore JS et al (2019) The long and short of the S-locus in Turnera (Passifloraceae). New Phytol 224: 1316–1329

Stavenga DG, Matsushita A, Arikawa K (2015) Combined pigmentary and structural effects tune wing scale coloration to color vision in the swallowtail butterfly *Papilio xuthus*. Zool Lett 1:1–10

Tuttle EM et al (2016) Divergence and functional degradation of a sex chromosome-like supergene. Curr Biol 26:344–350

Umebachi Y (1978) Red pigments in the wings of papilionid butterflies. Extraction and purification. Sci Rep Kanazawa Univ 23:119–128

Umebachi Y (1985) Papiliochrome, a new pigment group of butterfly. Zool Sci 2:163–174

Vane-Wright RI (1980) On the definition of mimicry. Biol J Linne Soc 13:1–6

Wallace AR (1865) On the phenomena of variation and geographical distribution as illustrated by the Papilionidae of the Malayan region. Trans Linn Soc (Lond) 25:1–71

Wang J et al (2013) A Y-like social chromosome causes alternative colony organization in fire ants. Nature 493:664–668

Westerman EL, Antonson N, Kreutzmann S, Peterson A, Pineda S, Kronforst MR, Olson-Manning CF (2019) Behaviour before beauty: signal weighting during mate selection in the butterfly *Papilio polytes*. Ethology 125:565–574

Yan Z et al (2020) Evolution of a supergene that regulates a trans-species social polymorphism. Nat Ecol Evol 4:240–249

Yoda S et al (2021) Genetic switch in UV response of mimicry-related pale-yellow colors in Batesian mimic butterfly, *Papilio polytes*. Sci Adv 7:eabd6475

Chapter 4
Spectrum of Sex Chromosomes in Mammals

Asato Kuroiwa

Abstract Placental mammals have two types of sex chromosomes, a relatively large X chromosome and a small Y chromosome. Sex is determined by a gene on the Y chromosome, *SRY*, and therefore, XY is male and XX is female. Although this XX/XY system is regarded as highly conserved in placental mammals, atypical sex chromosomes and sex determination mechanisms have been reported. Since species with such atypical chromosomes are non-model organisms, they are often not well-characterized. However, the application of advanced genome sequencing technologies has clearly revealed that diverse sex chromosomes have evolved in placental mammals. This chapter introduces the spectrum of sex chromosomes in placental mammals.

Keywords X chromosome · Y chromosome · Atypical sex chromosome · Evolution

4.1 Large X, Small Y

Sex chromosomes diverge rapidly in many lineages. Transitions in sex chromosomes are referred to as sex chromosome turnover. However, some groups have stable and conserved sex chromosomes. The most prominent example is placental mammals (referred to below as mammals), which have a highly conserved male heterogametic sex chromosome system, with XY males and XX females. There is no evidence for sex chromosome turnover in the long evolutionary history of the group, and the sex-determining mechanism triggered by the sex-determining region Y (*SRY*) gene on the Y chromosome is highly conserved (see Chap. 5). Among vertebrate groups, genomic differences between sexes are relatively large in mammals due to substantial differentiation between the X and Y chromosomes.

Mammalian X and Y chromosomes coevolved from an ordinary pair of autosomes in the common ancestor, first appearing about 300 million years ago

A. Kuroiwa (✉)
Division of Reproductive and Developmental Biology, Department of Biological Sciences, Faculty of Science, Hokkaido University, Sapporo, Hokkaido, Japan
e-mail: asatok@sci.hokudai.ac.jp

M. Tanaka, M. Tachibana (eds.), *Spectrum of Sex*,
https://doi.org/10.1007/978-981-19-5359-0_4

(Charlesworth 1991; Lahn and Page 1999; Ross et al. 2005). During evolution, X–Y recombination was suppressed and the portions of the X and Y chromosomes that do not recombine during male meiosis increased. The reduced efficacy of purifying selection in non-recombining regions led to the accumulation of deleterious mutations in Y-linked genes (Charlesworth and Charlesworth 2000). The subsequent inactivation and loss of Y-linked genes led to the genetic degeneration of the Y chromosome. The mammalian X chromosome is relatively large, making up about 5% of the haploid genome, and the size is highly conserved, with some exceptions (Ohno 1967). By contrast, the mammalian Y chromosome is extremely small and contains very few genes. Fewer than 5% of genes survived the massive mammalian Y chromosome degeneration (Bellott et al. 2014; Skinner et al. 2016).

The remaining Y chromosome genes acquired male-specific functions, including roles in sex determination (*SRY*) and spermatogenesis, and are consequently essential for maleness in mammals. It has been difficult to assemble the genome sequences of the Y chromosome owing to the excessive accumulation of repetitive sequences and transposable elements. However, advances in genome sequencing technology have enabled sequencing and comparative analyses of the Y chromosomes of many mammalian taxa. Recent genomic comparisons suggest that each lineage retained 9–16 ancient Y chromosome genes, and around 10 genes have been conserved in rodents. Five genes, *Sry*, *Ddx3y*, *Usp9y*, *Uty*, and *Zfy*, are common to all mammalian species with sequenced Y chromosomes (Table 4.1) (Bellott et al. 2014; Skinner et al. 2016; Cortez et al. 2014).

4.2 XO Males Without *SRY*

The Y chromosome is generally essential for males in mammals. In humans, XO individuals (45, XO, also known as Turner syndrome) show a female phenotype (although it is characterized by immature reproductive organs and functions). However, in some species, male sex is determined without a Y chromosome. Two species in the genus *Tokudaia* (Murinae, Rodentia) and one species in the genus *Ellobius* (Arvicolinae, Rodentia) show an XO/XO system and lack the *Sry* gene. Losses of the Y chromosome and *Sry* occurred independently in these species, as the genera *Ellobius* and *Tokudaia* belong to different subfamilies, Arvicolinae and Murinae, respectively.

4.3 XO/XO System in Spiny Rats, Genus *Tokudaia*

The spiny rats, which belong to the genus *Tokudaia*, are native to Japan. The genus includes three species, each indigenous to only a single island in southernmost Japan. The Okinawa spiny rat (*Tokudaia muenninki*), which lives on Okinawa-jima island, has an XX/XY sex chromosome system and a diploid chromosome

Table 4.1 Main genes on the X and Y chromosomes (Kuroiwa et al. 2010; Mulugeta et al. 2016; Murata et al. 2016)

M. musculus		*R. norvegicus*		*T. osimensis*		*T. muenninki*		*E. lutescens*		*E. talpinus*	
XX/XY		XX/XY		XO/XO		XX/XY		XO/XO		XX/XX	
X	Y	X	Y	X	Y-derived	X	Y	X	Y-derived	X	Y-derived
Zfx	*Zfy* + 2	*Zfx*	*Zfy* + 2	*Zfx*	*Zfy*	*Zfx*	*Zfy* (mc)	*Zfx*	*Zfy*	*Zfx*	*Zfy*
Uba1	*Uba1y*	*Uba1*	*Uba1y*	*Uba1*	–	*Uba1*	*Uba1y-ps* (mc)	*Uba1*	–	*Uba1*	–
Kdm5c	*Kdm5d*	*Kdm5c*	*Kdm5d*	*Kdm5c*	*Kdm5d*	*Kdm5c*	–	*Kdm5c*	–	*Kdm5c*	–
Eif2s3x	*Eif2s3y*	*Eif2s3x*	*Eif2s3y*	*Eif2s3x*	*Eif2s3y*	*Eif2s3x*	*Eif2s3y* (mc)	*Eif2s3x*	*Eif2s3y*	*Eif2s3x*	*Eif2s3y*
Kdm6a	*Uty*	*Kdm6a*	*Uty*	*Kdm6a*	*Uty*	*Kdm6a*	*Uty*	*Kdm6a*	–	*Kdm6a*	–
Tspyl2	*Tspy-ps*	*Tspyl2*	*Tspy-ps*	*Tspyl2*	*Tspy*	*Tspyl2*	*Tspy* (mc)	*Tspyl2*	–	*Tspyl2*	–
Ddx3x	*Ddx3y*	*Ddx3x*	*Ddx3y*	*Ddx3x*	*Ddx3y*	*Ddx3x*	*Ddx3y*	*Ddx3x*	–	*Ddx3x*	–
Usp9x	*Usp9y*	*Usp9x*	*Usp9y*	*Usp9x*	*Usp9y-ps*	*Usp9x*	*Usp9y* (mc)	*Usp9x*	*Usp9y*	*Usp9x*	–
Sox3	*Sry*	*Sox3*	*Sry*	*Sox3*	–	*Sox3*	*Sry* (mc)	*Sox3*	–	*Sox3*	–
Rbmx	*Rbmy* (mc)	*Rbmx*	*Rbmy* (mc)	*Rbmx*	*Rbmy*-ps	*Rbmx*	*Rbmy* (mc)	*Rbmx*	–	*Rbmx*	–

mc Multi-copy gene, ps: Pseudogene

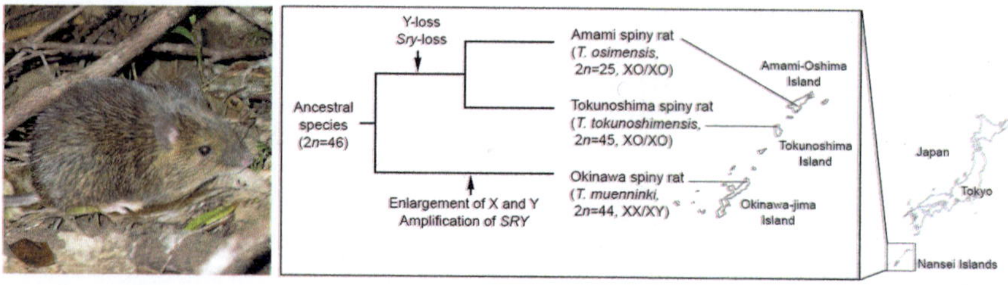

Fig. 4.1 Evolutionary events in the genus *Tokudaia*. (**a**) Photograph of *T. osimensis*. (**b**) Phylogenetic relationships and geographic distribution of species in the genus *Tokudaia*

number of $2n = 44$ (Tsuchiya et al. 1989; Murata et al. 2010). The Amami spiny rat (*Tokudaia osimensis*) and Tokunoshima spiny rat (*Tokudaia tokunoshimensis*), which live on Amami-Oshima island and Tokunoshima island, respectively, have XO/XO sex chromosome systems (i.e., they lack a Y chromosome). Therefore, *T. osimensis* and *T. tokunoshimensis* have odd-numbered diploid chromosome numbers, $2n = 25$ and $2n = 45$, respectively (Honda et al. 1977; Honda et al. 1978; Kobayashi et al. 2007). In addition, the genomes of these XO species lack the *SRY* gene (Murata et al. 2010; Soullier et al. 1998; Sutou et al. 2001).

Tokudaia is an interesting model of rapid and unique chromosomal evolution. A molecular phylogenetic analysis has indicated that *T. muenninki* was the first species in the genus to diverge (Murata et al. 2010), suggesting that the ancestral species of *Tokudaia* had a chromosome number of $2n = 46$ and had a Y chromosome and *SRY* (Fig. 4.1). Therefore, the Y and *Sry* were lost in the common lineage of *T. osimensis* and *T. tokunoshimensis* after they diverged from the ancestral *Tokudaia* species shared with *T. muenninki*. After the divergence of *T. osimensis* and *T. tokunoshimensis*, centric fusion and tandem fusion occurred in the *T. osimensis* lineage, resulting in a rapid decrease in chromosome number ($2n = 25$) (Nakamura et al. 2007). The single X chromosomes of *T. osimensis* and *T. tokunoshimensis* are submetacentric and subtelocentric, respectively (Honda et al. 1977; Honda et al. 1978; Kobayashi et al. 2007). The order of genes on the X chromosomes is conserved in these two species, whereas the positions of the X chromosome centromeres are different (Kobayashi et al. 2008). This observation indicates that centromere repositioning (CR) occurred in the *T. osimensis* lineage. CR is a recently discovered evolutionary phenomenon in which a new centromere emerges along a chromosome and the old one is inactivated without chromosomal rearrangement (Montefalcone et al. 1999). CR is not a common evolutionary event; however, it may be widespread in eukaryotes. CR has been reported in primates, cattle, marsupials, horses, birds, and rice (Kasai et al. 2003; Everts-van der Wind et al. 2004; Nagaki et al. 2004; Ventura et al. 2004; Carbone et al. 2006) and has been suggested to be widespread in eukaryotes. CR provides a potentially powerful evolutionary force for reproductive isolation and speciation (Rocchi et al. 2012).

4.4 Y Loss in the XO Spiny Rat

The Y chromosome has completely disappeared in the XO spiny rat; however, some genes were retained in the genome. A small region with several proto-Y-linked genes (i.e., genes originally located on the Y chromosome in the ancestral species) was translocated to the X chromosome. Cytogenetic and genome sequencing analyses have revealed that a 1.4 Mbp region on the Y chromosome was translocated to the distal region of the X chromosome long arm, including six genes, *KDM5D*, *DDX3Y*, *UTY*, *TSPY1*, *ZFY*, and *EIF2S3Y* (Table 4.1) (Kuroiwa et al. 2010; Arakawa et al. 2002). Two *KDM5D* copies (shorter and longer sequences) are located in a palindromic structure. *Usp9y* was identified as a pseudogene.

In mice, progeny can be generated from germ cells of males with only two Y chromosome genes, the testis-determining factor *Sry* and the spermatogonial proliferation factor *Eif2s3y* (Yamauchi et al. 2014). The XO species *Tokudaia* lacks *Sry*; therefore, a new sex-determining factor must have evolved in the species. Spermatogonial proliferation is regulated by *Eif2s3y* on the X chromosome in the developed testis. The X chromosome with a proto-Y region is shared by males and females, and therefore, females also have proto-Y genes. Surprisingly, these proto-Y genes are expressed in female tissues, especially in the ovary and brain (Kuroiwa et al. 2010; Ortega et al. 2019). However, the functions of the proto-Y genes in female tissues have not been established.

4.5 Avoidance of a Y Chromosome Loss in XY Spiny Rat

T. muenninki is the only species in the genus *Tokudaia* that has maintained the Y chromosome; as a consequence, it follows the general XY pattern of mammalian sex determination. However, the Y chromosome of *T. muenninki* has evolved in a unique manner, possibly because it was unstable in the common ancestor of the genus. In the ancestral lineage common to the two XO spiny rats, *T. osimensis* and *T. tokunoshimensis*, a few Y-linked genes escaped to the X chromosome, and the Y chromosome was subsequently lost. By contrast, in the ancestral population of *T. muenninki*, the X and Y chromosomes fused with a pair of autosomes and acquired neo-X and neo-Y regions (Murata et al. 2015; Murata et al. 2012) (Fig. 4.1). Consequently, the X and Y chromosomes of *T. muenninki* are unusually large. As a result of the autosomal fusions, the euchromatic regions of the X and Y chromosomes occupy 8% and 4% of the haploid genome, respectively, compared with 5% and less than 2% of the genome in other species (Murata et al. 2012).

The enlargement of the Y chromosome through fusion with an autosome extended the pseudoautosomal regions (PARs). PARs behave like an autosomal pair and recombine during male meiosis, and this is thought to play a critical role in spermatogenesis (Kauppi et al. 2011; Matsuda et al. 1992; Mohandas et al. 1992). XY male mice with an inverted X PAR, a Y PAR flanked at the distal end by the X

PAR boundary together with adjacent X-specific material, or both types of variant PARs produce XO progeny, leading to the production of unusual X and Y products and consequent sex chromosome loss (Burgoyne and Evans 2000). If the variant PAR was associated with Y chromosome loss in an ancestral lineage common to the two XO spiny rats, *T. osimensis* and *T. tokunoshimensis*, the Y chromosome would have been stabilized by the acquisition of new PARs in the ancestral lineage of *T. muenninki*, thus preventing the loss of the Y chromosome (Murata et al. 2012).

More than 70 copies of *SRY* with sequence variation are distributed on a long arm of the Y chromosome of *T. muenninki*; however, most of the sequences contain an internal stop codon and only a few copies retain the full-length coding sequence (Murata et al. 2010). The amino acid sequence of the HMG-box in *T. muenninki* SRY exhibits 92.5% identity with those of mouse and rat SRY; however, the 21st amino acid residue in the HMG-box domain (alanine) is replaced by serine in all copies of *T. muenninki* SRY (Murata et al. 2010). The amino acid at this position of SRY is alanine across other eutherian mammals and in other SOX family members, including SOX3, SOX8, SOX9, and SOX10. This position is located on the DNA-binding surface domain, which is important for the binding of SRY to a target DNA sequence (Phillips et al. 2004), suggesting that the transcriptional activity of *T. muenninki* SRY was weakened by this substitution. *T. muenninki* SRY is expressed in primate-derived COS7 cells transfected with a *T. muenninki* SRY expression vector (Ogata et al. 2019). However, the protein is not expressed in the gonads of transgenic mice carrying *T. muenninki* SRY. It has been thought that a long Q-rich domain plays essential roles in protein stabilization in mice; accordingly, the shorter Q-rich domain in *T. muenninki* SRY than in mouse SRY may explain why *T. muenninki* SRY is unstable in mouse cells. However, the stability of the SRY protein with a short Q-rich domain has been reported other rodent species (Zhao et al. 2017), suggesting that the instability may have other causes.

4.6 XCI in the Genus *Tokudaia*

To correct the dosage imbalance of X chromosome genes between XX females and XY males, mammals have evolved a unique mechanism. Females transcriptionally silence one of their two Xs to maintain equal X-linked gene expression levels to those of males in somatic cells (Lyon 1961). This dosage compensation mechanism is called X chromosome inactivation (XCI). XCI is achieved by *XIST* (X-inactive specific transcript) RNA, a long non-coding RNA expressed from the *XIST* gene located in the X inactivation center (XIC) region in all somatic cells of females. *XIST* RNA is expressed only from the inactive X (Xi) and not from the active X (Xa).

In the genus *Tokudaia*, XCI has also been studied. The X chromosome of the XX/XY species *T. muenninki* comprises an ancestral X region on the long arm (Xq), a large heterochromatin region (pericentromeric region), and a neo-X region on the short arm (Xp). RNA FISH experiments have revealed that *XIST* RNAs are distributed throughout the ancestral X (Xq) region of Xi, suggesting that XCI functions on the ancestral X (Zushi et al. 2017). *XIST* RNA was partially distributed on the neo-X (Xp). However, an RNA-seq analysis has shown that there is no compensation of genes located on the neo-X. By contrast, *XIST* RNA is not expressed in the XO species *T. osimensis* (Zushi et al. 2017). Mutation accumulation has been observed in the *XIST* gene in this species, consistent with the assumption that dosage compensation by XCI is not needed. These results show that *Tokudaia* species are good models for evolutionary analyses of XCI. In particular, the early stage of XCI observed in the neo-X of *T. muenninki* cannot be studied in most mammals.

4.7 Independent Losses of the Y Chromosome in the Mole Vole, Genus *Ellobius*

The mole vole is a subterranean animal belonging to the genus *Ellobius* (Arvicolinae, Cricetidae, Rodentia), which contains at least five species: *Ellobius fuscocapillus* (geographical distribution: Baluchistan, Afghanistan, Iran, and S. Turkmenistan), *Ellobius lutescens* (Armenia, Iran, Turkey, and Azerbaijan), the sister species *Ellobius tancrei* (Turkmenistan, Uzbekistan, Kazakhstan, and Xinjiang) and *Ellobius talpinus* (Ukraine, Crimea–Turkestan–Mongolia, Sinkiang, and N. Afghanistan), and *Ellobius alaicus* (Just et al. 2007). Of these, only *E. fuscocapillus* has an XX/XY karyotype, and *Sry* has been detected in the male genome of this species (Just et al. 1995). *E. lutescens*, which lives preferentially in semi-arid or grassland territories in riparian countries of the Caucasus Mountains, has an odd number of chromosomes ($2n = 17$, XO/XO) in both sexes (Coskun 2001; Vorontsov et al. 1980; Mulugeta et al. 2016). *E. tancrei*, *E. talpinus*, and *Ellobius alaicus* have XX/XX sex chromosome constitutions (Fredga and Lyapunova 1991; Veyrunes et al. 2005), and the autosome number varies due to Robertsonian fusion in *E. tancrei* and *E. talpinus* (Musser and Carleton 2005; Veyrunes et al. 2006).

The XO/XO and XX/XX species lack *Sry*, suggesting that the common ancestor of these *Ellobius* species possessed S*ry* and the gene was lost during or shortly after speciation. The X chromosome does not show sex-specific segregation in *E. lutescens* or *E. talpinus* (Vorontsov et al. 1980), suggesting that the sex-determining factor is located on an autosome. The genome sequence and assembly have been reported for male and female *E. lutescens* ($2n = 17$, XO/XO) and female *E. talpinus* ($2n = 54$, XX/XX) (Mulugeta et al. 2016). These results indicate that the Y chromosome was lost in two independent events in each lineage. Four proto-Y genes, *Zfy*, *Eif2s3y*, *Usp9y*, and *Ssty*, have been detected in male and female *E. lutescens*, three of which have also been detected in the female *E. talpinus* genome (Table 4.1).

Whole genome sequence data show that the gene content of the mammalian X chromosome is largely conserved in a single X chromosome of *E. lutescens*, including all genes known to control XCI in the XIC (Mulugeta et al. 2016). The whole XIC region shows over 90% identity between *E. lutescens* and *E. talpinus*, in which the XCI mechanism is most likely still intact (Fredga and Lyapunova 1991). These findings indicate that if XCI is dysfunctional, this functional failure happened recently and did not involve major genomic rearrangements of genes known to be involved in XCI.

4.8 African Pygmy Mouse, XY Females

The African pygmy mouse *Mus minutoides* has an atypical sex determination system. The pygmy mouse is one of the smallest African mammals (adult weight, 4–6 g) but one of the widest sub-Saharan distributions (Veyrunes et al. 2005). This species belongs to same genus of the house mouse, *Mus musculus*, but is assigned to a different subgenus, *Nannomys* (Veyrunes et al. 2005; Musser and Carleton 2005; Veyrunes et al. 2006). In this species, normal XX females and XY males exist in the population. In additional, individuals bearing a normal Y and a variant X (X*) develop as females (X*Y females), despite the presence of the Y chromosome and *Sry* (Veyrunes et al. 2010; Rahmoun et al. 2014). Up to 75% of females carry a Y chromosome and a peculiar feminizing X* chromosome and are fully fertile. A similar polygenetic system for sex determination has been described in the wood lemming *Myopus schisticolor*, with X, Y, and X* chromosomes.

In *M. minutoides*, the wild-type X and mutant X* can be cytogenetically distinguished owing to structural rearrangements (Veyrunes et al. 2010). They are fused to different autosomes (i.e., the X and Y to chromosome 1, and the X* to chromosome 16). The X and X* chromosomes are randomly inactivated in XX* females (Veyrunes and Perez 2018). No histone modification is observed as an indicator of XCI in the X* chromosome of X*Y females. In addition, X inactivation never spreads to autosomal regions in X and X*.

The genetic mechanism by which X* overrides the male sex-determining mechanism based on *SRY* has not been identified. In *M. minutoides*, *Sry* is expressed in embryonic and adult X*Y ovaries at higher levels than those in XY testes (Rahmoun et al. 2014), suggesting that female development in X*Y animals is unlikely due to a lack of *Sry* expression. Previous analyses of *Sry* in *M. minutoides* have identified no disruptive mutations in the HMG-box and part of the bridge domain and no differences in partial *Sry* sequences between XY males and X*Y females (Veyrunes et al. 2010; Veyrunes et al. 2013). However, gross defects in *Sry* are unlikely, given that the gene must retain its male sex-determining function in XY males. Accordingly, we hypothesized that the molecular function of the *Sry* pathway might be subtly altered in this species.

SRY proteins comprise an N-terminal HMG domain responsible for DNA binding and a large C-terminal Q-rich domain composed of 8 to 20 blocks of 2–13 glutamine residues interspersed by a short histidine-rich spacer sequence in the genus *Mus* (Albrecht and Eicher 1997; Coward et al. 1994). The HMG-box of SYR of *M. minutoides* and *Mus mattheyi*, a closely related species of *M. minutoides* without variant X (X*), is relatively conserved (Zhao et al. 2017).

However, there is sequence variation in the C-terminal Q-rich tract. SRY of *M. mattheyi* retains 11 glutamine blocks and histidine-rich spacer. By contrast, the C-terminus of *M. minutoides* SRY is further shortened and contains four blocks of only two glutamine residues and no histidine-rich spacer. These results indicate that the Q-rich tract of *M. minutoides* SRY evolved by the degeneration of a longer and more organized Q-rich tract present in a common ancestor. However, the stability of *M. minutoides* SRY is not affected by its degraded Q-rich tract.

4.9 Creeping Vole, XX Males

The creeping vole (*Microtus oregoni*) is one of very few mammals with atypical sex chromosomes. Classical cytogenetic analyses from the 1950s to the 1960s reported that females have paired sex chromosomes in the germ line ($2n = 18$, XX) but not in somatic cells ($2n = 17$, XO), whereas males have paired sex chromosomes in somatic cells ($2n = 18$, XY) but a single unpaired chromosome in the germ line ($2n = 17$, YO), the latter of which leads to the production of gametes bearing either a Y chromosome or no sex chromosome (Matthey 1956; Ohno et al. 1963).

In 2021, a highly contiguous male genome and transcriptome assembly were generated from short-read data for both sexes (Couger et al. 2021). Genome analysis shows that females have a single X chromosome, consistent with a previous karyotypic study; however, males have two types of X chromosome, X^M and X^P (Fig. 4.2). X^M is present in the somatic cells of both sexes but is maternally transmitted, and X^P is male-specific and is paternally transmitted. XIC generally occurs in female somatic cells with two X chromosomes. However, *Xist* RNA, which mainly functions in XCI, is expressed in male somatic cells but not in females in *M. oregoni.* Nine genes derived from the Y chromosome are present on X^M and X^P. Among these genes, *Ddx3y* and *Kdm5d* are expressed in both sexes, while *Uty, Ube1y,* and *Sry* are specific to males.

4.10 Sox9 Regulation in Atypical Sex Chromosomes

SRY proteins trigger cells of the supporting cell lineage to differentiate into Sertoli cells by regulating *Sox9* expression in undifferentiated gonads (Koopman et al. 1991). *Sox9* is a direct target of SRY and is critical for the differentiation of Sertoli cells. *Sox9* has many functions in various cell types and developmental stages (Jo et al. 2014). Its *cis*-regulatory region is spread over a gene desert of at least 2 Mb upstream of the coding sequence (Symon and Harley 2017). The SRY protein binds to the testis-specific enhancer upstream of *SOX9*. TESCO (TES [testis-specific enhancer of Sox9] COre) is an enhancer located 13 kb upstream of *Sox9* in mice (Sekido and Lovell-Badge 2008). SRY directly binds to the enhancer together with SF1 (nuclear receptor subfamily 5, group A, member 1; also referred to as NR5A1)

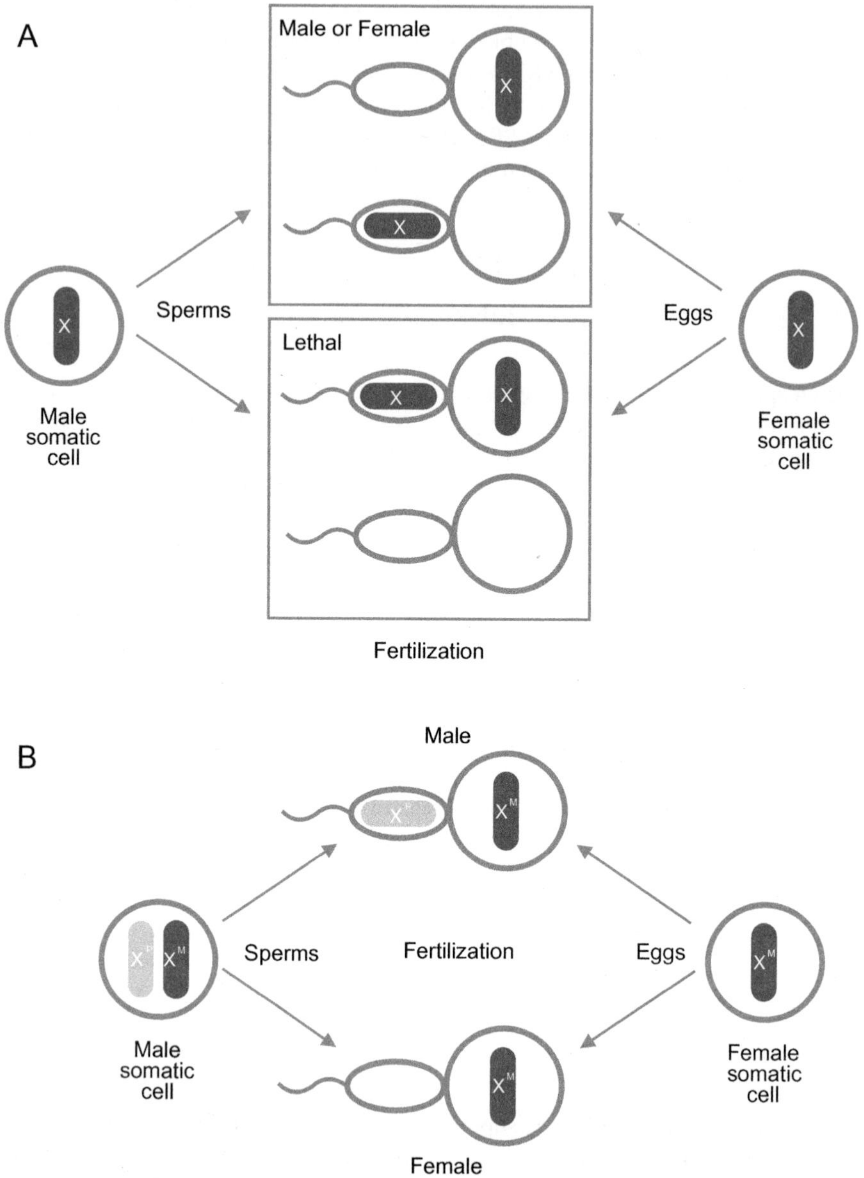

Fig. 4.2 Sex chromosome composition in the species which has atypical sex chromosomes. (**a**) XO/XO species in genus *Tokudai* and *Ellobius*. (**b**) Creeping vole, *M. oregoni* (Couger et al. 2021). XIC occurs on the X^P chromosome in male somatic cells

and synergistically upregulates the expression of *Sox9* (Sekido and Lovell-Badge 2008). FOXL2 (forkhead-box L2) also binds to the enhancer to suppress *Sox9* expression in developed ovaries (Uhlenhaut et al. 2009), and DAX1 (nuclear receptor subfamily 0, group B, member 1; also referred to as NR0B1) and the Wnt/beta-catenin pathway may antagonize the activation of TES by reducing the binding of SF1 to TES (Ludbrook et al. 2012; Bernard et al. 2012). These results suggest that TESCO represses *Sox9* in the development and/or maintenance of ovaries in mice.

A study in 2014 examined whether TESCO can act as a testis-specific enhancer in the genus *Tokudaia* (Kimura et al. 2014). TESCO sequences are highly conserved in the two XO species without *SRY* and a XX/XY species. The sequence of *Tokudaia* TESCO exhibited more than 83% identity with mouse TESCO. However, nucleotide substitution(s) were found in two of three SRY binding sites and in five of six SF1 binding sites. TESCO of all species showed low enhancer activity in a reporter gene assay. These findings indicate that the *SRY*-dependent sex-determining mechanism was lost in the common ancestor of the genus *Tokudaia* via binding site mutations. Enhancers other than TESCO are thought to regulate *SOX9* expression in *Tokudaia*.

In *Ellobius* species (Bagheri-Fam et al. 2012), TESCO contains an evolutionarily conserved region (ECR) of 180 bp. The ECRs of *E. lutescens*, *E. tancrei*, and *E. fuscocapillus* show high sequence identities (94–97%) and share a 14 bp deletion, removing a highly conserved SOX/TCF site. The deletion increases enhancer activity, as determined by a reporter gene assay. These findings indicate that the 14 bp deletion triggers the upregulation of *SOX9* in XX and XO gonads, leading to the destabilization of the XY/XX sex-determining mechanism in *Ellobius*.

SOX9 regulation via TESCO has also been examined in X*Y females of the African pygmy mouse *M. minutoides*. TESCO of *M. minutoides* has a sequence identity of approximately 93% with mouse TESCO (Matsuda et al. 1992). Sequence variation has been identified in TESCO at four SF1 binding sites, and these polymorphisms may impair SF1 binding. A reporter gene assay showed that the cotransfection of *M. minutoides* TESCO with mouse SF1 reduced mouse TESCO activity. Furthermore, *M. minutoides* TESCO exhibits low transcriptional activity by mouse SRY, and *M. minutoides* SRY fails to activate mouse TESCO. These results indicate that weakening the SRY/TESCO interaction may have facilitated the rise and spread of a variant X* chromosome carrying female-inducing modifier gene(s) (Zhao et al. 2017).

After the identification of TESCO, patients who are genetic males (*SRY*-positive 46, XY) and show male-to-female sex reversal with a 32.5 kb deletion located 600 kb upstream of *SOX9*, termed XY SR, have been reported (Kim et al. 2015). Furthermore, genetic females (*SRY*-negative 46, XX) show female-to-male sex reversal with a duplication or triplication of a 68 kb region 516–584 kb upstream of *SOX9*, termed XX SR (Kim et al. 2015; Benko et al. 2011). These findings indicate that several enhancers are essential for *SOX9* activation in early male gonads upstream of *SOX9*. In 2017, a functional analysis of TESCO was performed by CRISPR/Cas9 genome editing in mice (Gonen et al. 2018). Although the deletion of TESCO reduces *Sox9* expression levels in XY embryonic gonads to 60% relative to levels in wild-type gonads, sex reversal is not observed in TESCO knockout mice. Further chromatin accessibility analyses revealed that Enh13 located 565 kb 5′ from the transcription start site is an essential element for the initiation of testis development in mice (Gonen et al. 2018). In a comparative genome analysis of various mammals, the same region was identified as mXYSRa (Ogawa et al. 2018). Further analysis of *Sox9* regulation by the new enhancer may reveal an atypical sex-determining mechanism independent of the *Sry* gene.

References

Albrecht KH, Eicher EM (1997) DNA sequence analysis of Sry alleles (subgenus Mus) implicates misregulation as the cause of C57BL/6J-Y (POS) sex reversal and defines the SRY functional unit. Genetics 147:1267–1277

Arakawa Y, Nishida-Umehara C, Matsuda Y, Sutou S, Suzuki H (2002) X-chromosomal localization of mammalian Y-linked genes in two XO species of the Ryukyu spiny rat. Cytogenet Genome Res 99:303–309

Bagheri-Fam S, Sreenivasan R, Bernard P et al (2012) Sox9 gene regulation and the loss of the XY/XX sex-determining mechanism in the mole vole Ellobius lutescens. Chromosom Res 20: 191–199

Bellott DW, Hughes JF, Skaletsky H et al (2014) Mammalian Y chromosomes retain widely expressed dosage-sensitive regulators. Nature 508:494–499

Benko S, Gordon CT, Mallet D et al (2011) Disruption of a long distance regulatory region upstream of SOX9 in isolated disorders of sex development. J Med Genet 48:825–830

Bernard P, Ryan J, Sim H et al (2012) Wnt signaling in ovarian development inhibits Sf1 activation of Sox9 via the Tesco enhancer. Endocrinology 153:901–912

Burgoyne PS, Evans EP (2000) A high frequency of XO offspring from X(Paf)Y* male mice: evidence that the Paf mutation involves an inversion spanning the X PAR boundary. Cytogenet Cell Genet 91:57–61

Carbone L, Nergadze SG, Magnani E et al (2006) Evolutionary movement of centromeres in horse, donkey, and zebra. Genomics 87:777–782

Charlesworth B (1991) The evolution of sex chromosomes. Science 251:1030–1033

Charlesworth B, Charlesworth D (2000) The degeneration of Y chromosomes. Philos Trans R Soc Lond B Biol Sci 355:1563–1572

Cortez D, Marin R, Toledo-Flores D et al (2014) Origins and functional evolution of Y chromosomes across mammals. Nature 508:488–493

Coskun Y (2001) On distribution, morphology and biology of the mole vole, Ellobius lutescens Thomas, 1897 (Mammalia: Rodentia) in eastern Turkey. Zool Middle East 23:5–12

Couger MB, Roy SW, Anderson N et al (2021) Sex chromosome transformation and the origin of a male-specific X chromosome in the creeping vole. Science 372:592–600

Coward P, Nagai K, Chen D, Thomas HD, Nagamine CM, Lau YF (1994) Polymorphism of a CAG trinucleotide repeat within Sry correlates with B6.YDom sex reversal. Nat Genet 6:245–250

Everts-van der Wind A, Kata SR, Band MR, Rebeiz M (2004) A 1463 gene cattle-human comparative map with anchor points defined by human genome sequence coordinates. Genome Res 14:1424–1437

Fredga K, Lyapunova E (1991) Fertile males with two X chromosomes in Ellobius tancrei. Hereditas 115:86–97

Gonen N, Futtner CR, Wood S et al (2018) Sex reversal following deletion of a single distal enhancer of Sox9. Science 360:1469–1473

Honda T, Suzuki H, Itoh M (1977) An unusual sex chromosome constitution found in the Amami spinous country-rat, *Tokudaia osimensis osimensis*. Jpn J Genet 52:247–249

Honda T, Suzuki H, Itoh M, Hayashi K (1978) Karyotypical differences of the Amami spinous country-rats, *Tokudaia osimensis osimensis* obtained from two neighbouring islands. Jpn J Genet 53:297–299

Jo A, Denduluri S, Zhang B et al (2014) The versatile functions of Sox9 in development, stem cells, and human diseases. Genes Dis 1:149–161

Just W, Baumstark A, Süss A, Graphodatsky A, Rens W, Schäfer N, Bakloushinskaya I, Hameister H, Vogel W (2007) *Ellobius lutescens*: sex determination and sex chromosome. Sex Dev 1:211–221

Just W, Rau W, Vogel W, Akhverdian M, Fredga K, Graves JA, Lyapunova E (1995) Absence of Sry in species of the vole Ellobius. Nat Genet 11:117–118

Kasai F, Garcia C, Arruga MV, Ferguson-Smith MA (2003) Chromosome homology between chicken (Gallus gallus domesticus) and the red-legged partridge (Alectoris rufa); evidence of the occurrence of a neocentromere during evolution. Cytogenet Genome Res 102:326–330

Kauppi L, Barchi M, Baudat F, Romanienko PJ, Keeney S, Jasin M (2011) Distinct properties of the XY pseudoautosomal region crucial for male meiosis. Science 331:916–920

Kim G-J, Sock E, Buchberger A (2015) Copy number variation of two separate regulatory regions upstream of SOX9 causes isolated 46,XY or 46,XX disorder of sex development. Med Genet 52:240–247

Kimura R, Murata C, Kuroki Y, Kuroiwa A (2014) Mutations in the testis-specific enhancer of SOX9 in the SRY independent sex-determining mechanism in the genus Tokudaia. PLoS One 9: e108779

Kobayashi T, Yamada F, Hashimoto T, Abe S, Matsuda Y (2008) Centromere repositioning in the X chromosome of XO/XO mammals, Ryukyu spiny rat. Chromosome Res. 16:587–593

Kobayashi T, Yamada F, Hashimoto T, Abe S, Matsuda Y, Kuroiwa A (2007) Exceptional minute sex-specific region in the X0 mammal, Ryukyu spiny rat. Chromosome Res 15:175–187

Koopman P, Gubbay J, Vivian N, Goodfellow P, Lovell-Badge R (1991) Male development of chromosomally female mice transgenic for Sry. Nature 351:117–121

Kuroiwa A, Ishiguchi Y, Yamada F, Abe S, Matsuda Y (2010) The process of a Y-loss event in an XO/XO mammal, the Ryukyu spiny rat. Chromosoma 119:519–526

Lahn BT, Page DC (1999) Four evolutionary strata on the human X chromosome. Science 286: 964–967

Ludbrook LM, Bernard P, Bagheri-Fam S et al (2012) Excess DAX1 leads to XY ovotesticular disorder of sex development (DSD) in mice by inhibiting steroidogenic factor-1 (SF1) activation of the testis enhancer of SRYbox-9 (Sox9). Endocrinology 153:1948–1958

Lyon MF (1961) Gene action in the X-chromosome of the mouse (Mus musculus L.). Nature 190: 372–373

Matsuda Y, Moens PB, Chapman VM (1992) Deficiency of X and Y chromosomal pairing at meiotic prophase in spermatocytes of sterile interspecific hybrids between laboratory mice (*Mus domesticus*) and *Mus spretus*. Chromosoma 101:483–492

Matthey R (1956) Comparative cytology of Muridae; origin of Ellobii. Experientia 12:337–338

Mohandas TK, Speed RM, Passage MB, Yen PH, Chandley AC, Shapiro LJ (1992) Role of the pseudoautosomal region in sex-chromosome pairing during male meiosis: meiotic studies in a man with a deletion of distal Xp. Am J Hum Genet 51:526–533

Montefalcone G, Tempesta S, Rocchi M, Archidiacono N (1999) Centromere repositioning. Genome Res 12:1184–1188

Mulugeta E, Wassenaar E, Sleddens-Linkels E et al (2016) Genomes of Ellobius species provide insight into the evolutionary dynamics of mammalian sex chromosomes. Genome Res 26:1202–1210

Murata C, Kuroki Y, Imoto I, Kuroiwa A (2016) Ancestral Y-linked genes were maintained by translocation to the X and Y chromosomes fused to an autosomal pair in the Okinawa spiny rat Tokudaia muenninki. Chromosom Res 24:407–419

Murata C, Kuroki Y, Imoto I et al (2015) Initiation of recombination suppression and PAR formation during the early stages of neo-sex chromosome differentiation in the Okinawa spiny rat, Tokudaia muenninki. BMC Evol Biol 15:234

Murata C, Yamada F, Kawauchi N, Matsuda Y, Kuroiwa A (2010) Multiple copies of *SRY* on the large Y chromosome of the Okinawa spiny rat, *Tokudaia muenninki*. Chromosome Res 18:623–634

Murata C, Yamada F, Kawauchi N et al (2012) The Y chromosome of the Okinawa spiny rat, Tokudaia muenninki, was rescued through fusion with an autosome. Chromosom Res 20:111–125

Musser GG, Carleton MD (2005) Superfamily Muroidea. In: Wilson DE, Reeder DM (eds) Mammal species of the world, vol 2. John Hopkins University Press, Baltimore, MD, pp 894–1531

Nagaki K, Cheng Z, Ouyang S et al (2004) Sequencing of a rice centromere uncovers active genes. Nat Genet 36:138–145

Nakamura T, Kuroiwa A, Nishida-Umehara C et al (2007) Comparative chromosome painting map between two Ryukyu spiny rat species, Tokudaia osimensis and Tokudaia tokunoshimensis (Muridae, Rodentia). Chromosom Res 15:799–806

Ogata Y, Nishikata M, Kitada K et al (2019) Spiny rat SRY lacks a long Q-rich domain and is not stable in transgenic mice. Dev Dyn 248:784–794

Ogawa Y, Terao M, Hara S, Tamano M, Okayasu H, Kato T, Takada S (2018) Mapping of a responsible region for sex reversal upstream of Sox9 by production of mice with serial deletion in a genomic locus. Sci Rep 8:17514

Ohno S (1967) Sex chromosomes and sex-linked genes. Springer Verlag, Berlin

Ohno S, Jainchill J, Stenius C (1963) The creeping vole (Microtus Oregoni) as a gonosomic mosaic. I. The OY/XY constitution of the male. Cytogenetics 2:232–239

Ortega MT, Bivens NJ, Jogahara T et al (2019) Sexual dimorphism in brain transcriptomes of Amami spiny rats (Tokudaia osimensis): a rodent species where males lack the Y chromosome. BMC Genomics 20:87

Phillips NB, Nikolskaya T, Jancso-Radek A et al (2004) Sry-directed sex reversal in transgenic mice is robust with respect to enhanced DNA bending: comparison of human and murine HMG boxes. Biochemistry 43:7066–7081

Rahmoun M, Perez J, Saunders PA, Boizet-Bonhoure B, Wilhelm D, Poulat F, Veyrunes F (2014) Anatomical and molecular analyses of XY ovaries from the African pygmy mouse Mus minutoides. Sex Dev 8:356–363

Rocchi M, Archidiacono N, Schempp W, Capozzi O, Stanyon R (2012) Centromere repositioning in mammals. Heredity 108:59–67

Ross MT, Grafham DV, Coffey AJ et al (2005) The DNA sequence of the human X chromosome. Nature 434:325–337

Sekido R, Lovell-Badge R (2008) Sex determination involves synergistic action of SRY and SF1 on a specific Sox9 enhancer. Nature 453:930–934

Skinner BM, Sargent CA, Churcher C et al (2016) The pig X and Y chromosomes: structure, sequence, and evolution. Genome Res 26:130–139

Soullier S, Hanni C, Catzeflis F, Berta P, Laudet V (1998) Male sex determination in the spiny rat *Tokudaia osimensis* (Rodentia: Muridae) is not *Sry* dependent. Mamm Genome 9:590–592

Sutou S, Mitsui Y, Tsuchiya K (2001) Sex determination without the Y chromosome in two Japanese rodents *Tokudaia osimensis osimensis* and *Tokudaia osimensis* spp. Mamm Genome 12:17–21

Symon A, Harley V (2017) SOX9: a genomic view of tissue specific expression and action. Int J Biochem Cell Biol 87:18–22

Tsuchiya K, Wakana S, Suzuki H, Hattori S, Hayashi Y (1989) Taxonomic study of *Tokudaia* (Rodentia: Muridae): I. genetic differentiation. Memoirs. Nat Sci Museum, Tokyo 22:227–234

Uhlenhaut NH, Jakob S, Anlag K et al (2009) Somatic sex reprogramming of adult ovaries to testes by FOXL2 ablation. Cell 139:1130–1142

Ventura M, Weigl S, Carbone L et al (2004) Recurrent site for new centromere seeding. Genome Res 14:1696–1703

Veyrunes F, Britton-Davidian J, Robinson TJ, Calvet E, Denys C, Chevret P (2005) Molecular phylogeny of the African pygmy mice, subgenus Nannomys (Rodentia, Murinae, Mus): implications for chromosomal evolution. Mol Phylogenet Evol 36:358–369

Veyrunes F, Chevret P, Catalan J et al (2010) A novel sex determination system in a close relative of the house mouse. Proc Biol Sci 277:1049–1056

Veyrunes F, Dobigny G, Yang F, O'Brien PCM, Catalan J, Robinson TJ, Britton-Davidian J (2006) Phylogenomics of the genus Mus (Rodentia; Muridae): extensive genome repatterning is not restricted to the house mouse. Proc R Soc B 273:2925–2934

Veyrunes F, Perez J (2018) X inactivation in a mammal species with three sex chromosomes. Chromosoma 127:261–267

Veyrunes F, Perez J, Paintsil SNC, Fichet-Calvet E, Britton-Davidian J (2013) Insights into the evolutionary history of the X-linked sex reversal mutation in mus minutoides: clues from sequence analyses of the Y-linked Sry gene. Sex Dev 7:244–252

Vorontsov NN, Lyapunova E, Borissov YUM, Dovgal VE (1980) Sex chromosome variability in mammals. Genetica 52(53):361–372

Yamauchi Y, Riel JM, Stoytcheva Z, Ward MA (2014) Two Y genes can replace the entire Y chromosome for assisted reproduction in the mouse. Science 343:69–72

Zhao L, Quinn A, Ng ET, Veyrunes F, Koopman P (2017) Reduced activity of SRY and its target enhancer Sox9-TESCO in a mouse species with X*Y sex reversal. Sci Rep 7:41378

Zushi H, Murata C, Mizushima S, Nishida C, Kuroiwa A (2017) Unique XCI evolution in Tokudaia: initial XCI of the neo-X chromosome in Tokudaia muenninki and function loss of XIST in Tokudaia osimensis. Chromosoma 126:741–751

Chapter 5
The Evolutionary Aspects of the Mammalian Sex-Determining Gene *SRY*

Shingo Miyawaki and Makoto Tachibana

Abstract In eutherian mammals, gonadal sex is determined by the presence or absence of the *sex-determining region Y* (*SRY*) gene, which locates in the male-specific region of the Y chromosome. Although the process of sex determination triggered by *SRY* is conserved among animal species, its genome sequence is highly diverse. Thus, *SRY* represents a unique, dynamic, and rapid evolutionary process. Since its discovery 30 years ago, *SRY*, including mouse *Sry*, has been believed to be a single-exon gene. Recently, we identified a previously undiscovered second exon of mouse *Sry* and a corresponding novel transcript, the two-exon *Sry* (*Sry-T*). The discovery of *Sry-T* provided new insights into the genetic evolution of *Sry*. In this review, we discuss the genetic evolution of *Sry* sequences based on the discovery of *Sry-T*.

Keywords Sex development · *Sry* · Gene evolution

5.1 The Mammalian Sex Determination Gene *SRY*

The development of male and female sexes is essential for the survival of almost all animal species. Sex in mammals is determined by a combination of sex chromosomes. Given that an XX chromosome results in female and an XY chromosome results in male, it has been thought that there is a male factor (sex-determining factor) in the Y chromosome (McLaren 1988; Welshons and Russell 1959).

S. Miyawaki
Laboratory of Epigenome Dynamics, Graduate School of Frontier Biosciences, Osaka University, Suita, Osaka, Japan

Laboratory of Veterinary Surgery, Faculty of Applied Biological Sciences, Gifu University, Gifu, Gifu, Japan
e-mail: miyawaki@gifu-u.ac.jp

M. Tachibana (✉)
Laboratory of Epigenome Dynamics, Graduate School of Frontier Biosciences, Osaka University, Suita, Osaka, Japan
e-mail: tachiban@fbs.osaka-u.ac.jp

M. Tanaka, M. Tachibana (eds.), *Spectrum of Sex*,
https://doi.org/10.1007/978-981-19-5359-0_5

Sex-determining region Y (*SRY*) is a sex-determining gene in mammals (Gubbay et al. 1990; Sinclair et al. 1990; Koopman et al. 1991). Koopman et al. (1991) experimentally demonstrated that *Sry* alone is sufficient to trigger testis development in mice. In that experiment, insertion of a DNA fragment of the Y chromosome containing *Sry* and forcible expression of *Sry* resulted in testes formation in chromosomally female (XX-type) mice. This experiment also revealed that all of the sequences necessary for *Sry* expression are present within this 14.5-kb region. Human patients exhibiting XY gonadal dysgenesis and loss-of-function studies in mice, pig, and rabbit have demonstrated that *SRY/Sry* disruption causes sex reversal, such that chromosomal males (XY-type) are anatomically female (Berta et al. 1990; Jäger et al. 1990; Sinclair et al. 1990; Lovell-Badge and Robertson 1990; Wang et al. 2013; Kato et al. 2013; Song et al. 2017; Kurtz et al. 2021). These experiments suggested that SRY functions as the master regulator of male sex determination in all eutherian mammals (Waters et al. 2007).

5.2 Function of *Sry* in Sex Determination

Identification of the sex determination gene *SRY/Sry* led to an understanding of the mechanism of sex determination in mammals. Differentiation of Sertoli cells is the first step in male sex development in mammals (Albrecht and Eicher 2001; Koopman et al. 1990), with this process driven by *Sry* expression in the supporting cell lineage in the sexually undifferentiated gonad. Sertoli cells are essential for the coordination of testicular development and form the testis cords that encapsulate and direct germ cell differentiation as well as orchestrate the differentiation of other somatic cells, including Leydig cells, which represent a steroidogenic cell lineage (Habert et al. 2001). Sequential morphological and functional differentiation is subsequently induced under the influence of sex hormones (Fig. 5.1).

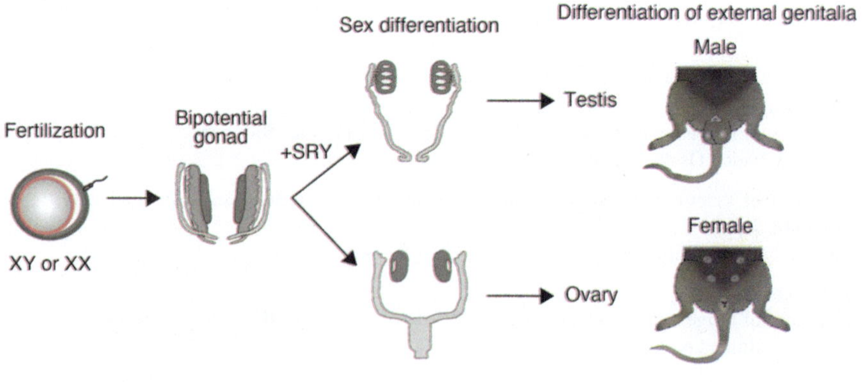

Fig. 5.1 Mammalian sex determination

In mice, *Sry* expression is strictly controlled in a cell-type- and temporal-specific manner. *Sry* expression starts at embryonic day 10.5 (E10.5) in pre-Sertoli cells of XY genital ridges, reaches a peak at E11.5, and declines by E12.5 (Hacker et al. 1995; Jeske et al. 1995; Koopman et al. 1990). If *Sry* expression is delayed by as little as a few hours, testis differentiation is disrupted or blocked completely in favor of ovarian development (Bullejos and Koopman 2001, 2005; Hiramatsu et al. 2009). *SRY*/*Sry* is more broadly expressed in animal species (human, rats, sheep, and rabbits) other than mice; however, recent studies using RNA sequencing (RNA-seq) analysis revealed that *SRY*/*Sry* also has a temporary expression pattern in humans and rats (Guo et al. 2021; Lecluze et al. 2020; Prokop et al. 2020). These observations indicate that sex determination in mammals is highly dependent on accurate transcription of *Sry*. Epigenetic regulation (i.e., histone modification and/or DNA methylation) is a mechanism that ensures accurate transcription and plays an important role in regulating *Sry* expression (Okashita and Tachibana 2021). In support of this view, complete or partial sex reversal has been observed in mice harboring knockouts of several epigenetic factors (Carré et al. 2018; Kuroki et al. 2013, 2017; Miyawaki and Tachibana 2019; Okashita et al. 2019).

A transgenic study demonstrated that ectopic expression of *SRY-box transcription factor 9* (*Sox9*), which belongs to the *Sry*-type high-mobility group box (Sox) family, can initiate the testis-determining pathway in XX mice (Vidal et al. 2001). In the developing testis, the critical function of *Sry* is to upregulate *Sox9* transcription (Sekido and Lovell-Badge 2008). SRY protein contains a high-mobility group (HMG) box DNA-binding domain that binds to regulatory elements upstream of *Sox9*. The testis-specific enhancer of *Sox9* (TES), a 3.2-kb element mapping to a 13-kb section at the 5′ end of the transcription start site of *Sox9*, and its 1.4-kb core region TESCO, activates *Sox9* expression *in cis* in Sertoli cells (Sekido and Lovell-Badge 2008). SRY and nuclear receptor subfamily 5 group A member 1 bind TESCO directly and activate *Sox9* expression. Recently, enhancer 13 (Enh13), a 557-bp element located at the 5′ end of the *Sox9* gene, was identified as a key enhancer of *Sox9* (Gonen et al. 2018). Mouse Enh13 contains a single consensus SRY-binding site, and targeted deletion of Enh13 reduced *Sox9* expression to a level equivalent to that in XX gonads and resulted in male-to-female sex reversal in mice (Gonen et al. 2018). These findings provided the first evidence that deletion of a noncoding genomic region causes sex reversal. Because most human XY females harbor mutations in the HMG box of SRY, its DNA-binding activity is essential for sex determination (Harley et al. 1994; Harley and Goodfellow 1994). Thus, future studies may reveal the existence of noncoding genomic regions that are necessary for sex determination in humans.

SOX9 activates a network of gene activity required for testis development while simultaneously impeding a network required for ovarian development (Warr and Greenfield 2012). Additionally, SRY represses the R-spondin1–Wnt/β-catenin signaling pathway that drives ovarian development (Bernard et al. 2008; Capel 2006; Lau and Li 2009; McElreavey et al. 1993). These two alternative gene regulatory networks confer the bipotential gonad its unique ability to differentiate into two morphologically and functionally distinct organs (Larney et al. 2014). In summary,

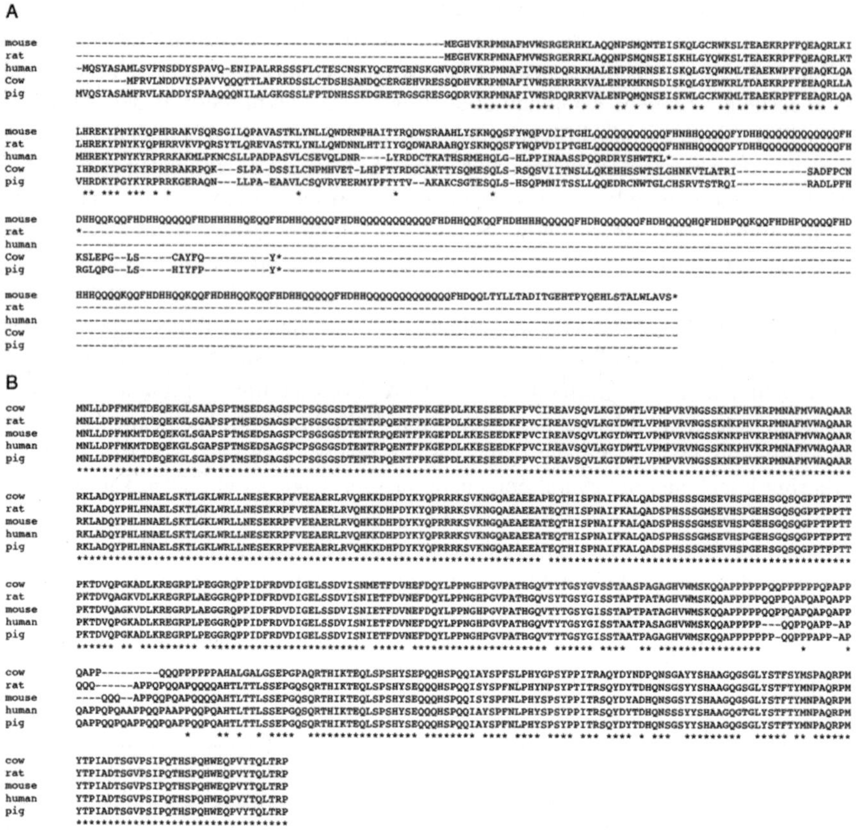

Fig. 5.2 Sequence comparison of SRY and SOX9. Amino acid sequences of SRY (**a**) and SOX9 (**b**) were aligned across the indicated animal species. Although the amino acid sequence of SRY is less conserved, that of SOX9 is highly conserved

the gene regulatory network for testis development is activated and that for ovarian development is repressed by the SRY–SOX9 axis. The amino acid sequence of SOX9 is highly conserved across animal species (Fig. 5.2), and these sex determination processes are consistent among different species (Nagahama et al. 2021). However, the sequence of *SRY* is diverse across species (Fig. 5.2).

5.3 Genetic Structure of *Sry*

Since the identification of the *SRY*/*Sry* gene in humans and mice, *SRY* has been identified in a variety of animal species. There is a high diversity of *SRY* sequences among animal species. SRY contains a highly conserved HMG DNA-binding domain and less-conserved N- and C-terminal domains (Tucker and Lundrigan

1993; Zhao and Koopman 2012). Mouse *Sry* has a unique CAG repeat sequence at the C-terminus (Bowles et al. 1999), and sequence comparison with rat and spiny rat *Sry* suggests that the CAG repeat sequence was not added to the C-terminus but inserted or extended just before the C-terminal sequence conserved in rodents (Ogata et al. 2019; Roy 2021). CAG repeats encode the polyglutamine (polyQ) amino acid sequence, which is essential for *Sry* transcription (Bowles et al. 1999). These findings suggest that the *Sry* sequence acquired CAGs to maintain transcriptional activity in rodents.

SRY/*Sry* is a single-copy gene in mice, humans, and horses (Behlke et al. 1993; Hacker et al. 1995; Janečka et al. 2018) but amplified in several species. For example, there are at least 11 copies of *Sry* in rats and two copies of *SRY* in dogs, rabbits, and pigs (Turner et al. 2007; Prokop et al. 2013; Geraldes et al. 2010; Skinner et al. 2016). Recent studies in rats showed that distinct copies of *Sry* are expressed (Prokop et al. 2020); however, it is unclear whether all of the *Sry* copies are required for sex determination or whether any of them act as true sex-determining genes. In a knockout study in pigs, a frameshift mutation in one copy of the *SRY* gene did not induce male-to-female sex reversal. These results suggest that expression from one *SRY* copy is sufficient for the development of male genitalia, or that only one of the two *SRY* copies is expressed, and the frameshift mutation was introduced into the unexpressed *SRY* copy (Kurtz et al. 2021).

Mouse *Sry* is embedded in a large palindromic sequence, which is also called an inverted repeat sequence. There is an almost indistinguishable 50-kb palindromic sequence flanking mouse *Sry* (Gubbay et al. 1992), with the palindromic nature of the *Sry* locus capable of potentially leading to the formation of a stem-loop structure. In this situation, juxtaposing the DNA breakpoint within the palindromic sequence might induce deletion of *Sry*, resulting in the generation of XY female mice (Gubbay et al. 1992). Until the recent establishment of genome-editing technology, the palindromic nature of the *Sry* locus made it difficult to manipulate endogenous *Sry* by conventional homologous recombination techniques.

Hacker et al. (1995) analyzed a mouse line termed 32.10 with 12 copies of L741 and concluded that *Sry* is a single-exon gene that encodes only one protein (SRY). Similarly, analysis of gonads during mouse development concluded that *Sry* is a single exon (Jeske et al. 1995). In humans and pigs, it was confirmed that the SRY protein is encoded by a single exon (Behlke et al. 1993; Daneau et al. 1996). Therefore, the conclusion that *SRY*/*Sry* is a single-exon gene was recognized as a confirmed fact that was not doubted for three decades. However, employment of comprehensive transcriptomics approaches identified a cryptic second exon within the flanking palindrome that is essential for male development (Miyawaki et al. 2020).

5.4 Discovery of Two-Exon *Sry* (*Sry-T*)

The genomic sequence of *SRY*, especially at the C-terminal coding sequence, is highly diverse. Various studies have discussed the sequence diversity of *SRY* (Larney et al. 2015; Roy 2021; Tucker and Lundrigan 1993; Zhao et al. 2014). Recently, we highlighted the previously unaddressed nature of the C-terminal sequences of mouse SRY (Miyawaki et al. 2020). In this section, we focus on the latest findings related to the discovery of a cryptic second exon of *Sry* and discuss the genetic evolution of *Sry* sequences.

5.4.1 Identification of Mouse Sry *Exon2*

First, we describe our discovery of two-exon SRY (SRY-T), which is the bona fide testis-determining factor in mice. During our previous study on regulation of *Sry* expression, we established a method to selectively collect gonadal somatic cells expressing *Sry* (Kuroki et al. 2013). Using this method, we conducted gene expression analysis (RNA-seq) of *Sry*-expressing cells, which led to the identification of an uncharacterized sequence transcribed from the 3′ palindromic sequence. We initially named this unknown sequence "*Srx*."

To characterize *Srx*, we performed comprehensive transcriptome analyses of embryonic gonadal somatic cells during the sex-determining period. Initially, we assumed that *Srx* would be a long noncoding RNA (lncRNA); however, analysis of the transcription start site using cap analysis of gene expression sequencing showed no transcription start signal at the 5′ end of *Srx*, suggesting that it might not be a lncRNA. We then performed long-read RNA-seq and found that this unknown transcript was spliced to the previously known sequence of *Sry* at the typical 5′ GT/3′ AG splice-site sequences with an internal polypyrimidine tract. These results clearly showed that *Srx* was not an independent transcript but was the previously unknown second exon of *Sry*. Sequencing data indicated that two-types of mRNA were transcribed from the *Sry* locus: the known single-exon type (*Sry-S*) and a novel two-exon type (*Sry-T*) (Fig. 5.3).

The SRY-T protein shares an HMG DNA-binding domain and polyQ sequences with the amino acid sequence of SRY-S. The 15 amino acids in the SRY-T C-terminus are encoded by the second exon, whereas the absence of splicing results in 18 different amino acids at the SRY-S C-terminus.

5.4.2 Sry-T *Is Essential for Testis Development*

To investigate the role of *Sry-T* in sex determination, we generated *Sry-T*-deficient mice in which *Sry* exon2 was deleted by genome editing using the CRISPR/Cas9

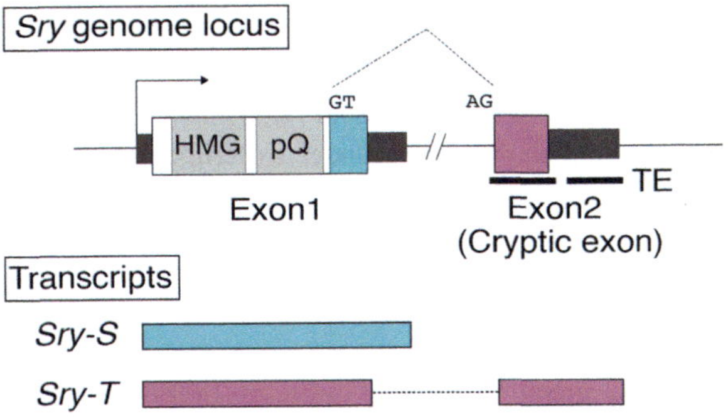

Fig. 5.3 Exon–intron structure of the mouse *Sry* locus. Transcripts specific for *Sry-S* (blue) and *Sry-T* (red) are shown along with the 5′ GT/3′ AG splicing motifs. TE: retrotransposon-derived sequence

Fig. 5.4 SRY-T is a necessary and sufficient sex determinant. *Sry-T*-deficient XY mice exhibit male-to-female sex reversal (left). *Sry-T*-expressing XX mice develop as males (right)

system. As a result, *Sry-T*-deficient mice showed a sex-reversal phenotype despite expressing *Sry-S*, indicating that *Sry-T* is essential for male sex determination. Furthermore, when *Sry-T* or *Sry-S* was expressed in XX mice under the same conditions, only mice expressing *Sry-T* underwent female-to-male sex reversal. These experiments demonstrated SRY-T as a necessary and sufficient sex determinant in vivo (Fig. 5.4).

5.4.3 Existence of a "Degron" in the SRY-S C-Terminus

As shown in previous studies, *Sry* transgenes encoding only SRY-S can induce testis development in XX mice under experimental conditions (Koopman et al. 1991; Bowles et al. 1999; Washburn et al. 2001; Zhao et al. 2014). However, our studies revealed that *Sry-S* alone could not induce male development under physiological conditions. To investigate the cause of the insufficiency of *Sry-S* for sex determination, we focused on the difference in the C-termini of SRY-S and SRY-T. We generated lentiviral constructs encoding blue fluorescent protein (BFP) and enhanced green fluorescent protein (EGFP) fused to either the C-terminus of SRY-S (termed S18) or SRY-T (termed T15), both of which were translated from the same transcript, and then transduced them into human embryonic kidney cell line 293T (HEK293T). The effect of S18 or T15 on protein stability was evaluated by examining the EGFP/BFP ratio. Although EGFP-T15 expression levels were similar to those of non-fused EGFP, EGFP-S18 expression levels were $<10\%$ those of non-fused EGFP. Further detailed investigation revealed that a protein-degrading sequence called a "degron" was present at the SRY-S C-terminus (Miyawaki et al. 2020). Because the amino acid at the second-to-last residue position is valine, the SRY-S degron was assumed to be a (V-2)-type degron (Koren et al. 2018). In (V-2)-type degrons, this valine residue is essential for degron activity. Replacement of the valine in the -2 site of endogenously expressed SRY-S with proline suppressed SRY-S degradation. Furthermore, the corresponding mice (*Sry-S*:V394P; *Sry*-exon2Δ mice) developed as males, even though they lacked SRY-T. These results showed that SRY-S is insufficient for male development under physiological conditions, because the C-terminal degron sequence makes the protein unstable.

5.4.4 Reconsidering Previous Research in Light of Sry-T Discovery

In the first experiment identifying *Sry* as a sex-determining gene in 1991, insertion of multiple copies of a DNA fragment (L741) encoding only *Sry-S* was successful in producing XX males. Of the eight XX individuals with transgenes, two mice changed their sex to male, whereas the others did not (Koopman et al. 1991). Other studies also reported variations in the ability to cause XX sex reversal on the part of the 14-kb *Sry* transgene lacking the second exon (Bowles et al. 1999; Zhao et al. 2014), which might reflect an inability to control the copy number and integration site of the transgene. Only mice expressing SRY-S in an amount sufficient to overcome its destabilization were able to become male. Furthermore, another study showed that a single copy of the L741 transgene at an autosomal locus (*Col1a1*) did not induce XX sex reversal (Quinn et al. 2014). In that article, the existence of an unknown DNA region necessary for males had been predicted.

Why was the second exon missed in previous studies? One reason might be attributable to the existence of the palindromic sequence. Specifically, the second exon was hidden by the palindrome sequences. As noted, there exists a 50-kb palindromic sequence flanking mouse *Sry* (Gubbay et al. 1992). When duplicate mapped reads are excluded from the mapping of sequence reads to the genome, the transcript corresponding to the second exon is not mapped. When duplicate mapping is allowed, this transcript appears on both sides of the palindrome sequences. Various other palindromic sequences are known to exist in the Y chromosome and reportedly contribute to its structural maintenance (Soh et al. 2014). To date, no functional sequences have been identified in the palindromic sequence. Given that the second exon of *Sry* is located within the palindromic sequence, this exon was identified as the first functional sequence existing within the palindromic sequence.

5.5 Evolution of the Y Chromosome and *Sry*

The discovery of *Sry* exon2 also provides new insights into the evolution of sex-determining genes. The Y chromosome, which contains *Sry*, lost its genes during the evolutionary process (Graves 2006; Lahn and Page 1999). It is believed that the X chromosome and Y chromosome share a common ancestor (Ohno 1967). After the ancestral Y chromosome acquired a male-determining gene, other male-advantage alleles accumulated, and recombination was suppressed. Mutations and deletions in non-recombinant regions resulted in rapid degradation of the proto-Y chromosome. As a result, unlike autosomes and the X chromosome, the Y chromosome is normally unable to recombine with homologous chromosomes; therefore, mutations created during the evolutionally process accumulate in the Y chromosome. Due to this accumulation of mutations, the Y chromosome is thought to be facing a crisis related to loss of function in various genes. Notably, in mammals, the Y chromosome appears susceptible to far more mutation, deletion, and insertion events than the rest of the genome (Lindblad-Toh et al. 2005; Makova and Li 2002).

The X chromosome gene *SRY-box transcription factor 3* (*SOX3*), which is believed to share a common ancestor with *SRY*, is a single-exon gene (Stevanovlć et al. 1993; Sutton et al. 2011). In humans and pigs, it has been confirmed that the *SRY* is single-exon genes (Behlke et al. 1993; Daneau et al. 1996). In the ancestor of rodents, *Sry* is also considered to be a single-exon gene; however, during its evolution, *Sry* might have mutated into a degron-coding gene, presumably leading to the loss of its male-determining function (Fig. 5.5). To avoid subsequent crises of species survival, distinct mechanisms could have been invoked between species.

We explain our hypothesis of how rats and mice avoided *Sry* dysfunction as follows. Rat *Sry* harbors of conserved DNA sequence similar to the mouse *Sry-S* sequence encoding the degron. However, because the degron sequence is not translated due to a preceding four-nucleotide insertion and stop codons, rat SRY is not presumed to be destabilized. Additionally, rat *Sry* has been duplicated through gene conversion at transposable elements on the Y chromosome (Prokop et al.

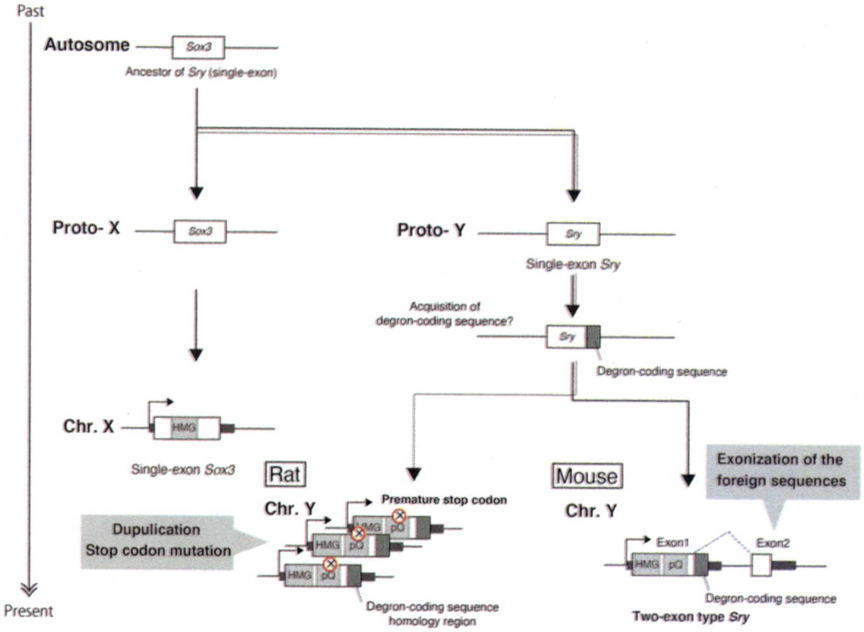

Fig. 5.5 Model for the evolution of the mouse and rat *Sry* locus

2013). At least three *Sry* genes (*Sry*1, *Sry*4A, and *Sry*3C) are expressed at the highest levels at E13, the time of testis differentiation, suggesting that multiple *Sry* genes potentially contribute to testis differentiation in rats (Prokop et al. 2020). These findings suggest that at least in rats, *Sry* might have evolved to avoid degron-mediated SRY dysfunction through gene duplication and early translation termination (Fig. 5.5).

In mice, we hypothesize that a different mechanism has evolved to avoid degron translation by creating a new exon (called exonization) in distal sequences and splicing to excise the degron. Interestingly, the second exon of *Sry* comprises a retrotransposon-derived sequence thought to originate from a retrovirus, suggesting that *Sry-T* was created by the exonization of a retrotransposon-derived sequence. In this case, exonization may have avoided translation of the degron-coding sequence and saved the ancestral *Sry*, which was in danger of losing its function due to the degron. This hypothesis suggests that virus-derived sequences might have evolved the host genes and saved the species from extinction. The discovery of *Sry-T* offer insight into a new aspect of the relationship between viruses and their host organisms. This protective mechanism represents an unusual and intriguing evolutionary mechanism that helps stop vulnerable Y chromosome genes from falling apart.

A previous study showed that although the sequence of the HMG box of SRY is conserved, the SRY C-terminus is rapidly evolving in mice and rats (Tucker and Lundrigan 1993). The hypothesis that the DNA sequence encoding the SRY C-terminus evolved to avoid degrons supports the existence of species-specific adaptive divergence through a process of positive Darwinian selection. From these perspectives, even if the Y chromosome undergoes degradation, *Sry* may persist through continuous evolution.

5.6 Consideration of the Evolution of *Sry* Exons from the Viewpoint of "Sex Spectrum"

The discovery of the second exon of *Sry* and the SRY-S degron offered insight into a part of the evolutionary process of sex-determining genes. The concept of "sex spectrum" explains that all cells/organs possess their own sex that can be located at any intermediate position between a typical male and a typical female. By acquisition of the SRY-S degron, the gonadal sex might have moved to the female side on the spectrum in the ancestor of rodents. However, in mice, the emergence of the second exon, which encodes degron-free SRY-T, moved the gonadal sex back to the male side. To support this concept, our genetic study using single-copy knock-in mouse lines that conditionally express each isoform of *Sry* showed that a copy of *Sry-T* was sufficient to activate the male pathway, whereas that of *Sry-S* was not. Alternatively, in rats, introducing the premature termination codon into *Sry* and increasing the *Sry* copy number might have evolved to move the gonadal sex back to the male side on the spectrum. These data suggest that not only the gene structure of *Sry* but also its sex-determining ability might have dynamically evolved in rodents. It may be worth examining whether the "cryptic" second exon or degron-coding sequence of *Sry* also exists in other mammals.

5.7 Conclusion and Future Directions

In the 30 years since *Sry* was discovered, no one questioned the fact that *Sry* is a single-exon gene. In contrast to this common knowledge, we showed that mouse *Sry* has a cryptic second exon that encodes previously unknown but functionally essential sex determinant, SRY-T. Our results provide insight into a critical neo-functionalization process that appears to be happening in real time; the gain-of-function replacement of SRY-S by SRY-T is currently ongoing in a mouse strain. In many strains of mice, the *Sry* sequence remains incompletely understood; therefore, further studies are warranted to comprehensively understand the evolutionary process of *Sry*.

Despite the high degree of conservation of the sex determination system triggered by SRY, its amino acid sequence shows dynamic evolution, with various species exhibiting novel protein domains, transcripts, and gene amplification. *Sry* is one of the few genes on the Y chromosome that is essential for survival and may be subject to different selection pressures for genetic evolution than autosomal and X chromosomal genes. Therefore, the function of male development may take the first priority, and preservation of the *Sry* DNA sequence may not be of critical importance. In general, the sequences of genes essential for survival tend to be highly conserved; however, despite *Sry* being essential for species survival, it shows sequence diversity among animal species. The diversity of *Sry* sequences in animal species may be the key to understanding the genetic evolution of *Sry*. As a result of these genetic evolutions, the spectrum-like diversity of SRY sequences may have been produced.

Approximately 10 years ago, it was proposed that the genes on the Y chromosome (including *Sry*) were in a state of degeneration and that males would eventually disappear (Bachtrog 2013; Graves 2006). Our findings have a profound influence on this notion.

Acknowledgments We are grateful to all members of the Tachibana laboratory for technical support. This work was supported by KAKENHI from the Japan Society for the Promotion of Science Grant Numbers 21K19221 (MT), 21H04769 (MT), 17H06424 (MT), 17H06423 (MT), and 21K19213 (SM).

References

Albrecht KH, Eicher EM (2001) Evidence that Sry is expressed in pre-Sertoli cells and Sertoli and granulosa cells have a common precursor. Dev Biol 240:92–107. https://doi.org/10.1006/dbio.2001.0438

Bachtrog D (2013) Y-chromosome evolution: emerging insights into processes of Y-chromosome degeneration. Nat Rev Genet 14:113–124. https://doi.org/10.1038/nrg3366

Behlke MA, Bogan JS, Beer-Romero P, Page DC (1993) Evidence that the SRY protein is encoded by a single exon on the human Y chromosome. Genomics 17(3):736–739. https://doi.org/10.1006/geno.1993.1395

Bernard P, Sim H, Knower K, Vilain E, Harley V (2008) Human SRY inhibits beta-catenin-mediated transcription. Int J Biochem Cell Biol 40:2889–2900. https://doi.org/10.1016/j.biocel.2008.06.006

Berta P, Hawkins JR, Sinclair AH, Taylor A, Griffiths BL, Goodfellow PN, Fellous M (1990) Genetic evidence equating SRY and the testis-determining factor. Nature 348:448–450. https://doi.org/10.1038/348448A0

Bowles J, Cooper L, Berkman J, Koopman P (1999) Sry requires a CAG repeat domain for male sex determination in Mus musculus. Nat Genet 22(4):405–408. https://doi.org/10.1038/11981

Bullejos M, Koopman P (2001) Spatially dynamic expression of Sry in mouse genital ridges. Dev Dyn 221:201–205. https://doi.org/10.1002/dvdy.1134

Bullejos M, Koopman P (2005) Delayed Sry and Sox9 expression in developing mouse gonads underlies B6-YDOM sex reversal. Dev Biol 278:473–481. https://doi.org/10.1016/j.ydbio.2004.11.030

Capel B (2006) R-spondin1 tips the balance in sex determination. Nat Genet 38:1233–1234. https://doi.org/10.1038/ng1106-1233

Carré GA, Siggers P, Xipolita M, Brindle P, Lutz B, Wells S, Greenfield A (2018) Loss of p300 and CBP disrupts histone acetylation at the mouse Sry promoter and causes XY gonadal sex reversal. Hum Mol Genet 27:190–198. https://doi.org/10.1093/hmg/ddx398

Daneau I, Ethier JF, Lussier JG, Silversides DW (1996) Porcine SRY gene locus and genital ridge expression. Biol Reprod 55:47–53. https://doi.org/10.1095/biolreprod55.1.47

Geraldes A, Rambo T, Wing RA, Ferrand N, Nachman MW (2010) Extensive gene conversion drives the concerted evolution of paralogous copies of the SRY gene in European rabbits. Mol Biol Evol 27:2437–2440. https://doi.org/10.1093/molbev/msq139

Gonen N, Futtner CR, Wood S, Garcia-Moreno SA, Salamone IM, Samson SC, Sekido R, Poulat F, Maatouk DM, Lovell-Badge R (2018) Sex reversal following deletion of a single distal enhancer of Sox9. Science 360:1469–1473. https://doi.org/10.1126/science.aas9408

Graves JAM (2006) Sex chromosome specialization and degeneration in mammals. Cell 124:901–914. https://doi.org/10.1016/j.cell.2006.02.024

Gubbay J, Collignon J, Koopman P, Capel B, Economou A, Münsterberg A, Vivian N, Goodfellow P, Lovell-Badge R (1990) A gene mapping to the sex-determining region of the mouse Y chromosome is a member of a novel family of embryonically expressed genes. Nature 346:245–250. https://doi.org/10.1038/346245a0

Gubbay J, Vivian N, Economou A, Jackson D, Goodfellow P, Lovell-Badge R (1992) Inverted repeat structure of the Sry locus in mice. Proc Natl Acad Sci U S A 89:7953–7957. https://doi.org/10.1073/pnas.89.17.7953

Guo J, Sosa E, Chitiashvili T, Nie X, Rojas EJ, Oliver E, Plath K, Hotaling JM, Stukenborg J-B, Clark AT, Cairns BR (2021) Single-cell analysis of the developing human testis reveals somatic niche cell specification and fetal germline stem cell establishment. Cell Stem Cell 28:764–778.e4. https://doi.org/10.1016/j.stem.2020.12.004

Habert R, Lejeune H, Saez JM (2001) Origin, differentiation and regulation of fetal and adult Leydig cells. Mol Cell Endocrinol 179:47–74. https://doi.org/10.1016/S0303-7207(01)00461-0

Hacker A, Capel B, Goodfellow P, Lovell-Badge R (1995) Expression of *Sry*, the mouse sex determining gene. Development 121:1603–1614. https://doi.org/10.1006/geno.1993.1395

Harley VR, Goodfellow PN (1994) The biochemical role of SRY in sex determination. Mol Reprod Dev 39:184–193. https://doi.org/10.1002/mrd.1080390211

Harley VR, Lovell-Badge R, Goodfellow PN (1994) Definition of a consensus DNA binding site for SRY. Nucleic Acids Res 22:1500–1501. https://doi.org/10.1093/nar/22.8.1500

Hiramatsu R, Matoba S, Kanai-Azuma M, Tsunekawa N, Katoh-Fukui Y, Kurohmaru M, Morohashi K-I, Wilhelm D, Koopman P, Kanai Y (2009) A critical time window of Sry action in gonadal sex determination in mice. Development 136:129–138. https://doi.org/10.1242/dev.029587

Jäger RJ, Anvret M, Hall K, Scherer G (1990) A human XY female with a frame shift mutation in the candidate testis-determining gene SRY. Nature 348:452–454. https://doi.org/10.1038/348452a0

Janečka JE, Davis BW, Ghosh S, Paria N, Das PJ, Orlando L, Schubert M, Nielsen MK, Stout TAE, Brashear W, Li G, Johnson CD, Metz RP, Al Zadjali AM, Love CC, Varner DD, Bellott DW, Murphy WJ, Chowdhary BP, Raudsepp T (2018) Horse Y chromosome assembly displays unique evolutionary features and putative stallion fertility genes. Nat Commun 9:2945. https://doi.org/10.1038/s41467-018-05290-6

Jeske YWA, Bowles J, Greenfield A, Koopman P (1995) Expression of a linear Sry transcript in the mouse genital ridge. Nat Genet 10:480–482. https://doi.org/10.1038/ng0895-480

Kato T, Miyata K, Sonobe M, Yamashita S, Tamano M, Miura K, Kanai Y, Miyamoto S, Sakuma T, Yamamoto T, Inui M, Kikusui T, Asahara H, Takada S (2013) Production of Sry knockout mouse using TALEN via oocyte injection. Sci Rep 3:1–8. https://doi.org/10.1038/srep03136

Koopman P, Münsterberg A, Capel B, Vivian N, Lovell-Badge R (1990) Expression of a candidate sex-determining gene during mouse testis differentiation. Nature 348:450–452. https://doi.org/10.1038/348450a0

Koopman P, Gubbay J, Vivian N, Goodfellow P, Lovell-Badge R (1991) Male development of chromosomally female mice transgenic for Sry. Nature 351(6322):117–121. https://doi.org/10.1038/351117a0

Koren I, Timms RT, Kula T, Xu Q, Li MZ, Elledge SJ (2018) The eukaryotic proteome is shaped by E3 ubiquitin ligases targeting C-terminal degrons. Cell 173:1622–1635.e14. https://doi.org/10.1016/j.cell.2018.04.028

Kuroki S, Matoba S, Akiyoshi M, Matsumura Y, Miyachi H, Mise N, Abe K, Ogura A, Wilhelm D, Koopman P, Nozaki M, Kanai Y, Shinkai Y, Tachibana M (2013) Epigenetic regulation of mouse sex determination by the histone demethylase Jmjd1a. Science 341:1106–1109. https://doi.org/10.1126/science.1239864

Kuroki S, Okashita N, Baba S, Maeda R, Miyawaki S, Yano M, Yamaguchi M, Kitano S, Miyachi H, Itoh A, Yoshida M, Tachibana M (2017) Rescuing the aberrant sex development of H3K9 demethylase Jmjd1a-deficient mice by modulating H3K9 methylation balance. PLoS Genet 13:e1007034. https://doi.org/10.1371/journal.pgen.1007034

Kurtz S, Lucas-Hahn A, Schlegelberger B, Göhring G, Niemann H, Mettenleiter TC, Petersen B (2021) Knockout of the HMG domain of the porcine SRY gene causes sex reversal in gene-edited pigs. Proc Natl Acad Sci U S A 118:e2008743118. https://doi.org/10.1073/pnas.2008743118

Lahn BT, Page DC (1999) Four evolutionary strata on the human X chromosome. Science 286:964–967. https://doi.org/10.1126/science.286.5441.964

Larney C, Bailey TL, Koopman P (2014) Switching on sex: transcriptional regulation of the testis-determining gene Sry. Development 141:2195–2205. https://doi.org/10.1242/dev.107052

Larney C, Bailey TL, Koopman P (2015) Conservation analysis of sequences flanking the testis-determining gene Sry in 17 mammalian species gametogenesis and fertilization. BMC Dev Biol 15:1–14. https://doi.org/10.1186/s12861-015-0085-6

Lau Y-FC, Li Y (2009) The human and mouse sex-determining SRY genes repress the Rspol/beta-catenin signaling. J Genet Genomics 36:193–202. https://doi.org/10.1016/S1673-8527(08)60107-1

Lecluze E, Rolland AD, Filis P, Evrard B, Leverrier-Penna S, Maamar MB, Coiffec I, Lavoué V, Fowler PA, Mazaud-Guittot S, Jégou B, Chalmel F (2020) Dynamics of the transcriptional landscape during human fetal testis and ovary development. Hum Reprod 35:1099–1119. https://doi.org/10.1093/humrep/deaa041

Lindblad-Toh K, Wade CM, Mikkelsen TS, Karlsson EK, Jaffe DB, Kamal M, Clamp M, Chang JL, Kulbokas EJ, Zody MC, Mauceli E, Xie X, Breen M, Wayne RK, Ostrander EA, Ponting CP, Galibert F, Smith DR, DeJong PJ, Kirkness E, Alvarez P, Biagi T, Brockman W, Butler J, Chin C-W, Cook A, Cuff J, Daly MJ, DeCaprio D, Gnerre S, Grabherr M, Kellis M, Kleber M, Bardeleben C, Goodstadt L, Heger A, Hitte C, Kim L, Koepfli K-P, Parker HG, Pollinger JP, Searle SMJ, Sutter NB, Thomas R, Webber C, Baldwin J, Abebe A, Abouelleil A, Aftuck L, Ait-Zahra M, Aldredge T, Allen N, An P, Anderson S, Antoine C, Arachchi H, Aslam A, Ayotte L, Bachantsang P, Barry A, Bayul T, Benamara M, Berlin A, Bessette D, Blitshteyn B, Bloom T, Blye J, Boguslavskiy L, Bonnet C, Boukhgalter B, Brown A, Cahill P, Calixte N, Camarata J, Cheshatsang Y, Chu J, Citroen M, Collymore A, Cooke P, Dawoe T, Daza R, Decktor K, DeGray S, Dhargay N, Dooley K, Dooley K, Dorje P, Dorjee K, Dorris L, Duffey N, Dupes A, Egbiremolen O, Elong R, Falk J, Farina A, Faro S, Ferguson D, Ferreira P, Fisher S, FitzGerald M, Foley K, Foley C, Franke A, Friedrich D, Gage D, Garber M, Gearin G, Giannoukos G, Goode T, Goyette A, Graham J, Grandbois E, Gyaltsen K, Hafez N, Hagopian D, Hagos B, Hall J, Healy C, Hegarty R, Honan T, Horn A, Houde N, Hughes L, Hunnicutt L, Husby M, Jester B, Jones C, Kamat A, Kanga B, Kells C, Khazanovich D, Kieu AC, Kisner P, Kumar M, Lance K, Landers T, Lara M, Lee W, Leger J-P, Lennon N, Leuper L, LeVine S, Liu J, Liu X, Lokyitsang Y, Lokyitsang T, Lui A, Macdonald J, Major J, Marabella R, Maru K, Matthews C, McDonough S, Mehta T, Meldrim J, Melnikov A, Meneus L, Mihalev A, Mihova T, Miller K, Mittelman R, Mlenga V, Mulrain L, Munson G, Navidi A, Naylor J, Nguyen T, Nguyen N, Nguyen C, Nguyen T, Nicol R, Norbu N, Norbu C, Novod N, Nyima T, Olandt P, O'Neill B, O'Neill K, Osman S, Oyono L, Patti C, Perrin D, Phunkhang P, Pierre F, Priest M, Rachupka A, Raghuraman S, Rameau R, Ray V, Raymond C, Rege F, Rise C, Rogers J, Rogov P, Sahalie J, Settipalli S, Sharpe T, Shea T, Sheehan M, Sherpa N, Shi J,

Shih D, Sloan J, Smith C, Sparrow T, Stalker J, Stange-Thomann N, Stavropoulos S, Stone C, Stone S, Sykes S, Tchuinga P, Tenzing P, Tesfaye S, Thoulutsang D, Thoulutsang Y, Topham K, Topping I, Tsamla T, Vassiliev H, Venkataraman V, Vo A, Wangchuk T, Wangdi T, Weiand M, Wilkinson J, Wilson A, Yadav S, Yang S, Yang X, Young G, Yu Q, Zainoun J, Zembek L, Zimmer A, Lander ES (2005) Genome sequence, comparative analysis and haplotype structure of the domestic dog. Nature 438:803–819. https://doi.org/10.1038/nature04338

Lovell-Badge R, Robertson E (1990) XY female mice resulting from a heritable mutation in the primary testis-determining gene, Tdy. Development 109:635–646. https://doi.org/10.1242/dev.109.3.635

Makova KD, Li W-H (2002) Strong male-driven evolution of DNA sequences in humans and apes. Nature 416:624–626. https://doi.org/10.1038/416624a

McElreavey K, Vilain E, Abbas N, Herskowitz I, Fellous M (1993) A regulatory cascade hypothesis for mammalian sex determination: SRY represses a negative regulator of male development. Proc Natl Acad Sci U S A 90:3368–3372. https://doi.org/10.1073/pnas.90.8.3368

McLaren A (1988) Sex determination in mammals. Trends Genet 4:153–157. https://doi.org/10.1016/0168-9525(88)90020-0

Miyawaki S, Tachibana M (2019) Role of epigenetic regulation in mammalian sex determination. Curr Top Dev Biol 134:195–221. https://doi.org/10.1016/bs.ctdb.2019.01.008

Miyawaki S, Kuroki S, Ryo Maed R, Okashita N, Koopman P, Tachibana M (2020) The mouse Sry locus harbors a cryptic exon that is essential for male sex determination. Science 370(6512):121–124. https://doi.org/10.1126/science.abb6430

Nagahama Y, Chakraborty T, Paul-Prasanth B, Ohta K, Nakamura M (2021) Sex determination, gonadal sex differentiation, and plasticity in vertebrate species. Physiol Rev 101(3):1237–1308. https://doi.org/10.1152/physrev.00044.2019

Ogata Y, Nishikata M, Kitada K, Mizushima S, Jogahara T, Kuroiwa A (2019) Spiny rat SRY lacks a long Q-rich domain and is not stable in transgenic mice. Dev Dyn 248(9):784–794. https://doi.org/10.1002/dvdy.73

Ohno S (1967) Sex chromosomes and sex-linked genes. In: Labhart A (ed) et al Monographs on endocrinology, Springer, New York. ISBN: 978-3-642-88178-7

Okashita N, Tachibana M (2021) Transcriptional regulation of the Y-linked mammalian testis-determining gene SRY. Sex Dev 15(5–6):351–359. https://doi.org/10.1159/000519217

Okashita N, Kuroki S, Maeda R, Tachibana M (2019) TET2 catalyzes active DNA demethylation of the Sry promoter and enhances its expression. Sci Rep 9:13462. https://doi.org/10.1038/s41598-019-50058-7

Prokop JW, Underwood AC, Turner ME, Miller N, Pietrzak D, Scott S, Smith C, Milsted A (2013) Analysis of Sry duplications on the Rattus norvegicus Y-chromosome. BMC Genomics 14:792. https://doi.org/10.1186/1471-2164-14-792

Prokop JW, Chhetri SB, van Veen JE, Chen X, Underwood AC, Uhl K, Dwinell MR, Geurts AM, Correa SM, Arnold AP (2020) Transcriptional analysis of the multiple Sry genes and developmental program at the onset of testis differentiation in the rat. Biol Sex Differ 11:28. https://doi.org/10.1186/s13293-020-00305-8

Quinn A, Kashimada K, Davidson TL, Ng ET, Chawengsaksophak K, Bowles J, Koopman P (2014) A site-specific, single-copy transgenesis strategy to identify 5′ regulatory sequences of the mouse testis-determining gene Sry. PLoS One 9(4):e94813. https://doi.org/10.1371/journal.pone.0094813

Roy SW (2021) Noncoding RNA, intragenomic conflict, and rodent SRY evolution. Trends Genet 37:102–104. https://doi.org/10.1016/j.tig.2020.11.004

Sekido R, Lovell-Badge R (2008) Sex determination involves synergistic action of SRY and SF1 on a specific Sox9 enhancer. Nature 453:930–934. https://doi.org/10.1038/nature06944

Sinclair AH, Berta P, Palmer MS, Hawkins JR, Griffiths BL, Smith MJ, Foster JW, Frischauf AM, Lovell-Badge R, Goodfellow PN (1990) A gene from the human sex-determining region encodes a protein with homology to a conserved DNA-binding motif. Nature 346:240–244. https://doi.org/10.1038/346240a0

Skinner BM, Sargent CA, Churcher C, Hunt T, Herrero J, Loveland JE, Dunn M, Louzada S, Fu B, Chow W, Gilbert J, Austin-Guest S, Beal K, Carvalho-Silva D, Cheng W, Gordon D, Grafham D, Hardy M, Harley J, Hauser H, Howden P, Howe K, Lachani K, Ellis PJI, Kelly D, Kerry G, Kerwin J, Ng BL, Threadgold G, Wileman T, Wood JMD, Yang F, Harrow J, Affara NA, Tyler-Smith C (2016) The pig X and Y chromosomes: structure, sequence, and evolution. Genome Res 26:130–139. https://doi.org/10.1101/gr.188839.114

Soh YQS, Alföldi J, Pyntikova T, Brown LG, Graves T, Minx PJ, Fulton RS, Kremitzki C, Koutseva N, Mueller JL, Rozen S, Hughes JF, Owens E, Womack JE, Murphy WJ, Cao Q, De Jong P, Warren WC, Wilson RK, Skaletsky H, Page DC (2014) Sequencing the mouse Y chromosome reveals convergent gene acquisition and amplification on both sex chromosomes. Cell 159:800–813. https://doi.org/10.1016/j.cell.2014.09.052

Song Y, Liu T, Wang Y, Deng J, Chen M, Yuan L, Lu Y, Xu Y, Yao H, Li Z, Lai L, Song Y, Liu T, Wang Y, Deng J, Chen M, Yuan L, Lu Y, Xu Y, Yao H, Li Z, Lai L (2017) Mutation of the Sp1 binding site in the 5′ flanking region of SRY causes sex reversal in rabbits. Oncotarget 5:1–8. https://doi.org/10.18632/oncotarget.16979

Stevanovlć M, Lovell-badge R, Collignon J, Goodfellow PN (1993) SOX3 is an X-linked gene related to SRY. Hum Mol Genet 2(12):2013–2018. https://doi.org/10.1093/hmg/2.12.2013

Sutton E, Hughes J, White S, Sekido R, Tan J, Arboleda V, Rogers N, Knower K, Rowley L, Eyre H, Rizzoti K, McAninch D, Goncalves J, Slee J, Turbitt E, Bruno D, Bengtsson H, Harley V, Vilain E, Sinclair A, Lovell-Badge R, Thomas P (2011) Identification of SOX3 as an XX male sex reversal gene in mice and humans. J Clin Invest 121(1):328–341. https://doi.org/10.1172/JCI42580

Tucker PK, Lundrigan BL (1993) Rapid evolution of the sex determining locus in Old World mice and rats. Nature 364:715–717. https://doi.org/10.1038/364715a0

Turner ME, Martin C, Martins AS, Dunmire J, Farkas J, Ely DL, Milsted A (2007) Genomic and expression analysis of multiple Sry loci from a single Rattus norvegicus Y chromosome. BMC Genet 8:1–11. https://doi.org/10.1186/1471-2156-8-11

Vidal VPI, Chaboissier MC, De Rooij DG, Schedl A (2001) Sox9 induces testis development in XX transgenic mice. Nat Genet 28:216–217. https://doi.org/10.1038/90046

Wang H, Hu Y-C, Markoulaki S, Welstead GG, Cheng AW, Shivalila CS, Pyntikova T, Dadon DB, Voytas DF, Bogdanove AJ, Page DC, Jaenisch R (2013) TALEN-mediated editing of the mouse Y chromosome. Nat Biotechnol 31:530–532. https://doi.org/10.1038/nbt.2595

Warr N, Greenfield A (2012) The molecular and cellular basis of gonadal sex reversal in mice and humans. Wiley Interdiscip Rev Dev Biol 1:559–577. https://doi.org/10.1002/wdev.42

Washburn LL, Albrecht KH, Eicher EM (2001) C57BL/6J-T-associated sex reversal in mice is caused by reduced expression of a Mus domesticus Sry allele. Genetics 158:1675–1681

Waters PD, Wallis MC, Graves JAM (2007) Mammalian sex—origin and evolution of the Y chromosome and SRY. Semin Cell Dev Biol 18:389–400. https://doi.org/10.1016/j.semcdb.2007.02.007

Welshons WJ, Russell LB (1959) The Y-chromosome as the bearer of male determining factors in the mouse. Proc Natl Acad Sci U S A 45:560–566. https://doi.org/10.1073/pnas.45.4.560

Zhao L, Koopman P (2012) SRY protein function in sex determination: thinking outside the box. Chromosom Res 20:153–162. https://doi.org/10.1007/s10577-011-9256-x

Zhao L, Ng ET, Davidson TL, Longmuss E, Urschitz J, Elston M, Moisyadi S, Bowles J, Koopman P (2014) Structure-function analysis of mouse Sry reveals dual essential roles of the C-terminal polyglutamine tract in sex determination. Proc Natl Acad Sci U S A 111(32):11768–11773. https://doi.org/10.1073/pnas.1400666111

Part II
Endocrine Regulation of the Sex Spectrum

Chapter 6
Revisiting the Role of Steroid Hormones in Gonadal Fate Determination

Kiyoshi Kikuchi and Takashi Koyama

Abstract Early in the twentieth century, it was posited that sexual differentiation in vertebrates is under the control of hormones. Referred to as the hormone theory of sex differentiation, it gave rise to the specific hypothesis that sex steroids control gonadal fate decisions. Although this hypothesis is no longer supported for placental mammals, there seems to be some uncertainty in the case of nonmammalian vertebrates, especially fish. Based on the success in generating complete sex reversals of medaka fish by exogenous steroid administration in the 1950s, Yamamoto hypothesized that sex steroids were endogenous sex inducers in fish embryos. While this model has provided a framework for research in fish developmental biology and physiology, the validity of the hypothesis has been brought into question, since experimental evidence has been lacking. However, the recent identification of master sex-determining (MSD) genes in some fish clearly supports part of this hypothesis, since estrogens play a role in determining gonadal fate. This commentary aims to give a brief overview of the developments related to the steroid theory and focuses on the role of sex steroids in gonadal sex determination and early sex differentiation in nonmammalian species, with a special emphasis on fish. The recent partial validation of Yamamoto's model should nevertheless lead to a better understanding of the evolution of sex-determining cascades in vertebrates and encourage the exploration of the target genes of estrogen signaling leading to gonadal fate specification.

Keywords Steroid hormones · Gonadal sex determination · Fish · Nonmammalian vertebrates

K. Kikuchi (✉)
Fisheries Laboratory, University of Tokyo, Hamamatsu, Japan
e-mail: akikuchi@mail.ecc.u-tokyo.ac.jp

T. Koyama
Graduate School of Fisheries and Environmental Sciences, Nagasaki University, Nagasaki, Japan
e-mail: takashi.k1201@gmail.com

M. Tanaka, M. Tachibana (eds.), *Spectrum of Sex*,
https://doi.org/10.1007/978-981-19-5359-0_6

6.1 Introduction

The idea that hormones can affect sexual dimorphism and sex determination can be traced back to the mid-nineteenth century (Berthold 1849; Josso 2008). However, more easily traceable history of this theory begins with Bouin and Ancel (1903), who suggested that the secretions of the fetal pig testes were responsible for the male characteristics in adults (termed the interstitial theory at that time) (Bouin and Ancel 1903; Scheib 1983; Josso 2008). However, there was scathing criticism of this theory (Klein and Bouin 1970), which was revived only in 1916 by Lillie, and independently by Keller and Tandler, all of whom studied freemartins, genetic female (XX) cattle that are infertile with partially masculinized phenotypes in gonads, reproductive tracts, etc. (Bouin and Ancel 1903; Lillie 1916; Scheib 1983). The hormonal theory inspired a long series of experiments such as grafting experiments in amphibians, initiated by Burns (1925) and pursued by Humphrey and Witschi (Burns 1961; Jost 1971; Josso 2008). In addition, the availability of purified sex hormones in the mid-1930s led to numerous administration experiments, first in birds and followed by amphibians and fish (for a review of experiments in fish, see Piferrer 2001, and for those in other vertebrates, see Burns 1961), which gave rise to the hypothesis attributed to the hormonal theory that sex steroids control gonadal fate determination. Although this idea is no longer supported for placental mammals, it appears to still have some traction in the case of nonmammalian vertebrates, especially fish.

After a series of experiments in the 1950s and 1960s with medaka fish (*Oryzias latipes*), in which exogenous sex steroids were administered and functional sex reversal was first obtained in fish (Yamamoto 1953, 1958, 1959a, b, 1961), Yamamoto hypothesized that the sex steroids were the endogenous sex inducers that determined gonadal fate in fish (Yamamoto 1962, 1969). Since then, numerous studies have experimented with steroid treatments in fish (e.g., Piferrer 2001; Guiguen et al. 2010). While Yamamoto's hypothesis provided an important framework for research that has continued until today, it has not been without controversy since many have questioned the validity of the hypothesis (Devlin and Nagahama 2002; Piferrer and Guiguen 2008; Guiguen et al. 2010; Li et al. 2019; Nagahama et al. 2021). As a result, there have been modifications to Yamamoto's framework (as detailed in Piferrer and Guiguen 2008; Guiguen et al. 2010), which can be summarized as follows.

1. Bogart (1987) proposed a theoretical model where the balance between androgens and estrogens, rather than their absolute amount, at a critical period during development, was the determinant of gonadal sex in vertebrates. This model has since been shown to not be the case in mammals (reviewed in Cutting et al. 2013).
2. Contrary to Bogart's modification of the Yamamoto hypothesis, it appears that changes in androgen levels (leading to changes in the balance between androgens and estrogens) may not play a critical role during early testicular differentiation after all, at least in the case of fish (Nagahama 2005). Rather, an estrogen-centric model was proposed, in which estrogens would be required for ovarian

differentiation, while their absence would be required for testicular differentiation (Kobayashi and Nagahama 2009; Guiguen et al. 2010).

3. In fish, 11-ketotestosterone (11-KT) rather than testosterone is one of the most potent steroids for male sex determination and/or differentiation (Piferrer et al. 1993; reviewed in Borg 1994; Tokarz et al. 2015).

In the following, I will provide a brief overview of the history of the hormonal theory of sex differentiation, while elaborating on some salient research selected from older studies in addition to the latest ones.

6.2 A Short History of the Hormonal Theory of Sex Differentiation

6.2.1 Hormones and Gonadal Sex Determination

The "hormone theory" has been thought to encompass hypotheses relating to hormonal control of several specific phenomena, such as gonadal sex determination, early gonadal differentiation, development of the sex cords and genital structure, and secondary sexual characteristics in adults (Burns 1961; Jost 1971). Among those, this commentary primarily focuses on models related to the gonadal sex determination by which the bipotential embryonic gonad commits to either the ovarian or testicular fate (Capel 2017). The subsequent process is often termed gonadal sex differentiation in which the molecular and cellular processes lead to the formation of morphologically and functionally distinct gonads, either the ovaries or testes (Devlin and Nagahama 2002; Adolfi et al. 2021). It should be noted that the conceptual distinction between the two processes, determination and differentiation, is not made in the old literature, and even today it is often difficult to clearly identify the transition between the two processes except for well-studied species such as mice.

6.2.2 Freemartins

Freemartin is a genetic female cattle with infertile ovaries and partially masculinized phenotype, typically as a result of being born as a twin to a male. At birth, freemartin ovaries are hypoplastic and depleted of germ cells, and they sometimes contain seminiferous tubules that would be formed in testes (Josso 2008; Harikae et al. 2012). In most cattle twins, the blood vessels between the two placentas become connected, thus creating a shared circulation (Fig. 6.1). In the early nineteenth century, it was hypothesized that masculinization was due to the transmission of a hormone produced by the gonads of the male twin to the female twin via the shared circulation (Keller and Tandler 1916; Lillie 1916; reviewed in Burns 1961; Padula 2005). Several hormones such as testosterone and anti-Müllerian hormone have been

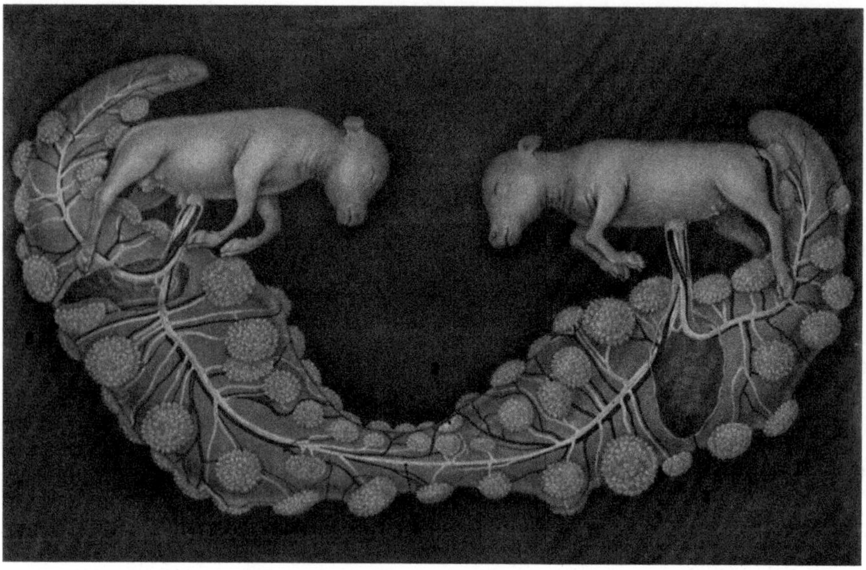

Fig. 6.1 Placentas connected between two cattle twins (Lillie 1916). The male is on the left and a freemartin on the right. (Reprinted with permission of Wiley-Liss, a subsidiary of John Wiley & Sons)

implicated, but the causative factor has been elusive (Padula 2005; Miura et al. 2019).

6.2.3 *Parabiosis and Gonadal Transplantation in Amphibians*

Parabiosis experiments conducted in the 1920s and 1930s on amphibian larvae (reviewed by Jost 1971) appeared to support the hormone theory (Fig. 6.2). In these experiments, the two embryos were conjoined in the manner of Siamese twins, providing an opportunity to test the possibility of hormonal action through a common circulation. Although the degree and direction (from the ovary to the testis or the reverse) varied greatly depending on the species under study and various experimental conditions, partially sex-reversed gonads were observed (reviewed in Burns 1961).

The parabiosis method was soon improved by the development of a technique by which the prospective gonad-forming tissue from a donor embryo was transplanted to a host (Humphrey 1928) (Fig. 6.3). This method has yielded results that have in general confirmed and extended those obtained by parabiosis. Remarkably, Humphrey (1945) obtained a fertile male by sex-reversing the prospective ovary (the left gonad) through orthotopic transplantation of the donor testis primordium to the right

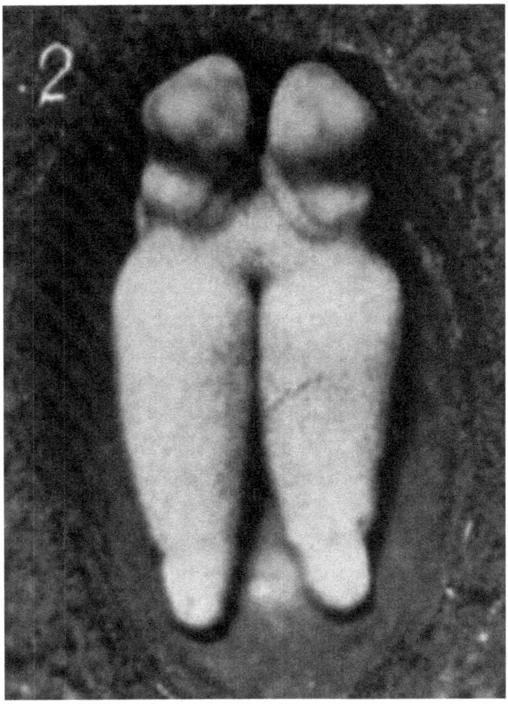

Fig. 6.2 A ventral view of two *Amblystoma* embryos joined in parabiosis (Burns 1925). (Reprinted with permission of Wiley-Liss, a subsidiary of John Wiley & Sons)

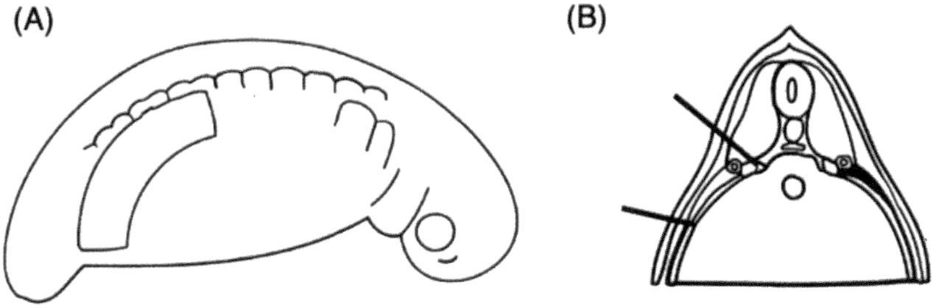

Fig. 6.3 Humphrey's orthotopic transplantation. (**a**) Position of the gonad- and mesonephros-forming area in the embryo (framed). This primordium was excised and replaced into the corresponding part of the donor embryo. (**b**) Cross-section of the host at a later stage shows the position of the implanted tissues (between thick lines) on the left. The Wolffian duct is lateral to the mass of primordial germ cells, and above the gonad- and mesonephros-forming regions. (Illustration modified from Humphrey (1928) and Burns (1961))

side of host embryos and confirmed the functionality of host-derived sperm and the sex chromosome constitution by means of crossing experiments.

6.2.4 Exogenous Sex Steroids

The hypothesized effects of diffusible substances on gonadal sex determination appeared to gain further traction by the sex reversals produced after 1935 through exogenous sex steroid administration in nonmammalian species (reviewed in Burns 1961; Bull et al. 1988). Male-to-female sex reversals occurred after treatment with estrogens in fish, amphibians, reptiles, and birds (reviewed in Piferrer 2001; Pieau and Dorizzi 2004; Lance 2009; Nakamura 2010; Flament 2016). While androgen treatments resulted in sex reversal from female to male in fish and some amphibians, it was far less effective and prone to paradoxical results (e.g., feminizing effect of androgens) in reptiles and birds (Burns 1961; Pieau et al. 1999; Ganesh and Raman 1995). Exposure to exogenous estrogens prior to gonadal sex differentiation can cause almost complete gonadal sex reversal in male marsupials as well (Burns 1955; Coveney et al. 2001; Pask et al. 2010). However, the embryonic gonads in placental mammals were found to be resistant to exogenous sex steroid (reviewed in Burns 1961), although estrogen appears to have a role in maintaining the postnatal ovary (Britt and Findlay 2002). As a result of these findings, it came to be considered that the hormonal theory on primary sex determination was only applicable to nonmammalian vertebrates (Burns 1961; Cutting et al. 2013).

Note that there appears to be a gradual evolutionary gain of resistance to exogenous estrogen in gonadal sex determination or differentiation, from fish to mammals (Cutting et al. 2013). It has been hypothesized that mammals discarded the use of steroid hormones in sex determination and/or early gonadal differentiation with the evolution of placenta and intrauterine development to resist the influence of maternal estrogen (Wolf 1999; Cutting et al. 2013; Capel 2017).

6.2.5 Fish

The earliest experiments that administered sex steroid hormones to fish to control their gonadal sex can be traced to 1937 (Berkowitz 1937; Castelnuovo 1937; Padoa 1937). However, intersex fish were produced in the early experiments (reviewed in Piferrer 2001). Yamamoto (1953) appears to have been the first to achieve complete (functional) sex reversal in which male-to-female reversal of medaka fish was induced by oral administration of estrogen. In this work, the genetic sex of the fish and the fertility were experimentally confirmed. Based on a series of experiments that followed (e.g., Yamamoto 1958, 1959a, b), Yamamoto (1962, 1969) hypothesized that endogenous estrogens are female-inducing substances (female inducers) that drive ovarian differentiation during the sex determination period in fish, and androgens are male-inducing ones (male inducers) that trigger testicular differentiation (Yamamoto 1962, 1969) (Fig. 6.4).

The term, "inducers," appears to have been adapted by Yamamoto from that of Witschi (1914) who had postulated that, in amphibians, the inducing substances, named medullarin and corticin, behaved like the hypothetical-inducing substances in

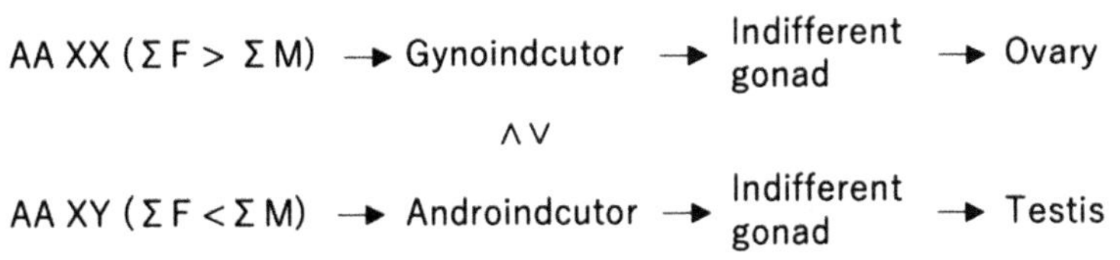

Fig. 6.4 Yamamoto's scheme for the steroid theory of sex inducers. Yamamoto assumed a polygenetic system in which sex is determined mainly by sex-determining genes on the sex chromosomes (X and Y) with influences of many modifiers on autosomes (A), based on (Winge 1934). The genetic information is realized through the action of sex steroids (gynoinductor and androinductor) on indifferent gonads. (Illustration modified from Yamamoto (1969))

the Spemann organizer (Spemann and Hilde 1924), being transmitted by diffusion and having localized effects to induce the formation of either female or male gonads (Witschi 1914; Burns 1961; Gorbman 1979). Interestingly, Witschi consistently assumed the sex inducers to be proteins, rather than sex steroids.

6.3 Are Steroids Natural Inducers of Gonadal Sex Differentiation?

6.3.1 Controversies

Since Yamamoto's experiments, a large number of studies in many fish species have been conducted on the influence of steroid hormones on sex determination and/or differentiation (for review Hunter and Donaldson 1983; Pandian and Sheela 1995). In most cases, the results have shown that when fish are treated with the appropriate dose of sex steroids during the critical time window of sex determination, exogenous estrogen and androgen are able to induce female and male sex reversals, respectively (Piferrer 2001).

While these studies appear to support Yamamoto's model, the effects of exogenous steroid treatments do not necessarily reflect the physiological role of endogenous steroids (e.g., Stewart et al. 2020). Thus, the actual role of endogenous steroids in sex determination, in other words, whether sex steroids are endogenous sex inducers ("direct sex determiners" or "natural sex inducers") has continued to be debated.

6.3.2 Androgen in Model Fish

The controversy regarding the role of androgens during sex determination has been well covered elsewhere (Piferrer and Guiguen 2008; Guiguen et al. 2010; Devlin and Nagahama 2002; Ijiri et al. 2008; Fernandino et al. 2012; Yamaguchi et al. 2010), and therefore, we only briefly mention it here.

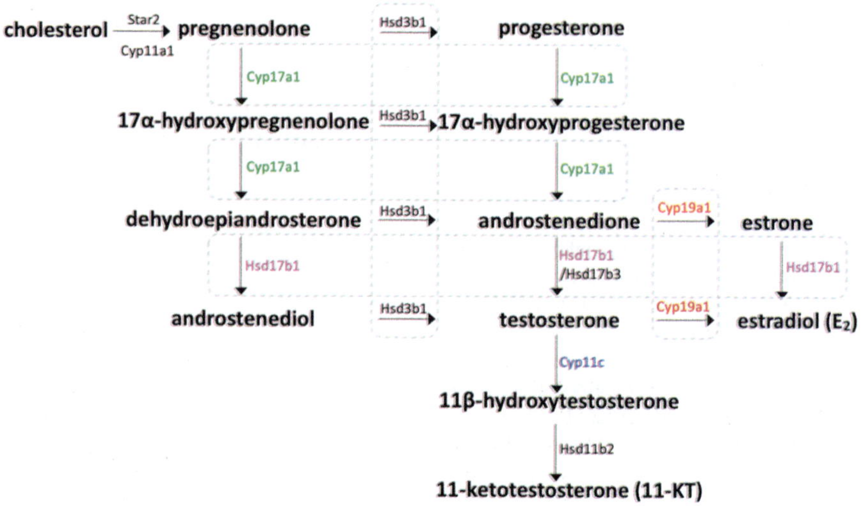

Fig. 6.5 Postulated major pathway of steroidogenesis in teleost fishes. Key enzymes are shown next to arrows at each step of biosynthesis according to (Tokarz et al. 2015; Koyama et al. 2019; Nagahama et al. 2021). Note that all pathways and enzymes are not shown. *Cyp* cytochrome P450, *Hsd* hydroxysteroid dehydrogenase. Cyp17a1 (P450c17-I), shown in green, is required for producing both estrogens and androgens. Cyp11c, shown in blue, is involved in synthesis of 11-ketotestosterone (11-KT) that is one of the most potent androgens in fish (Piferrer et al. 1993). Cyp19a1 (aromatase), shown in red, is required for the synthesis of estradiol (E_2), a major estrogen in ovarian differentiation. Hsd17b1, shown in magenta, is also required for estradiol production. It is speculated that male Seriola homozygous for the hypomorphic allele of *Hsd17b* (the Z-type Hsd17b) can synthesize testosterone/11-KT via postlarval expression of *Hsd17b3* gene (Koyama et al. 2019)

Yamamoto's hypothesis predicts that the level of androgens is higher in the prospective testes during sex determination, as is that of estrogens in the prospective ovaries. However, there have not been many studies on their levels in fish embryonic gonads, likely due to the technical difficulty in measuring steroid levels in the very small fish gonads during the early developmental stages (Guiguen et al. 2010). In contrast to the paucity of studies of the early stages, it has been shown in many fish species that the gonads exhibit contrasting properties in the synthesis of sex steroids during the differentiating stage or later, with marked androgen (specifically 11-KT, a potent androgen in fish) synthesis in the differentiating testes and significant estrogen production in the differentiating ovaries (Guiguen et al. 2018).

While these observations are consistent, if not in direct agreement, with the steroid hypothesis, some studies on gene expression related to androgen production appear to question the role of androgens during sex determination. For example, expression analysis of the *Cyp11c/Cyp11b* gene that encodes a steroid enzyme involved in 11-KT production in the rainbow trout (*Oncorhynchus mykiss*) and Nile tilapia (*Oreochromis niloticus*) (shown in green in Fig. 6.5) suggested that their expression in the male gonad could not be detected before the first sign of morphological gonadal differentiation (Vizziano et al. 2007; Ijiri et al. 2008; Kobayashi and Nagahama 2009). The implication of these results was that androgens are not yet actively produced at this stage in these species.

More decisive evidence came from the experiments that blocked estrogen synthesis with aromatase (Cyp19a) inhibitors, originally conducted with the aim of clarifying the role of estrogen (Piferrer et al. 1994; Guiguen et al. 2010). Cyp19a1 is a key enzyme in the biosynthesis of estrogens (Nagahama and Yamashita 2008; Lubzens et al. 2010) (shown in red in Fig. 6.5). In these experiments, complete female-to-male sex reversals were obtained in many fish species including the rainbow trout and Nile tilapia (Guiguen et al. 1999; Kitano et al. 2000; Rashid et al. 2007; Kobayashi and Nagahama 2009) (but see an exception below). Given that the presumed levels of androgens are very low at the sex determination period, the results suggested that the mere absence of estrogens is sufficient to drive testicular differentiation, and androgens are not inducers for the development of testes (Kobayashi and Nagahama 2009). Summarizing these results, Guiguen et al. (2010) proposed an estrogen-centric model in which estrogens are required for ovarian differentiation, while their absence would be sufficient to trigger testicular differentiation. The dispensability of androgens in the early testicular differentiation was further confirmed in medaka, tilapia, and zebrafish (*Danio rerio*) by mutant analysis in which the genes required for the production of testosterone and/or 11-KT (*Cyp17a1* or *Cyp11c*) were disrupted (Fig. 6.5) (Sato et al. 2008; Zhang et al. 2020; Zheng et al. 2020).

6.3.3 Androgen in Fish Undergoing Temperature-Induced Masculinization

Apart from their role in sex determination in the absence of stress mentioned above, androgens have recently been suspected to be endogenous drivers for testicular differentiation in temperature-induced masculinization (Fernandino et al. 2012). For example, in the pejerrey (*Odontesthes bonariensis*) and Japanese flatfish (*Paralichthys olivaceus*), larvae exposed to masculinizing temperature during a critical period of sex determination tend to develop testes rather than ovaries, with increased levels of whole-body cortisol (Hattori et al. 2009; Yamaguchi et al. 2010). It was hypothesized that the masculinizing effect could be due to either the concomitant synthesis of 11-KT with cortisol inactivation by Hsd11b (shown in blue in Fig. 6.6) (Fernandino et al. 2012) or the suppressing role of cortisol on the aromatase (Cyp19a1a) gene expression (Yamaguchi et al. 2010) (shown in red in Fig. 6.6).

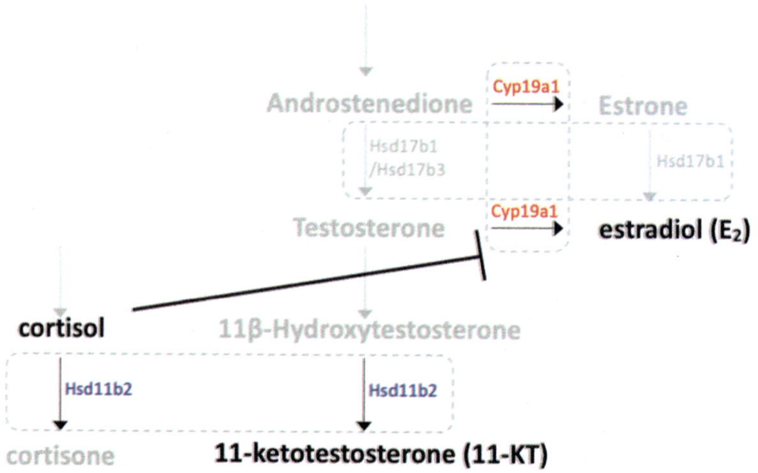

Fig. 6.6 Simplified pathways of biosynthesis of glucocorticoids and sex steroids in fish and two proposed models for temperature-induced masculinization. Yamaguchi et al. (2010) proposed that masculinization is due to the direct suppression of cortisol induced by the temperature stress on the aromatase/Cyp19a1 (red) gene expression (shown as "-|"). On the other hand, Fernandino et al. (2012) postulated that masculinization is due to the concomitant synthesis of 11-KT by Hsd11b2 (blue) whose expression is increased via the inactivation process of cortisol induced by the temperature stress

6.3.4 Androgen in Fish Undergoing in Sequential Sex-Change

In sequential hermaphroditic fishes (sex-changing fishes), it is now generally considered that a rapid decrease of estrogens is important for the initiation of gonadal sex change from females to males, while androgens play roles in the maintenance of the testicular function (Ortega-Recalde et al. 2020; Li et al. 2019; Nagahama et al. 2021). However, the endogenous androgen (11-KT) is suspected to be the potential trigger of the gonadal sex change in some protogynous fishes that exhibit female-to-male sex change. For example, an increase in 11-KT levels without a rapid decrease in estradiol levels was observed in an early transition phase from female to male in the Honeycomb grouper (*Epinephelus merra*) (Bhandari et al. 2003; Alam et al. 2006). Moreover, a recent study revealed that many grouper species commonly contain androgen-producing cells in the tunica of the ovaries, which are likely involved in the increase in 11-KT in the early transition phase, and hence in the onset of the sex change (Murata et al. 2021).

6.3.5 Androgen in a Frog Species

Outside of teleost fish, the endogenous androgen (testosterone) is suspected to play a key role in the early steps of gonadal sex differentiation in a species of amphibians, the Japanese wrinkled frog (*Glandirana* (formerly *Rana*) *rugosa*) (Miura 2007; Oike et al. 2017). In this species, the androgen receptor (*Ar*) gene resides on both sex chromosomes (W and Z) along with *Sox3* and *Sf-1* (Miura 2017), but the expression of the W-type *Ar* allele is barely observed (Oike et al. 2017). Overexpression and knockdown of *Ar* in this frog resulted in the formation of ovotestis, suggesting that *Ar* expression is required fore normal testis development (Oike et al. 2017).

6.3.6 An Inconvenient Truth: Estrogen in Medaka Fish

In contrast to the case of androgens, the importance of estrogens in the gonadal fate determination or the early ovarian differentiation in fish has been well accepted (Piferrer and Guiguen 2008; Guiguen et al. 2010; Devlin and Nagahama 2002). However, Yamamoto's model was criticized soon after it was proposed, ironically based on studies using medaka. Specifically, it was reported that the steroid-producing cells were clearly observed in medaka only after the initiation of sexual dimorphic differentiation in the ovary as well as in the testis (Satoh 1974; Kanamori et al. 1985), implying that sex steroids are not involved in determining the gonadal fate in this species.

6.3.7 Medaka Depleted in Both Androgens and Estrogens

Experimental evidence against the hypothesis was obtained from genetic studies using medaka. As already touched on in the section on androgen, Cyp17a1 (P450c17-I) is involved in the steroidogenic pathway that produces both estrogens and androgens (shown in green in Fig. 6.5) (Devlin and Nagahama 2002; Zhou et al. 2007; Sato et al. 2008). Therefore, the synthesis of estrogens as well as androgens is expected to be severely impaired when the function of Cyp17a1 is lost. However, in the mutant medaka with a female genotype (XX), early oogenesis and folliculogenesis appeared not to be affected (Sato et al. 2008). Later, the mutant developed a gonad with both oocytes and spermatozoa, suggesting that estrogens have a role in maintaining the ovarian fate in later stages.

6.3.8 *Estrogen in Medaka*

Evidence more directly related to estrogen was obtained when a pharmacological inhibitor of aromatase/Cyp19a1 was administrated to medaka to block estrogen synthesis (shown in red in Fig. 6.5). Repeated oral administration of the aromatase inhibitor to hatching larva revealed that the inhibitor treatment did not affect early oogenesis in genetically female medaka, and spermatogenesis was occasionally observed later in the ovaries (Suzuki et al. 2004). A similar phenotype in oogenesis was also obtained in a mutant medaka deficient in ovarian aromatase (Cyp19a1): the gonads first differentiated into ovaries with follicles, and the ovaries subsequently degenerated, followed by the appearance of testicular tissues (Nakamoto et al. 2018). These results suggest that endogenous estrogens are likely not essential for either sex determination or early oogenesis/folliculogenesis in female medaka but are required for the maintenance of ovarian fate at a later stage. Of note, the master sex-determining (MSD) gene in medaka was identified as *Dmy* (a duplicated copy of *Dmrt1*) (Matsuda et al. 2002; Nanda et al. 2002). However, the pathway linking this gene to steroid synthesis remains elusive.

6.3.9 *Estrogen in Nile Tilapia and the Rainbow Trout*

Contrary to the medaka studies, however, evidence supporting at least the requirement of estrogen for early stages of gonadal development, if not gonadal sex determination itself, has been accumulating in other fish such as the Nile tilapia (Nakamura et al. 2003) and rainbow trout (Vizziano et al. 2007). For example, *Cyp19a1*-positive cells became detectable before morphological sex differentiation in the prospective ovary of the Nile tilapia and rainbow trout (Nakamura et al. 1998; Vizziano et al. 2007). Gene expression profiling during the critical period of sex differentiation in both species confirmed that key genes for estrogen synthesis including *Cyp19a1* have been expressed before ovarian morphological differentiation (Vizziano et al. 2007; Ijiri et al. 2008). Oral administration of an aromatase inhibitor resulted in female-to-male sex reversal in 100% of the rainbow trout and 75.3% of the Nile tilapia (Guiguen et al. 1999). In *Cyp19a1*-deficient tilapia with a female genotype (XX), the gonads differentiated into testes rather than ovaries (Zhang et al. 2017). Remarkably, a sign of sex reversal in this mutant was observed at a very early stage of gonadal development when a sexually dimorphic difference in the germ cell number appeared. This histological differentiation precedes other morphological differentiation such as the formation of a testicular efferent duct or ovarian cavity in the Nile tilapia (Kobayashi and Nagahama 2009).

Taken together, these results suggest that estrogens are needed in the very early stages of ovarian differentiation in these species (Fig. 6.7). Furthermore, a recent study on the MSD gene, *sdY*, in the rainbow trout sheds light on a link between its MSD gene and estrogen synthesis (Bertho et al. 2018). The rainbow trout *sdY*

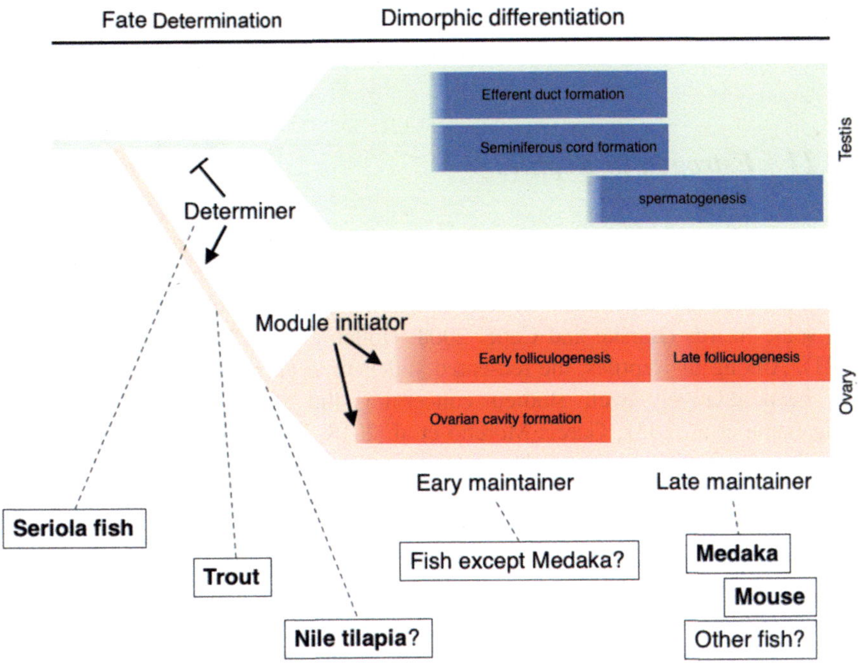

Fig. 6.7 The role of estrogens in the sex determination and differentiation period varies among fish species

encodes a truncated form of an immune-related protein and directly modulates *Cyp19a1* expression in cooperation with Foxl2 and Nr5a1, resulting in depletion of estrogens in a male genotype (XY) (Bertho et al. 2018) (Fig. 6.7).

6.3.10 Estrogen in Birds

Outside of teleost fish, it is well recognized that estrogen plays a major role in the early stages of ovarian differentiation in birds (Lambeth et al. 2013), while a very likely candidate MSD gene has been *DMRT1* that resides on the Z chromosome but not on the W chromosome (Smith et al. 2009). Estrogen treatment of genetic male (ZZ genotype) embryos results in testis-to-ovary sex reversal, and inhibition of estrogen synthesis in genetic female (ZW) embryos promotes ovary-to-testis sex reversal (Elbrecht and Smith 1992; Burke and Henry 1999). A recent genome editing experiment in chicken (*Gallus domesticus*) showed that loss of one copy of *DMRT1* in genetic males ($ZZ^{DMRT1+/DMRT1-}$) caused testis-to-ovary sex reversal, confirming that *DMRT1* is the key switch gene necessary for testis development (Ioannidis et al. 2021). In addition, the experiment with an aromatase inhibitor revealed that the production of estrogen is also a key factor in primary sex determination in chicken by suppressing the testis pathway activated by DMRT1 (Ioannidis et al. 2021).

6.3.11 Estrogen in Reptiles

Sex determination in reptiles is very complicated, ranging from temperature-dependent sex determination, genetic sex determination, and their coexistence, in which temperature-dependent sex determination prevails in crocodiles, most turtles, and some lizards (Barske and Capel 2008; Holleley et al. 2016; Nagahama et al. 2021). Among the reptiles, the red-eared slider turtle (*Trachemys scripta elegans*) has been relatively well studied concerning the sex determination process (Nagahama et al. 2021; Garcia-Moreno et al. 2018; Yao and Capel 2005). In this species, eggs incubated at a male-producing temperature with exogenous estrogen developed into females, whereas administration of an aromatase inhibitor to an egg incubating at female-producing temperature resulted in male development (Crews and Bergeron 1994; Wibbels and Crews 1994; Lance 2009; Pieau and Dorizzi 2004). Moreover, *Cyp19a1* expression is higher in embryonic gonads at the female-producing temperature than at the male-producing temperature (Czerwinski et al. 2016), suggesting that estrogens play a role in sex determination or subsequent differentiation. While recent studies in this species suggest that DNA methylation of the *Dmrt1* promoter might be key to determining gonadal fate by temperature (Ge et al. 2017), the direct link between Dmrt1 and estrogen production or the precise role of estrogens in sex determination is still unknown.

6.3.12 Estrogen in Amphibians

In many amphibians, gonadal differentiation is responsive to steroid manipulation, but the extent of the response varies among species (Nagahama et al. 2021; Flament 2016). Among amphibians, two species, *Xenopus laevis* and *G. rugosa*, have been well studied concerning the mechanism of sex determination and sex differentiation. Female-biased expression of the *Cyp19a1* gene in the gonads is observed just after the sex determination step or during early gonadal differentiation in both species (Maruo et al. 2008; Mawaribuchi et al. 2014). While their MSD genes or candidates have been known: *Dmw* in *X. laevis* and *Ar*, *Sox3* or *Sf-1* in *G. rugosa* (Yoshimoto et al. 2008; Miura 2017), the link between estrogen production and their MSD genes or the precise role of estrogens in sex determination is still unknown.

6.4 Insights from Master Sex-Determining Genes

6.4.1 Limitations of Loss-of-Function Experiments

As discussed above, pharmacological and genetic inhibition of the endogenous enzymes that catalyze sex steroids resulted in sex-reversal phenotypes in several species, providing strong evidence to support the requirement of estrogens to trigger ovarian differentiation, promote early differentiation, or maintain ovarian identity. However, loss-of-function experiments alone are not sufficient to distinguish the role of estrogens among them because estrogens are often essential for maintaining gonadal identity in nonmammalian vertebrates (e.g., Paul-Prasanth et al. 2013; Takatsu et al. 2013). Thus, late developmental phenotypes in loss of function experiments, i.e., gonadal sex reversal, can obscure early developmental phenotypes. Furthermore, in the case of teleost fish, the existence of two paralogs of the steroidogenic enzyme gene in the genome due to an ancient genome duplication does not permit a simple interpretation of the genetic inhibition experiments (Tenugu et al. 2021; Zhou et al. 2021; Force et al. 1999; Jaillon et al. 2004).

6.4.2 Master Sex-Determining Genes

Alternative evidence demonstrating the steroid hypothesis could be obtained if there is a species whose gonadal sex is determined by the master control gene that affects sex steroid levels. Although the search for such species has been previously considered to be impractical due to the difficulty for the identification of MSD genes in nonmodel organisms, recent advances in DNA sequencing technology and its commoditization enabled us to identify MSD genes (or strong candidates) in many vertebrate species, in particular, in fish, by using genome-wide association studies (e.g., Table 1 in Nagahama et al. 2021; Pan et al. 2021).

As mentioned in the previous section for some examples, more than a dozen different MSD genes have been identified in nonmammalian vertebrates to date. However, the majority of them encode Dmrt1, the Sox-family proteins, or proteins involved in the TGF-β pathway (e.g., Matsuda et al. 2002; Smith et al. 2009; Hattori et al. 2012; Kamiya et al. 2012; Takehana et al. 2014; reviewed in Kikuchi and Hamaguchi 2013; Bachtrog et al. 2014; Nagahama et al. 2021; Pan et al. 2021). Thus, the direct link between MSD genes and sex steroid levels has been unclear until very recently, although it can reasonably be surmised that steroidogenesis acts downstream of these MSD genes.

In 2018, a link between the MSD gene and sex steroid levels was reported in rainbow trout, as mentioned above (Bertho et al. 2018). An even more direct link was reported in four *Seriola* fish (the California yellowtail *S. dorsalis*, the greater amberjack *S. dumerili*, the Japanese amberjack *S. quinqueradiata*, and the yellowtail kingfish *S. lalandi*) in 2018 and 2019 (Purcell et al. 2018; Koyama et al. 2019).

In three Seriola fish including the greater amberjack, the Japanese amberjack, and the yellowtail kingfish, a Z-specific single nucleotide missense substitution in the coding region of *Hsd17b1* has been identified as the sole polymorphism associated with phenotypic sex (Koyama et al. 2019). Biochemical and molecular dynamics analyses revealed that Z-type Hsd17b1 protein is a hypomorphic form attenuating production of estradiol relative to the allelic product from the W chromosome (shown in magenta in Fig. 6.5), by disrupting the hydrogen bond network between the steroid and the enzyme's catalytic residues. Therefore, fish homozygous for the hypomorphic allele ($ZZ^{Hsd17bz/Hsd17bz}$) develop into males due to the depletion of estrogens (especially estradiol) during the sex determination period (Koyama et al. 2019). In the California yellowtail, a W-specific 61-base deletion upstream of the *Hsd17b1* gene has been identified. It was hypothesized that the deletion disrupts a putative silencer motif of *Hsd17b1* and thereby enhances estrogen production, leading to ovarian development in a female genotype ($ZW^{+/del}$) (Purcell et al. 2018).

Besides, in two species of *Trachinotus* fish (*T. ovatus* and *T. anak*) as well, allelic variation at the *Hsd17b1* locus, an intronic polymorphism in the splicing site this time, has been shown again to be associated with the phenotypic sex (Fan et al. 2021; Guo et al. 2021). Furthermore, in two species of tuna fish (*Thunnus orientalis* and *T. maccoyii*), the male-specific *Sult1st6Y* gene encoding an estrogen sulfotransferase appears to trigger testicular differentiation through inactivation of estrogens by sulfation in genetic males ($XY^{-/Sult1st6Y}$) (Nakamura et al. 2021).

In addition, an allelic variation of an estrogen signaling-related gene, *Bcar1*, has been proposed as the candidate MSD allele in the channel catfish (*Ictalurus punctatus*) (Bao et al. 2019). Since human BCAR1 interacts with estrogen receptor α (ERα) in human breast carcinoma cells (Cabodi et al. 2004) and the channel catfish *Bcar1* resides on the Y chromosome, it was assumed that the male-biased expression of this gene during the sex determination period resulted in an inhibition of estrogen receptor signaling and drives testicular development in catfish with a male genotype ($XY^{Bcar1x/Bcar1y}$).

6.5 Conclusions and Future Perspective

Together, these genetic studies clearly indicate that estrogens can be endogenous sex inducers by triggering ovarian differentiation in some species of fish, but not in others. In other words, the role of estrogens in cell fate specification and maintenance at the early gonadal differentiation likely vary among species even within teleost fish (Fig. 6.7). The situation in many species may fall between two extremes: (1) Seriola fish in which estrogen production is the initial trigger for sex determination, and (2) medaka fish in which estrogens are required neither for gonadal sex determination nor early ovarian differentiation (Fig. 6.7).

While it is now evident that the whole process of gonadal sex determination can lie downstream of estrogen activity in some fishes, there are still many unanswered questions about the detailed molecular processes that lead from estrogens to the

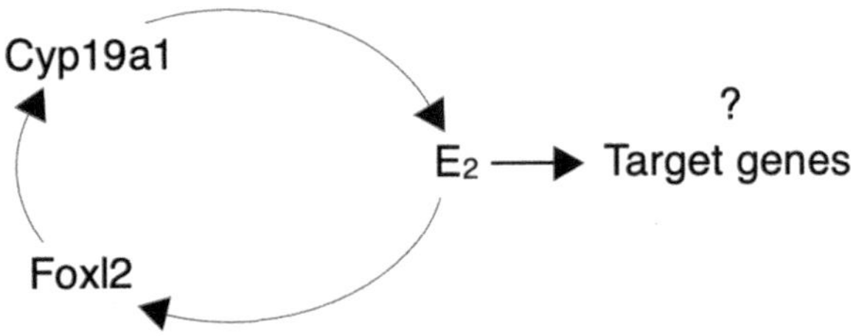

Fig. 6.8 Target genes of estrogens responsible for the gonadal fate determination and its maintenance. The action of estrogens is likely maintained by a positive feedback loop of Cyp19a, estrogens (represented by E_2), and Foxl2 in fish (Wang et al. 2007; Bertho et al. 2018). However, target genes of estrogens during sex determination/differentiation have been yet to be determined

gonadal fate determination. For example, although the action of estrogens should be mediated by estrogen receptors (ERs) (Nagahama et al. 2021) and likely maintained by a positive feedback loop of Cyp19a, estrogens, and Foxl2 in fish (Wang et al. 2007; Bertho et al. 2018) (Fig. 6.8), the mode of mediation of estrogen's effects and their target genes during sex determination/differentiation have been elusive. Indeed, despite a large number of studies reporting transcriptome analysis of differentiating ovaries and testes in fish (e.g., Sreenivasan et al. 2008; Tao et al. 2013; Koyama et al. 2019), direct target genes of ERs leading to ovarian development have yet to be identified (Fig. 6.8).

Given that an enormous number of ER target genes have been reported in human breast cancer cells and mouse mammary gland (Carroll et al. 2006; Palaniappan et al. 2019), it is reasonable to speculate that estrogens control many ER target genes during the period of sex determination and early steps of ovarian differentiation in nonmammalian vertebrates. Consistent with this expectation, of 337 female-biased genes in the differentiating gonads of the American alligator *Alligator mississippiensis*, 116 genes are in predicted estrogen-responsive (ER1-targeted) genomic regions (Rice et al. 2017). However, recent studies in zebrafish raised the intriguing possibility that the number of target genes may be limited. By generating double mutants of *cyp19a1* and *dmrt1*, Wu et al. (2020) showed that disruption of *dmrt1* rescues an initiation defect of folliculogenesis observed in the *cyp19a1a* mutant. In this double mutant, ovarian follicles developed normally up to the previtellogenic stage. This result suggests that estrogens are dispensable for early folliculogenesis in the absence of dmrt1 function (Wu et al. 2020), and thus, the ER target genes may be restricted to dmrt1-activated genes or *dmrt1* itself (Fig. 6.9). A similar dispensability of estrogen in early ovarian differentiation in the absence of DMRT1 function was recently reported in chicken (Ioannidis et al. 2021), though this study focused on earlier stages than those of folliculogenesis.

From the above discussion, the conclusions that could be drawn are as follows.

1. Although Yamamoto proposed that androgens can act as endogenous testes inducers in fish, it is not likely the case in many fishes except for fish that undergo temperature-induced masculinization or sequential sex change. Testes are likely

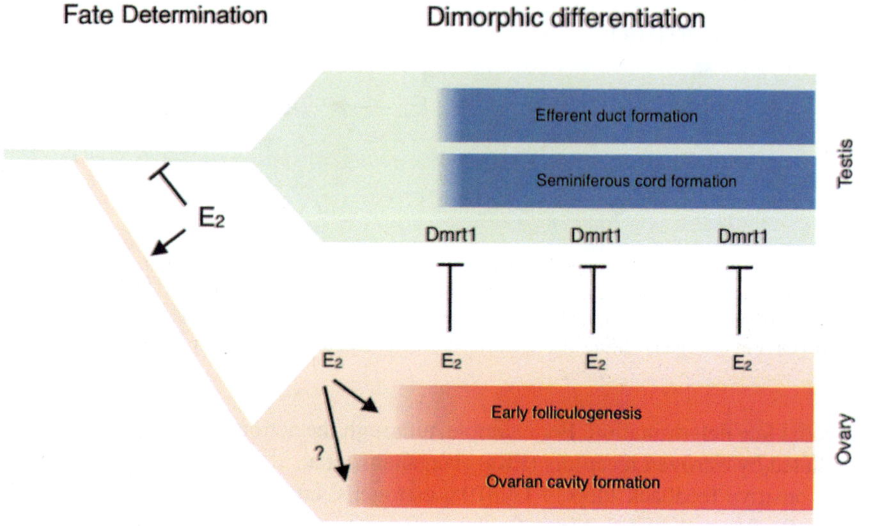

Fig. 6.9 A model for the role of estrogens in the sex determination and early sex differentiation. The role of estrogens (represented by E_2) as a gonadal sex determiner would be to activate the ovarian, while repressing the testicular pathway, or both during the sex determination period. To maintain the ovarian fate, continuous repression of the Dmrt1-activated pathway by E_2 is required. E_2 may play no active role in the developmental module of early folliculogenesis except for the initiation, contrary to a traditional view in nonmammalian vertebrates. The role of E_2 at the sex determination period can be performed by other MSD genes. Furthermore, the role of E_2 in the subsequent differentiation can be performed by other unknown factors

the default gonads in many fishes as far as sex steroid signaling is concerned: the undifferentiated gonads autonomously develop into testes in the absence of sex steroids. With some caution, the default state of gonads with respect to MSD gene in a species can be changeable with the replacement of their MSD gene. The role of androgens in sex determination remains to be elucidated in amphibians and reptiles.

2. As Yamamoto predicted, estrogens (namely estradiol) are most likely natural inducers that trigger ovarian differentiation in some fishes such as *Seriola* (Fig. 6.7). However, this role of estrogen to turn on or off the ovarian pathway can be performed by other MSD genes such as *Dmrt1*, *Sox*, and *Tgfb*-related genes. Currently, no direct link between these MSD genes and estrogens has been shown except for *sdY* in rainbow trout. It would be interesting to see whether such divergent MSD signaling converged into estrogen signaling or if estrogens only play a part in their downstream pathway in other species. From this view point, the sex determination mechanism of sharks would be also interesting. The precise role of estrogens in sex determination remains to be elucidated in amphibians and reptiles.
3. In a simplified model (Fig. 6.9), the role of estrogens as the gonadal sex determiner would be achieved by either activating the ovarian pathway or repressing the

testicular pathway during the sex determination period, and continuous repression of the pro-testis factors such as Dmrt1 during the early differentiation period. Estrogens may play no active role in the developmental module of the early folliculogenesis, and thus, the number of genes directed by estrogens in sex determination and early ovarian development could be far fewer than previously thought. It should be emphasized that direct target genes of the estrogen signaling leading to ovarian development have yet to be identified (Fig. 6.8).

References

Adolfi MC, Herpin A, Schartl M (2021) The replaceable master of sex determination: bottom-up hypothesis revisited. Philos Trans R Soc B Biol Sci 376:20200090

Alam MA, Bhandari RK, Kobayashi Y, Nakamura S, Soyano K et al (2006) Changes in androgen-producing cell size and circulating 11-ketotestosterone level during female-male sex change of honeycomb grouper Epinephelus merra. Mol Reprod Dev 73:206–214

Bachtrog D, Mank JE, Peichel CL, Kirkpatrick M, Otto SP et al (2014) Sex determination: why so many ways of doing it? PLoS Biol 12:e1001899

Bao L, Tian C, Liu S, Zhang Y, Elaswad A et al (2019) The Y chromosome sequence of the channel catfish suggests novel sex determination mechanisms in teleost fish. BMC Biol 17:1–16

Barske LA, Capel B (2008) Blurring the edges in vertebrate sex determination. Curr Opin Genet Dev 18:499–505

Berkowitz P (1937) Effect of estrogenic substances in Lebistes reticulatus (guppy). Proc Soc Exp Biol Med 36:416–418

Bertho S, Herpin A, Branthonne A, Jouanno E, Yano A et al (2018) The unusual rainbow trout sex determination gene hijacked the canonical vertebrate gonadal differentiation pathway. Proc Natl Acad Sci U S A 115:12781–12786

Berthold AA (1849) Transplantation der Hoden. Archiv für Anatomie, Physiologie und Wissenschaftliche Medici 16:42–46

Bhandari RK, Komuro H, Nakamura S, Higa M, Nakamura M (2003) Gonadal restructuring and correlative steroid hormone profiles during natural sex change in protogynous honeycomb grouper (Epinephelus merra). Zool Sci 20:1399–1404

Bogart MH (1987) Sex determination: a hypothesis based on steroid ratios. J Theor Biol 128:349–357

Borg B (1994) Androgens in teleost fishes. Comp Biochem Physiol C: Pharmacol Toxicol Endocrinol 109:219–245

Bouin P, Ancel P (1903) Sur la signification de la glande interstitielle du testicule embryonnaire. C R Soc Biol 55:1682

Britt K, Findlay J (2002) Estrogen actions in the ovary revisited. J Endocrinol 175:269–276

Bull J, Gutzke W, Crews D (1988) Sex reversal by estradiol in three reptilian orders. Gen Comp Endocrinol 70:425–428

Burke W, Henry M (1999) Gonadal development and growth of chickens and turkeys hatched from eggs injected with an aromatase inhibitor. Poult Sci 78:1019–1033

Burns RK (1925) The sex of parabiotic twins in Amphibia. J Exp Zool 42:31–89

Burns RK (1955) Experimental reversal of sex in the gonads of the opossum didelphis Virginiana. Proc Natl Acad Sci U S A 41:669

Burns RK (1961) Role of hormones in the differentiation of sex. In: Young WC (ed) Sex and internal secretions I. Williams & Wilkins, Baltimore, pp 76–160

Cabodi S, Moro L, Baj G, Smeriglio M, Di Stefano P et al (2004) p130Cas interacts with estrogen receptor α and modulates non-genomic estrogen signaling in breast cancer cells. J Cell Sci 117: 1603–1611

Capel B (2017) Vertebrate sex determination: evolutionary plasticity of a fundamental switch. Nat Rev Genet 18:675–689

Carroll JS, Meyer CA, Song J, Li W, Geistlinger TR et al (2006) Genome-wide analysis of estrogen receptor binding sites. Nat Genet 38:1289–1297

Castelnuovo G (1937) Effetti di alcuni ormoni sulla maturazione delle carpe. Riv Biol 23:365–372

Coveney D, Shaw G, Renfree MB (2001) Estrogen-induced gonadal sex reversal in the tammar wallaby. Biol Reprod 65:613–621

Crews D, Bergeron J (1994) Role of reductase and aromatase in sex determination in the red-eared slider (Trachemys scripta), a turtle with temperature-dependent sex determination. J Endocrinol 143:279–289

Cutting A, Chue J, Smith CA (2013) Just how conserved is vertebrate sex determination? Dev Dyn 242:380–387

Czerwinski M, Natarajan A, Barske L, Looger LL, Capel B (2016) A timecourse analysis of systemic and gonadal effects of temperature on sexual development of the red-eared slider turtle Trachemys scripta elegans. Dev Biol 420:166–177

Devlin RH, Nagahama Y (2002) Sex determination and sex differentiation in fish: an overview of genetic, physiological, and environmental influences. Aquaculture 208:191–364

Elbrecht A, Smith RG (1992) Aromatase enzyme activity and sex determination in chickens. Science 255:467–470

Fan B, Xie D, Li Y, Wang X, Qi X et al (2021) A single intronic single nucleotide polymorphism in splicing site of steroidogenic enzyme hsd17b1 is associated with phenotypic sex in oyster pompano, Trachinotus anak. Proc R Soc B 288:20212245

Fernandino JI, Hattori RS, Kishii A, Strüssmann CA, Somoza GM (2012) The cortisol and androgen pathways cross talk in high temperature-induced masculinization: the 11-β-hydroxysteroid dehydrogenase as a key enzyme. Endocrinology 153:6003–6011

Flament S (2016) Sex reversal in amphibians. Sex Dev 10:267–278

Force A, Lynch M, Pickett FB, Amores A, Yan Y-L et al (1999) Preservation of duplicate genes by complementary, degenerative mutations. Genetics 151:1531–1545

Ganesh S, Raman R (1995) Sex reversal by testosterone and not by estradiol or temperature in Calotes versicolor, the lizard lacking sex chromosomes. J Exp Zool 271:139–144

Garcia-Moreno SA, Plebanek MP, Capel B (2018) Epigenetic regulation of male fate commitment from an initially bipotential system. Mol Cell Endocrinol 468:19–30

Ge C, Ye J, Zhang H, Zhang Y, Sun W et al (2017) Dmrt1 induces the male pathway in a turtle species with temperature-dependent sex determination. Development 144:2222–2233

Gorbman A (1979) Emil Witschi and the problem of vertebrate sexual differentiation. Am Zool 19(4):1261–1270

Guiguen Y, Baroiller JF, Ricordel MJ, Iseki K, McMeel OM et al (1999) Involvement of estrogens in the process of sex differentiation in two fish species: the rainbow trout (Oncorhynchus mykiss) and a tilapia (Oreochromis niloticus). Mol Reprod Dev 54:154–162

Guiguen Y, Fostier A, Piferrer F, Chang C-F (2010) Ovarian aromatase and estrogens: a pivotal role for gonadal sex differentiation and sex change in fish. Gen Comp Endocrinol 165:352–366

Guiguen Y, Fostier A, Herpin A (2018) Sex determination and differentiation in fish: genetic, genomic, and endocrine aspects. In: Sex control in aquaculture, pp 35–63

Guo L, Yang J-W, Liu B, Malara D, Battaglia P et al (2021) Turnover of sex defining mutation provides an insight into evolution of sex chromosomes in the golden pompano (Trachinotus ovatus). Authorea. https://doi.org/10.22541/au.163706527.77995768/v1

Harikae K, Tsunekawa N, Hiramatsu R, Toda S, Kurohmaru M et al (2012) Evidence for almost complete sex-reversal in bovine freemartin gonads: formation of seminiferous tubule-like structures and transdifferentiation into typical testicular cell types. J Reprod Dev 58(6):654–660

Hattori RS, Fernandino JI, Kishii A, Kimura H, Kinno T et al (2009) Cortisol-induced masculinization: does thermal stress affect gonadal fate in pejerrey, a teleost fish with temperature-dependent sex determination? PLoS One 4:e6548

Hattori RS, Murai Y, Oura M, Masuda S, Majhi SK et al (2012) A Y-linked anti-Müllerian hormone duplication takes over a critical role in sex determination. Proc Natl Acad Sci U S A 109:2955–2959

Holleley CE, Sarre SD, O'Meally D, Georges A (2016) Sex reversal in reptiles: reproductive oddity or powerful driver of evolutionary change? Sex Dev 10:279–287

Humphrey RR (1928) Sex differentiation in gonads developed from transplants of the intermediate mesoderm of Amblystoma. Biol Bull 55:317–338

Humphrey RR (1945) Sex determination in ambystomid salamanders: a study of the progeny of females experimentally converted into males. Am J Anat 76:33–66

Hunter GA, Donaldson EM (1983) 5. Hormonal sex control and its application to fish culture. In: Fish physiology. Elsevier, pp 223–303

Ijiri S, Kaneko H, Kobayashi T, Wang DS, Sakai F et al (2008) Sexual dimorphic expression of genes in gonads during early differentiation of a teleost fish, the Nile tilapia Oreochromis niloticus. Biol Reprod 78:333–341

Ioannidis J, Taylor G, Zhao D, Liu L, Idoko-Akoh A et al (2021) Primary sex determination in birds depends on DMRT1 dosage, but gonadal sex does not determine adult secondary sex characteristics. Proc Natl Acad Sci U S A 118(10):e2020909118

Jaillon O, Aury J-M, Brunet F, Petit J-L, Stange-Thomann N et al (2004) Genome duplication in the teleost fish Tetraodon nigroviridis reveals the early vertebrate proto-karyotype. Nature 431:946–957

Josso N (2008) Professor Alfred Jost: the builder of modern sex differentiation. Sex Dev 2:55–63

Jost A (1971) Embryonic sexual differentiation (morphology, physiologie, abnormalities). Williams and Wilkins Company, Baltimore

Kamiya T, Kai W, Tasumi S, Oka A, Matsunaga T et al (2012) A trans-species missense SNP in Amhr2 is associated with sex determination in the tiger pufferfish, Takifugu rubripes (fugu). PLoS Genet 8:e1002798

Kanamori A, Nagahama Y, Nobuo E (1985) Development of the tissue architecture in the gonads of the medaka *Oryzias latipes*. Zool Sci 2:695–706

Keller K, Tandler I (1916) Über das Verhalten der Eihäute bei der Zwillingsträchtigkeit des Rindes

Kikuchi K, Hamaguchi S (2013) Novel sex-determining genes in fish and sex chromosome evolution. Dev Dyn 242:339–353

Kitano T, Takamune K, Nagahama Y, Abe SI (2000) Aromatase inhibitor and 17-α-methyltestosterone cause sex reversal from genetical females to phenotypic males and suppression of P450 aromatase gene expression in Japanese flounder (Paralichthys olivaceus). Mol Reprod Dev 56:1–5

Klein M, Bouin PA (1970) Complete dictionary of scientific biography. https://www.encyclopedia.com. Accessed 1 Mar 2022

Kobayashi T, Nagahama Y (2009) Molecular aspects of gonadal differentiation in a teleost fish, the Nile tilapia. Sex Dev 3:108–117

Koyama T, Nakamoto M, Morishima K, Yamashita R, Yamashita T et al (2019) A SNP in a steroidogenic enzyme is associated with phenotypic sex in Seriola fishes. Curr Biol 29:1901–1909.e1908

Lambeth LS, Cummins DM, Doran TJ, Sinclair AH, Smith CA (2013) Overexpression of aromatase alone is sufficient for ovarian development in genetically male chicken embryos. PLoS One 8: e68362

Lance VA (2009) Is regulation of aromatase expression in reptiles the key to understanding temperature-dependent sex determination? J Exp Zool A Ecol Genet Physiol 311:314–322

Li M, Sun L, Wang D (2019) Roles of estrogens in fish sexual plasticity and sex differentiation. Gen Comp Endocrinol 277:9–16

Lillie FR (1916) The theory of the free-martin. Science 43:611–613

Lubzens E, Young G, Bobe J, Cerdà J (2010) Oogenesis in teleosts: how fish eggs are formed. Gen Comp Endocrinol 165:367–389

Maruo K, Suda M, Yokoyama S, Oshima Y, Nakamura M (2008) Steroidogenic gene expression during sex determination in the frog Rana rugosa. Gen Comp Endocrinol 158:87–94

Matsuda M, Nagahama Y, Shinomiya A, Sato T, Matsuda C et al (2002) DMY is a Y-specific DM-domain gene required for male development in the medaka fish. Nature 417:559–563

Mawaribuchi S, Ikeda N, Fujitani K, Ito Y, Onuma Y et al (2014) Cell-mass structures expressing the aromatase gene Cyp19a1 lead to ovarian cavities in Xenopus laevis. Endocrinology 155: 3996–4005

Miura I (2007) An evolutionary witness: the frog rana rugosa underwent change of heterogametic sex from XY male to ZW female. Sex Dev 1:323–331

Miura I (2017) Sex determination and sex chromosomes in Amphibia. Sex Dev 11:298–306

Miura K, Harikae K, Nakaguchi M, Imaimatsu K, Hiramatsu R et al (2019) Molecular and genetic characterization of partial masculinization in embryonic ovaries grafted into male nude mice. PLoS One 14:e0212367

Murata R, Nozu R, Mushirobira Y, Amagai T, Fushimi J et al (2021) Testicular-inducing steroidogenic cells trigger sex change in groupers. Sci Rep 11:11117

Nagahama Y (2005) Molecular mechanisms of sex determination and gonadal sex differentiation in fish. Fish Physiol Biochem 31:105–109

Nagahama Y, Yamashita M (2008) Regulation of oocyte maturation in fish. Dev Growth Differ 50: S195–S219

Nagahama Y, Chakraborty T, Paul-Prasanth B, Ohta K, Nakamura M (2021) Sex determination, gonadal sex differentiation, and plasticity in vertebrate species. Physiol Rev 101:1237–1308

Nakamoto M, Shibata Y, Ohno K, Usami T, Kamei Y et al (2018) Ovarian aromatase loss-of-function mutant medaka undergo ovary degeneration and partial female-to-male sex reversal after puberty. Mol Cell Endocrinol 460:104–122

Nakamura M (2010) The mechanism of sex determination in vertebrates—are sex steroids the key-factor? J Exp Zool A Ecol Genet Physiol 313:381–398

Nakamura M, Kobayashi T, Chang XT, Nagahama Y (1998) Gonadal sex differentiation in teleost fish. J Exp Zool 281:362–372

Nakamura M, Bhandari RK, Higa M (2003) The role estrogens play in sex differentiation and sex changes of fish. Fish Physiol Biochem 28:113–117

Nakamura Y, Higuchi K, Kumon K, Yasuike M, Takashi T et al (2021) Prediction of the sex-associated genomic region in tunas (Thunnus fishes). Int J Genomics 2021:7226353

Nanda I, Kondo M, Hornung U, Asakawa S, Winkler C et al (2002) A duplicated copy of DMRT1 in the sex-determining region of the Y chromosome of the medaka, Oryzias latipes. Proc Natl Acad Sci U S A 99:11778–11783

Oike A, Kodama M, Yasumasu S, Yamamoto T, Nakamura Y et al (2017) Participation of androgen and its receptor in sex determination of an amphibian species. PLoS One 12:e0178067

Ortega-Recalde O, Goikoetxea A, Hore TA, Todd EV, Gemmell NJ (2020) The genetics and epigenetics of sex change in fish. Annu Rev Anim Biosci 8:47–69

Padoa E (1937) Differenziazione e inversione sessuale (femminizzazione) di avanotti di trota (Salmo irideus) trattati con ormone follicolare. Monitore zoologico italiano 48:195–203

Padula A (2005) The freemartin syndrome: an update. Anim Reprod Sci 87:93–109

Palaniappan M, Nguyen L, Grimm SL, Xi Y, Xia Z et al (2019) The genomic landscape of estrogen receptor α binding sites in mouse mammary gland. PLoS One 14:e0220311

Pan Q, Kay T, Depincé A, Adolfi M, Schartl M et al (2021) Evolution of master sex determiners: TGF-β signalling pathways at regulatory crossroads. Philos Trans R Soc B 376:20200091

Pandian TJ, Sheela SG (1995) Hormonal induction of sex reversal in fish. Aquaculture 138:1–22

Pask AJ, Calatayud NE, Shaw G, Wood WM, Renfree MB (2010) Oestrogen blocks the nuclear entry of SOX9 in the developing gonad of a marsupial mammal. BMC Biol 8:1–11

Paul-Prasanth B, Bhandari R, Kobayashi T, Horiguchi R, Kobayashi Y et al (2013) Estrogen oversees the maintenance of the female genetic program in terminally differentiated gonochorists. Sci Rep 3:1–11

Pieau C, Dorizzi M (2004) Oestrogens and temperature-dependent sex determination in reptiles: all is in the gonads. J Endocrinol 181:367–377
Pieau C, Dorizzi M, Richard-Mercier N (1999) Temperature-dependent sex determination and gonadal differentiation in reptiles. Cell Mol Life Sci 55:887–900
Piferrer F (2001) Endocrine sex control strategies for the feminization of teleost fish. Aquaculture 197:229–281
Piferrer F, Guiguen Y (2008) Fish gonadogenesis. Part II: molecular biology and genomics of sex differentiation. Rev Fish Sci 16:35–55
Piferrer F, Baker IJ, Donaldson EM (1993) Effects of natural, synthetic, aromatizable, and nonaromatizable androgens in inducing male sex differentiation in genotypic female chinook salmon (Oncorhynchus tshawytscha). Gen Comp Endocrinol 91:59–65
Piferrer F, Zanuy S, Carrillo M, Solar II, Devlin RH et al (1994) Brief treatment with an aromatase inhibitor during sex differentiation causes chromosomally female salmon to develop as normal, functional males. J Exp Zool 270:255–262
Purcell CM, Seetharam AS, Snodgrass O, Ortega-García S, Hyde JR et al (2018) Insights into teleost sex determination from the Seriola dorsalis genome assembly. BMC Genomics 19:1–11
Rashid H, Kitano H, Lee KH, Nii S, Shigematsu T et al (2007) Fugu (Takifugu rubripes) sexual differentiation: CYP19 regulation and aromatase inhibitor induced testicular development. Sex Dev 1:311–322
Rice ES, Kohno S, John JS, Pham S, Howard J et al (2017) Improved genome assembly of American alligator genome reveals conserved architecture of estrogen signaling. Genome Res 27:686–696
Sato T, Suzuki A, Shibata N, Sakaizumi M, Hamaguchi S (2008) The novel mutant scl of the medaka fish, Oryzias latipes, shows no secondary sex characters. Zool Sci 25:299–306
Satoh N (1974) An ultrastructural study of sex differentiation in the teleost Oryzias latipes. J Embryol Exp Morphol 32:195–215
Scheib D (1983) Effects and role of estrogens in avian gonadal differentiation. Differentiation 23 (Suppl):S87–S92
Smith CA, Roeszler KN, Ohnesorg T, Cummins DM, Farlie PG et al (2009) The avian Z-linked gene DMRT1 is required for male sex determination in the chicken. Nature 461:267–271
Spemann H, Mangold H (1924) Über die Induktion von Embryonalanlagen durch Implantation artfremder Organisatoren. Roux's Archiv für mikroskopische Anatomie und Entwicklungsmechanik. 100:599–638. (On the Induction of Embryonic Primordia by Implantation of Organizers from Different Species. Roux'Archives for' Microscopic Anatomy and Developmental Mechanics 100: 599-638)
Sreenivasan R, Cai M, Bartfai R, Wang X, Christoffels A et al (2008) Transcriptomic analyses reveal novel genes with sexually dimorphic expression in the zebrafish gonad and brain. PLoS One 3:e1791
Stewart MK, Mattiske DM, Pask AJ (2020) Exogenous oestrogen impacts cell fate decision in the developing gonads: a potential cause of declining human reproductive health. Int J Mol Sci 21: 8377
Suzuki A, Tanaka M, Shibata N, Nagahama Y (2004) Expression of aromatase mRNA and effects of aromatase inhibitor during ovarian development in the medaka, Oryzias latipes. J Exp Zool A Comp Exp Biol 301:266–273
Takatsu K, Miyaoku K, Roy SR, Murono Y, Sago T et al (2013) Induction of female-to-male sex change in adult zebrafish by aromatase inhibitor treatment. Sci Rep 3:1–7
Takehana Y, Matsuda M, Myosho T, Suster ML, Kawakami K et al (2014) Co-option of Sox3 as the male-determining factor on the Y chromosome in the fish Oryzias dancena. Nat Commun 5:1–10
Tao W, Yuan J, Zhou L, Sun L, Sun Y et al (2013) Characterization of gonadal transcriptomes from Nile tilapia (Oreochromis niloticus) reveals differentially expressed genes. PLoS One 8:e63604
Tenugu S, Pranoty A, Mamta S-K, Senthilkumaran B (2021) Development and organisation of gonadal steroidogenesis in bony fishes—a review. Aquac Fish 6:223–246

Tokarz J, Möller G, de Angelis MH, Adamski J (2015) Steroids in teleost fishes: a functional point of view. Steroids 103:123–144

Vizziano D, Randuineau G, Baron D, Cauty C, Guiguen Y (2007) Characterization of early molecular sex differentiation in rainbow trout, Oncorhynchus mykiss. Dev Dyn 236:2198–2206

Wang D-S, Kobayashi T, Zhou L-Y, Paul-Prasanth B, Ijiri S et al (2007) Foxl2 up-regulates aromatase gene transcription in a female-specific manner by binding to the promoter as well as interacting with ad4 binding protein/steroidogenic factor 1. Mol Endocrinol 21:712–725

Wibbels T, Crews D (1994) Putative aromatase inhibitor induces male sex determination in a female unisexual lizard and in a turtle with temperature-dependent sex determination. J Endocrinol 141:295–299

Winge Ö (1934) The experimental alteration of sex chromosomes into autosomes and vice versa, as illustrated by Lebistes. Lab Carlsberg Ser Physiol 21:1–47

Witschi E (1914) Experimentelle Untersuchungen über die Entwicklungsgeschichte der Keimdrüsen von Rana temporaria. Arch Mikrosk Anat 85:9–113

Wolf U (1999) Reorganization of the sex-determining pathway with the evolution of placentation. Hum Genet 105:288–292

Wu K, Song W, Zhang Z, Ge W (2020) Disruption of dmrt1 rescues the all-male phenotype of the cyp19a1a mutant in zebrafish–a novel insight into the roles of aromatase/estrogens in gonadal differentiation and early folliculogenesis. Development 147:dev182758

Yamaguchi T, Yoshinaga N, Yazawa T, Gen K, Kitano T (2010) Cortisol is involved in temperature-dependent sex determination in the Japanese flounder. Endocrinology 151:3900–3908

Yamamoto T (1953) Artificially induced sex-reversal in genotypic males of the medaka (Oryzias latipes). J Exp Zool 123:571–594

Yamamoto T (1958) Artificial induction of functional sex-reversal in genotypic females of the medaka (Oryzias latipes). J Exp Zool 137:227–263

Yamamoto T (1959a) A further study on induction of functional sex reversal in genotypic males of the medaka (Oryzias latipes) and progenies of sex reversals. Genetics 44:739

Yamamoto T (1959b) The effects of estrone dosage level upon the percentage of sex-reversals in genetic male (XY) of the medaka (Oryzias latipes). J Exp Zool 141:133–153

Yamamoto T (1961) Progenies of sex-reversal females mated with sex-reversal males in the medaka, Oryzias latipes. J Exp Zool 146:163–179

Yamamoto T (1962) Harmonic factors affecting gonadal sex differentiation in fish. Gen Comp Endocrinol 1:341–345

Yamamoto T (1969) Sex differentiation. In: Fish physiology. Elsevier, pp 117–175

Yao HH, Capel B (2005) Temperature, genes, and sex: a comparative view of sex determination in Trachemys scripta and Mus musculus. J Biochem 138:5–12

Yoshimoto S, Okada E, Umemoto H, Tamura K, Uno Y et al (2008) A W-linked DM-domain gene, DM-W, participates in primary ovary development in Xenopus laevis. Proc Natl Acad Sci U S A 105:2469–2474

Zhang X, Li M, Ma H, Liu X, Shi H et al (2017) Mutation of foxl2 or cyp19a1a results in female to male sex reversal in XX Nile tilapia. Endocrinology 158:2634–2647

Zhang Q, Ye D, Wang H, Wang Y, Hu W et al (2020) Zebrafish cyp11c1 knockout reveals the roles of 11-ketotestosterone and cortisol in sexual development and reproduction. Endocrinology 161

Zheng Q, Xiao H, Shi H, Wang T, Sun L et al (2020) Loss of Cyp11c1 causes delayed spermatogenesis due to the absence of 11-ketotestosterone. J Endocrinol 244:487–499

Zhou L-Y, Wang D-S, Kobayashi T, Yano A, Paul-Prasanth B et al (2007) A novel type of P450c17 lacking the lyase activity is responsible for C21-steroid biosynthesis in the fish ovary and head kidney. Endocrinology 148:4282–4291

Zhou L, Li M, Wang D (2021) Role of sex steroids in fish sex determination and differentiation as revealed by gene editing. Gen Comp Endocrinol 313:113893

Chapter 7
Sex Steroid Regulation of Male- and Female-Typical Mating Behaviors in Teleost Fish

Kataaki Okubo, Yuji Nishiike, Thomas Fleming, Yukiko Kikuchi, and Towako Hiraki-Kajiyama

Abstract Males and females exhibit innate sex-specific mating behaviors, which are established developmentally. In mammals and birds, phenotypic sex differences in mating behaviors are stable and essentially irreversible, because the underlying neural substrates are irreversibly sex-differentiated prior to puberty due to the effects of gonadal steroids and the sex chromosome complement. In contrast, experimental manipulation of the hormonal milieu of teleost fish in adulthood effectively reverses male and female mating behaviors, illustrating the lifelong sexual lability in their underlying neural substrates. Consistent with this, evidence is accumulating that, in teleosts, both early gonadal steroids and the sex chromosome complement have little effect on the development of sex-typical mating behaviors; instead, recent work in medaka (*Oryzias latipes*) demonstrates that mutual antagonism between estradiol-17β signaling through an estrogen receptor subtype Esr2b and 11-ketotestosterone signaling through androgen receptor in adulthood, rather than during development, mediates sex-typical mating behaviors in a reversible and transient manner. Further evidence is provided that the pronounced sexual dimorphism and adult steroid-dependent lability in the expression of Esr2b and downstream effectors, including neuropeptide B, in the telencephalic and preoptic nuclei underlie the neural basis of induction, maintenance, and reversal of male and female mating behaviors in teleosts.

Keywords Brain · Mating behavior · Sex steroid · Sexual spectrum · Teleost

K. Okubo (✉) · Y. Nishiike · T. Fleming · Y. Kikuchi
Department of Aquatic Bioscience, Graduate School of Agricultural and Life Sciences, The University of Tokyo, Bunkyo, Tokyo, Japan
e-mail: okubo@marine.fs.a.u-tokyo.ac.jp; nishiike@g.ecc.u-tokyo.ac.jp; fleming@g.ecc.u-tokyo.ac.jp

T. Hiraki-Kajiyama
Department of Aquatic Bioscience, Graduate School of Agricultural and Life Sciences, The University of Tokyo, Bunkyo, Tokyo, Japan

Laboratory for Systems Molecular Ethology, RIKEN Center for Brain Science, Wako, Saitama, Japan
e-mail: towako.kajiyama@riken.jp

M. Tanaka, M. Tachibana (eds.), *Spectrum of Sex*,
https://doi.org/10.1007/978-981-19-5359-0_7

7.1 Introduction

From invertebrates to humans, males and females of a given species display differences in a wide range of physiological and behavioral traits. Perhaps the most obvious differences are found in traits directly relevant to reproduction, including mating behaviors. Males commonly perform elaborate courtship displays to attract females for mating, while female behavior is primarily an evaluation of male suitability. These differences result from differential development and activation of the underlying neural substrates in males versus females (Yang and Shah 2014, 2016; Chen and Hong 2018).

In vertebrates, sex-specific mating behaviors and the underlying neural substrates are highly dependent on the milieu of gonadal hormones, particularly sex steroids, which is typical of each sex (McCarthy et al. 2017; Jennings and de Lecea 2020). In adult males, high and persistent levels of circulating androgens activate the neural substrates to facilitate the expression of male-typical mating behaviors, whereas in adult females, cyclic elevations in circulating estrogens, progestins, and prostaglandins activate the neural substrates to achieve female-typical mating behaviors. While this mechanism seems to be fundamental to the expression of sex-typical mating behaviors in all vertebrate taxa, large variations in the impact and mode of action of gonadal hormones are apparent across taxa.

This review briefly summarizes current knowledge on how gonadal hormones regulate male and female mating behaviors in teleost fish, which are unique among vertebrates in that their sex-specific phenotypes, including behaviors, are quite labile and can even be reversed as adults (Okubo et al. 2019; Nagahama et al. 2021). Considering the framework of this book, which regards sex as a spectrum rather than a binary, teleosts can transition from one end of the spectrum to the other throughout their lifetime. This, combined with knowledge from other vertebrates, highlights the mechanistic underpinnings of the permanence/lability of sex-specific mating behaviors, or in other words, the settlement and transition of these behaviors on the sexual spectrum.

7.2 Sex Differences in Mating Behaviors Are Stable and Essentially Irreversible in Mammals and Birds

While the pattern of gonadal hormone secretion in adulthood is certainly a proximate factor that influences sex-typical mating behaviors, the overall picture of the mechanism underlying these behaviors is not that simple. Reversal of sex-typical mating behaviors in adult mammals and birds does not generally occur. Even after ovariectomy and androgen administration to alter their sex steroid milieu to that typical of males, females do not exhibit male-typical mating behaviors to the same extent as males. Similarly, adult males of these taxa, if castrated and treated with estrogens and progestins, fail to show female-typical behaviors. These observations indicate

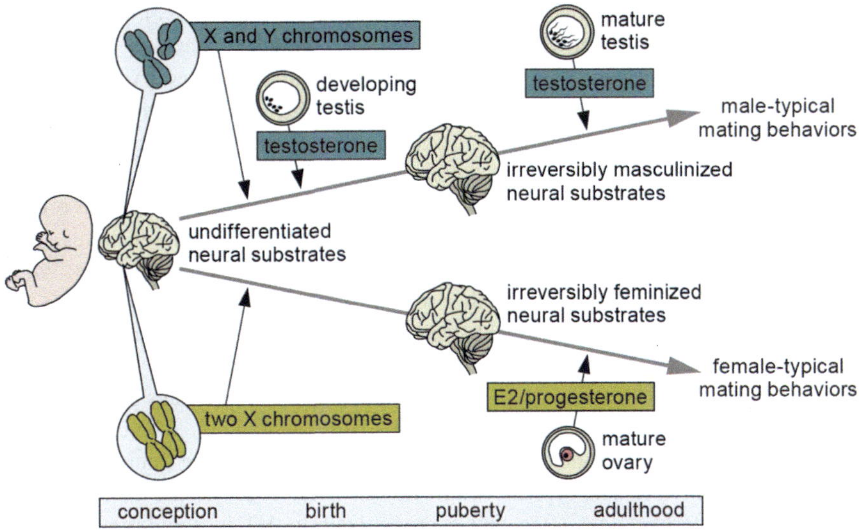

Fig. 7.1 Mechanisms underlying the development of sex-typical mating behaviors in mammals. The neural substrates that mediate mating behaviors are irreversibly and constitutively sex-differentiated prior to puberty, depending on early developmental exposure to gonadal steroids (testicular testosterone) and the sex chromosome complement of brain cells (XX vs. XY chromosomes). The sex-differentiated neural substrates are then activated in the adult by increased steroid secretion from mature gonads in a sex-specific manner (testosterone from the testes and estradiol-17β (E2) and progesterone from the ovaries) for manifestation of sex-typical behaviors

that differential expression of mating behaviors by males and females is essentially permanent and irreversible in mammals and birds.

More than half a century of research on these taxa (mostly in rodents) has revealed the mechanisms underlying the permanent differentiation of the male and female brain (McCarthy and Arnold 2011; McCarthy et al. 2017; McCarthy 2020; Balthazart 2019) (Fig. 7.1). In rodents, the fetal testes of males secrete large amounts of testosterone, whereas the fetal ovaries of females remain hormonally quiescent. Once in the developing male brain, testicular testosterone is largely converted by cytochrome p450 aromatase into estradiol-17β (E2), which then acts to masculinize (induce male-typical phenotypes of, or direct toward the male end of the sexual spectrum) and defeminize (prevent female-typical phenotypes of, or direct away from the female end of the spectrum) the neural substrates that later mediate sex-specific behaviors. In the absence of testosterone during development, as in the female brain, the neural substrates are feminized (female-typical phenotypes are induced or directed toward the female end of the sexual spectrum) and demasculinized (male-typical phenotypes are prevented or directed away from the male end of the spectrum), although exposure to E2 during the prepubertal period is required for full feminization (McCarthy et al. 2017; Balthazart 2019). A similar mechanism is present in primates, although testosterone acts directly to masculinize the brain, without prior conversion to E2 (Bao and Swaab 2011). In quail, in contrast to rodents, the fetal female ovaries secrete E2, which acts on the developing brain to feminize and demasculinize the neural substrates, while the absence of fetal E2 secretion masculinizes and defeminizes them as occurs in males (Adkins-Regan 2009; Maekawa et al. 2014; Balthazart 2019). In spite of such variations among species, these steroidal effects early in development, traditionally referred to as "organizational effects", represent a common mechanism for sexual differentiation of brain and behavior in mammals and birds. Importantly, the early steroid effects are basically irreversible and permanently differentiate the neural substrates. This allows for the manifestation of behaviors typical of one sex later in life, while concurrently preventing that of the other sex.

The sex-differentiated neural substrates are then activated in adulthood by increasing sex steroid secretion from mature gonads (McCarthy and Arnold 2011; McCarthy et al. 2017). In adult males, testicular testosterone activates the masculinized neural substrates to allow male-typical mating behaviors, whereas in adult females, E2 and progesterone arising from the ovaries activate the feminized neural substrates to allow female-typical behaviors (in male rodents, the bulk of adult testicular testosterone acts after being aromatized to E2 in the brain, as fetal testosterone does, and thus E2 is primarily responsible for mating behaviors, as in females). These effects of sex steroids in adulthood are traditionally referred to as "activational effects" and differ from early steroid effects in that they are reversible and transient in nature, thus existing only when steroids are continuously present.

Although gonadal sex steroids are the dominant drivers of most sexually differentiated behaviors, studies over the past few decades have established the contribution of the sex chromosome complement (XX vs. XY in mammals, or ZZ vs. ZW in birds) which directly and constitutively affects the sexual differentiation of some behaviors, independent of steroidal effects (Arnold 2017, 2020). A notable example comes from zebra finches, where the sex chromosome complement contributes to the masculine development of the song system, a series of interconnected nuclei that are essential for learning and producing courtship songs, a male-biased behavioral trait (Agate et al. 2003). These effects, referred to as "sex chromosome effects", have also been shown to contribute to parental and aggressive behaviors and social interactions (McCarthy and Arnold 2011), although the proximate mechanisms of these effects remain elusive.

Collectively, sex-specific behaviors and the underlying neural substrates are shaped by (1) the effects of gonadal sex steroids early in life, which cause irreversible and enduring sex differences in the developing neural substrates; (2) the effects of gonadal sex steroids in adulthood, which activate the differentiated neural substrates to drive sex-specific behaviors in a reversible and transient manner; and (3) the effects of sex chromosome complement, where direct sex-specific actions of sex chromosome genes within the brain induce sex differences in some behavioral phenotypes (Fig. 7.1). Mammals and birds exhibit stable sex differences in mating behaviors, most probably because their underlying neural substrates are irreversibly and constitutively sex-differentiated prior to puberty under the influence of early gonadal steroids and sex chromosome complement.

7.3 Mating Behaviors in Teleosts Are Highly Sexually Labile Across Their Lifetime

Although differential expression of mating behaviors by males and females is essentially fixed in mammals and birds, some studies, albeit few, have challenged the permanence of these behaviors and suggest that a certain degree of sexual lability is retained into adulthood. For example, female rodents can exhibit mating behaviors typical of males, though less frequently, either in response to testosterone or E2 treatment in adulthood (Edwards and Burge 1971; Södersten 1972). In addition, the medial preoptic nucleus of the quail brain, which reportedly is involved in male mating behaviors, is about 1.4 times larger in males than in females, and this sex difference relies to some extent on the adult steroid milieu (Adkins-Regan 2009; Balthazart et al. 2010; Balthazart 2019). In agreement with these findings, female quail treated with testosterone as adults display some male-specific courtship behaviors, such as crowing and strutting (Ball et al. 2014).

Notably, and interestingly, sexual lability in behavioral phenotypes is seen much more frequently and thoroughly in teleost fish, in which experimental manipulation of the hormonal milieu of adult males and females effectively reverses sex-typical behaviors. For example, adult male and female goldfish (*Carassius auratus*) receiving acute treatment with prostaglandin F2α (PGF2α) or androgens, respectively, display mating behaviors typical of the opposite sex (Stacey and Kyle 1983; Stacey and Kobayashi 1996; Ghosal and Sorensen 2016). Similarly, female stickleback (*Gasterosteus aculeatus*) and medaka (*Oryzias latipes*) respectively exhibit nest-building behavior or courtship displays that are typical of males, when treated with androgens as adults (Wai and Hoar 1963; Nishiike et al. 2021). Moreover, the frequency of courtship displays induced in androgen-treated female medaka can be as high as in males (Nishiike et al. 2021).

Teleosts are even more unique in that many species spontaneously undergo phenotypic sex reversal, involving both anatomical and behavioral changes, depending on physiological, environmental, and social cues (Godwin 2010; Liu et al. 2017; Capel 2017). Even in teleost species that do not normally sex-reverse in adult life, such as tilapia (*Oreochromis niloticus*), medaka, African cichlid (*Astatotilapia burtoni*), and zebrafish (*Danio rerio*), morphological, and often behavioral, sex reversal can be induced by chronically exposing adult females to an aromatase inhibitor to block the conversion of androgens to estrogens (Paul-Prasanth et al. 2013; Takatsu et al. 2013; Sun et al. 2014; Göppert et al. 2016). These facts strongly imply that brain and behavior of teleosts, regardless of whether the species normally switch sex or not, are sexually labile throughout their lifetime and can even be reversed as adults (Okubo et al. 2019).

7.4 Sex-Typical Mating Behaviors in Teleosts Are Largely Dependent on the Adult Steroid Milieu

The enduring adult sexual lability in mating behaviors in teleost fish suggests that their neural substrates undergo a unique process of sexual differentiation from those of mammals and birds. It is plausible to assume that gonadal sex steroids early in life cause only minimal, if any, irreversible sex differences in the neural substrates, or that these differences can readily be reversed by the prevailing sex steroid milieu in adulthood. The former supposition is supported by the observation that neither testosterone nor E2 levels are appreciably elevated in medaka embryos (Iwamatsu et al. 2005, 2006). In addition, the levels of aromatase expression in the developing medaka brain are much lower than those in adult fish, with no transient increase or sex difference (Okubo et al. 2011). This is in sharp contrast to what occurs in developing rodent brain, where high levels of aromatase activity are transiently induced in a male-biased manner to help masculinize and defeminize the neural substrates (Okubo et al. 2019).

Presumably, the sex chromosome complement also plays a minor role, if any, in sexual differentiation of the neural substrates in teleosts, because the permanent and invariable effects that the constitutive presence of sex chromosomes should produce are inconsistent with the sexual lability of teleosts. In line with this notion, in medaka and tongue sole (*Cynoglossus semilaevis*), sex-reversed males and females (i.e., males with the sex chromosome constitution XX or ZW and females with XY or ZZ chromosomes) appear to be as fertile and behaviorally active in mating as normal males and females (Volff et al. 2007; Paul-Prasanth et al. 2013; Chen et al. 2014), although there has been no quantitative examination of behavior or neural function of sex-reversed fish. The sex chromosomes of teleosts have arisen only recently and independently many times in different families, genera, or species (many teleost species do not have sex chromosomes, and sex is instead determined by environmental signals) (Marshall Graves and Peichel 2010; Kikuchi and Hamaguchi 2013; Gammerdinger and Kocher 2018). Since their sex chromosomes are still in the early stages of differentiation, the chromosome pairs are not morphologically distinguishable and are almost identical with only one or a few different loci. As an extreme example, in some fish species including pufferfish (*Takifugu rubripes*) and amberjacks (*Seriola dumerili*), the sex chromosome pairs essentially differ by only a single nucleotide (Kamiya et al. 2012; Koyama et al. 2019). The lack of extensive genetic differences between males and females probably limits the potential impact of the sex chromosome complement on sex-typical behaviors and the underlying neural substrates in teleosts (Okubo et al. 2019).

Assuming that neither early gonadal steroids nor the sex chromosome complement has much effect on sexual differentiation of the neural substrates underlying sex-typical mating behaviors, it is highly likely that gonadal steroids in adulthood play a decisive role in teleosts. The reversible and transient nature of the adult steroidal effects agrees with the marked sexual lability of behavioral phenotypes in teleosts. The reversal of sex-typical mating behaviors upon manipulation of the adult hormonal milieu described above exactly reflects such a nature, and suggests that these effects are largely, if not solely, responsible for sex-typical mating behaviors. In teleosts, gonadal steroids in adult life may serve to both differentiate the neural substrates and activate them to actuate sex-typical mating behaviors in a reversible and transient manner. It appears that the neural substrates for sex-typical mating behaviors in teleosts remain undifferentiated until the onset of puberty, after which they are differentiated and simultaneously activated depending on the sex-specific steroid milieu in the adult.

7.5 Teleosts Have a Unique Adult Steroid Milieu Compared to Mammals and Birds

The adult steroid milieu of teleost fish differs from that of mammals and birds in several important respects (Figs. 7.2 and 7.3). The mature teleost testes secrete copious quantities of 11-ketotestosterone (11KT) in addition to testosterone into the circulation, which eventually gain access to the brain. 11KT is generally more effective than testosterone in stimulating the development of secondary sexual characteristics, spermatogenesis, and mating behaviors in teleosts (Borg 1994). Because of this and the fact that the circulating levels of 11KT, unlike testosterone (see below), are much higher in males than in females, 11KT is regarded as the primary testicular androgen in teleosts (Devlin and Nagahama 2002). Importantly, 11KT also differs from testosterone in that it cannot be aromatized into estrogens, and all of its actions are mediated by the androgen receptor (AR) (Borg 1994). This mode of action is also distinct from that of 5α-dihydrotestosterone (DHT), a potent

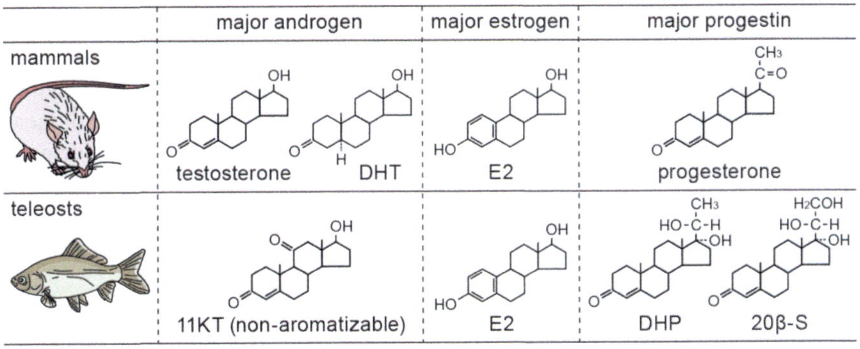

Fig. 7.2 Comparison of major sex steroids in mammals and teleosts. The major androgen in teleosts is 11-ketotestosterone (11KT; which cannot be aromatizable), and not testosterone or 5α-dihydrotestosterone (DHT) as in mammals. In teleosts, testosterone serves as a precursor to 11KT and estradiol-17β (E2), rather than as an androgenic steroid. In addition, the major progestin in teleosts is not progesterone as in mammals, but 17,20β-dihydroxy-4-pregnen-3-one (DHP), or in some species, 17,20β,21-trihydroxy-4-pregnen-3-one (20β-S). The major estrogen in both mammals and teleosts is E2

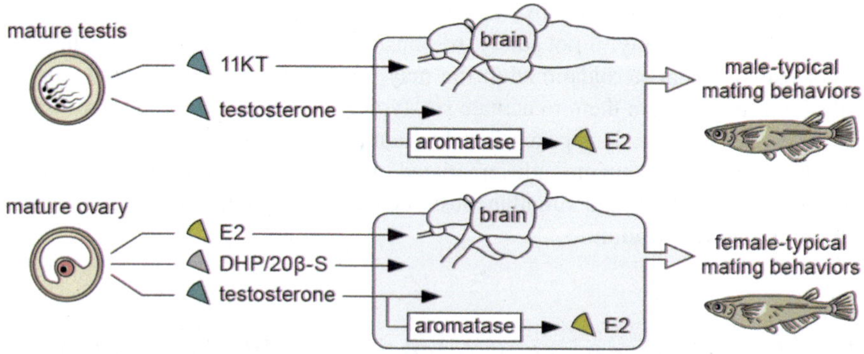

Fig. 7.3 The adult sex steroid milieu in teleosts. The mature testis primarily secretes 11-ketotestosterone (11KT) and testosterone, whereas the mature ovary mainly secretes estradiol-17β (E2), 17,20β-dihydroxy-4-pregnen-3-one (DHP) (or 17,20β,21-trihydroxy-4-pregnen-3-one (20β-S) in some species), and testosterone. Sex-typical mating behaviors in teleosts largely depend on the sex-specific adult steroid milieu. Of note, teleosts have much higher brain aromatase activity than any other vertebrates. Hence, circulating testosterone, once in the brain, is converted to E2, resulting in a significant accumulation of E2 even in the male brain

androgen in mammals and birds, which cannot be aromatized into estrogens, but whose metabolite, 3β-androstanediol, binds to the estrogen receptor (ESR, also known as ER) and exerts estrogenic effects (Handa et al. 2009).

The mature ovaries of teleosts secrete large amounts of E2 and progestin into the circulation, as do those of mammals and birds; however, the major progestin in teleosts is not progesterone as in mammals and birds, but 17,20β-dihydroxy-4-pregnen-3-one (DHP), or in some species, 17,20β,21-trihydroxy-4-pregnen-3-one (20β-S) (Scott et al. 2010). Strikingly, teleost ovaries also produce substantial amounts of testosterone, and its circulating levels in adult females are comparable to and sometimes higher than the levels found in adult males (Katz and Eckstein 1974; Borg 1994). Testosterone therefore likely functions as a precursor to both the major androgen 11KT and the major estrogen E2, rather than as an androgenic steroid, in teleosts. DHT is also produced in both the testes and ovaries of teleosts; however, its circulating levels are typically lower than those of testosterone and 11KT and are not significantly male-biased, leaving the role of DHT in teleosts largely unexplored (Margiotta-Casaluci et al. 2013; Martyniuk et al. 2013; Nishiike et al. 2021; Yazawa et al. 2021).

Another salient feature of the sex steroid milieu in teleosts is that their adult brain, irrespective of sex, has more than 100-fold greater aromatase activity than the adult mammalian and avian brains, and is thus a major site of E2 production along with the ovary (Pasmanik and Callard 1985). In rodents, brain aromatase activity reaches its peak during fetal life and declines gradually thereafter to adult levels, representing only a small percentage of the fetal peak; the remaining activity in adulthood is, however, still important for mating behaviors (Tobet et al. 1985; Beyer et al. 1993; Lephart 1996) (Fig. 7.4). In contrast, brain aromatase expression and activity in

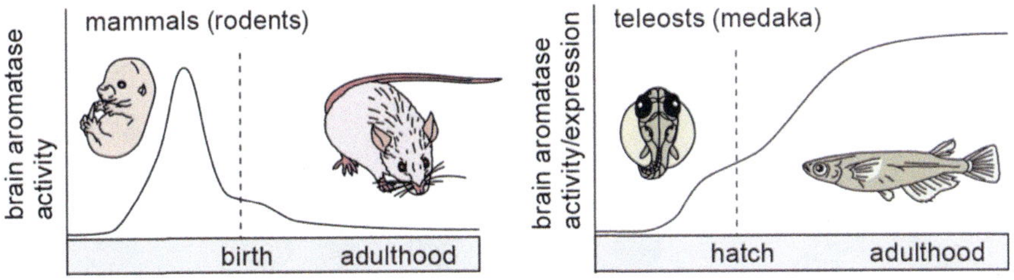

Fig. 7.4 Lifetime changes in brain aromatase activity/expression in mammals and teleosts. In mammals (rodents), brain aromatase activity peaks during fetal development and gradually decreases toward adulthood. In contrast, in teleosts (medaka), brain aromatase activity/expression increases with age and sexual maturation, reaching levels in adulthood comparable to or higher than the fetal peak in rodents

teleosts increases with age and sexual maturation and reaches levels comparable to or higher than the fetal peak in rodents during adult life (Le Page et al. 2010; Okubo et al. 2011) (Fig. 7.4). Due to this very high level of aromatase activity, a substantial fraction of the circulating testosterone that reaches the adult brain is converted to E2, resulting in high levels of E2 even in the male brain (though not as much as in the female).

7.6 Duplicated Steroid Receptor Subtypes in Teleosts Have Largely Nonredundant Roles in Mediating Sex-Typical Mating Behaviors

What then is the mechanism of action of adult sex steroids in establishing and reversing sex-typical mating behaviors in teleosts? The action of sex steroids is primarily mediated via binding to specific intracellular receptors, including AR, ESR, and progestin receptor (PGR, also known as PR), which function as ligand-dependent transcription factors to regulate the expression of downstream genes. While tetrapods have a single AR subtype and two ESR subtypes (ESR1 and ESR2), most teleosts have two AR subtypes (Ara and Arb) and three ESR subtypes (Esr1, Esr2a, and Esr2b), except for some species, including zebrafish, which have only a single AR (Ikeuchi et al. 1999; Hawkins et al. 2000; Douard et al. 2008) (Fig. 7.5). Ara and Arb are duplicates that arose because of a whole-genome duplication event in the teleost lineage and are co-orthologous to tetrapod AR (Ogino et al. 2018). Similarly, Esr2a and Esr2b are co-orthologs of tetrapod ESR2, arising from the teleost-specific genome duplication (Ogino et al. 2018). In contrast to AR and ESR, both tetrapods and teleosts have only a single subtype of PGR.

Duplicated genes often diverge in function, partitioning the multiple functions of their single ancestral gene (Force et al. 1999; Lynch and Force 2000; Postlethwait et al. 2004). Studies of mice and several teleost species rendered deficient in AR and

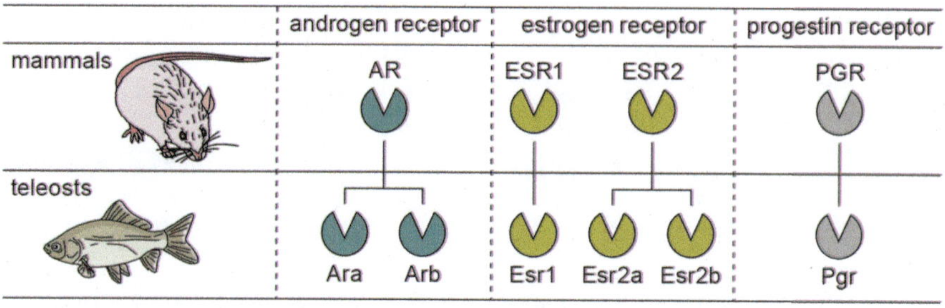

Fig. 7.5 Comparison of intracellular steroid receptor subtypes in mammals and teleosts. While mammals have a single androgen receptor (AR) subtype and two ESR subtypes (ESR1 and ESR2), most teleosts have two AR subtypes (Ara and Arb) and three ESR subtypes (Esr1, Esr2a, and Esr2b). Orthologous subtypes are indicated by connecting lines. Teleosts have more subtypes of these receptors as a consequence of a whole-genome duplication event early in their evolution. In contrast to AR and ESR, teleosts have only a single progestin receptor (PGR) subtype, as do mammals

ESR subtypes have revealed that they indeed have largely nonredundant roles in mediating mating behaviors. In male rodents, testicular testosterone, once in the brain, is aromatized to E2, which then acts through ESR1 to masculinize and activate the neural substrates to actuate male-typical mating behaviors and, in addition, through ESR2 to defeminize the neural substrates (McCarthy and Arnold 2011; McCarthy et al. 2017). The most compelling evidence for this is the finding that male mice lacking *Esr1* show greatly reduced male-typical mating behaviors (Ogawa et al. 1997, 1999), whereas those lacking *Esr2* show no overt defect in these behaviors, but display lordosis, a characteristic mating posture of female rodents, when castrated and given estrogens and progesterone as adults (Kudwa et al. 2005). Testosterone also acts directly on AR in the male rodent brain without being aromatized to E2. In male mice, however, dysfunction of AR in the brain leads to only a modest reduction in mating behaviors (Juntti et al. 2010), while loss of both ESR subtypes completely abolishes them (Ogawa et al. 2000). It thus seems that AR-mediated signaling is not essential for the masculinization of mating behaviors, but rather regulates the extent of behavioral displays in males (Juntti et al. 2010).

In contrast, in teleosts, where nonaromatizable 11KT is the primary androgen, studies with different agonists and antagonists have repeatedly shown that AR-mediated 11KT signaling is critical for male-typical mating behaviors (e.g., Stacey and Kobayashi 1996; van Breukelen 2013; Alward et al. 2019). This is confirmed by recent findings in zebrafish that *ar*-deficient males (zebrafish have only a single AR) are fertile, but court females less vigorously (Yong et al. 2017) and that males deficient in a steroidogenic enzyme *cyp17a1* show reduced androgen levels and impaired mating behaviors, which can be rescued by administration of 11KT (Zhai et al. 2018; Shu et al. 2020). In addition, in African cichlid, loss of *ara*, but not *arb*, results in males with diminished mating behaviors, indicating that *ara* mediates the effects of 11KT on male mating behaviors (Alward et al. 2020). In contrast, the involvement of ESR in male mating behaviors remains unclear in teleosts. Male medaka lacking *esr2b* show no defects in mating behaviors, suggesting that this ESR subtype does not play a role in eliciting male-typical mating behaviors (Nishiike et al. 2021); however, no attempt has been made to determine whether *esr1* and *esr2a* have any such role. Considering that large amounts of E2 are produced locally by aromatase in the male brain of teleosts, ESR may play some role, albeit minor, in male mating behaviors in teleosts as it does in rodents. In fact, it has been reported that treating adult male guppies (*Poecilia reticulata*) with an aromatase inhibitor attenuates their courtship activities (Hallgren et al. 2006). However, the aromatase inhibitor has no such effect in African cichlid, and further testing with other species would be necessary to draw any conclusions (Huffman et al. 2013).

Then, what about female-typical mating behaviors? In female rodents, E2 activates the feminized neural substrates to induce female-typical mating behaviors mainly through Esr1. This is evidenced by the finding that female mice lacking *Esr1* are not receptive to male courtship, while those lacking *Esr2* mate normally (Ogawa et al. 1996, 1998, 1999; Rissman et al. 1997). Recent work in medaka has shown that female-typical mating behaviors in teleosts also require ESR-mediated E2 signaling (Nishiike et al. 2021). However, unlike in rodents, the effect of E2 in teleosts appears to be mediated primarily by Esr2b (a teleost ortholog of ESR2), not Esr1; female medaka deficient in *esr2b* are not sexually receptive to courting males, despite retaining normal ovarian function with an unaltered sex steroid milieu (Nishiike et al. 2021), whereas *esr1* and *esr2a* deficiency do not prevent normal mating behaviors in females (Tohyama et al. 2017; Kayo et al. 2019). Even more strikingly, *esr2b*-deficient female medaka are not only unreceptive to male courtship, but often court other females. This finding indicates that Esr2b is the major determinant of sex-typical mating behaviors, playing a decisive role in demasculinization as well as feminization of these behaviors. This is in marked contrast to rodents, in which neither E2/ESR1 nor E2/ESR2 signaling is apparently involved in demasculinization of mating behaviors (rather they serve critical roles in the masculinization and defeminization, respectively, as mentioned above). It should be noted, however, that recent evidence suggests that, in quail, fetal E2 demasculinizes the neural substrates of mating behaviors primarily through Esr2 (Court et al. 2020). The extent to which the demasculinizing effect of E2/ESR2 signaling is prevalent in vertebrates is an important question that warrants further investigation.

7.7 E2/Esr2b and 11KT/AR Signaling Act Antagonistically to Regulate Mating Behaviors in Teleosts

The study of *esr2b*-deficient female medaka also revealed that they develop male-typical mating behaviors due to the AR-mediated action of small amounts of 11KT (or, less likely, DHT) secreted by the ovary (Nishiike et al. 2021). Although the ovary of normal females also secretes small amounts of 11KT, they do not show

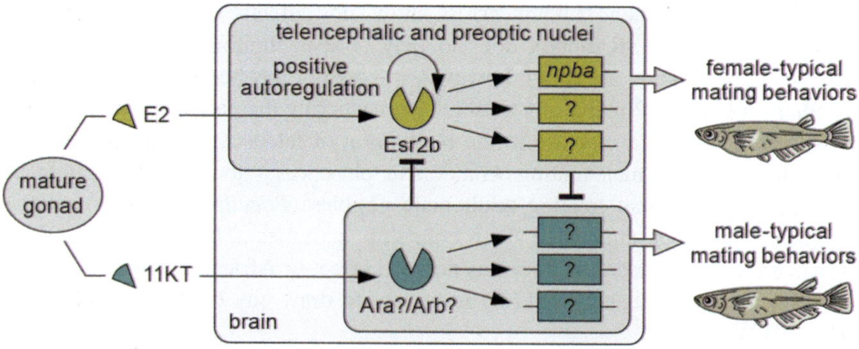

Fig. 7.6 Regulation of teleost mating behaviors by gonadal steroid signaling in adulthood, as revealed by studies in medaka. Estradiol-17β (E2) signaling through an estrogen receptor subtype Esr2b is essential for the induction of female-typical, and the suppression of male-typical, mating behaviors, acting antagonistically to androgen receptor (Ara and/or Arb)-mediated 11-ketotestosterone (11KT) signaling. Presumably, the positive autoregulation of Esr2b expression in the telencephalic and preoptic nuclei by E2, followed by the activation of downstream effector genes such as *npba*, feminizes and demasculinizes mating behaviors, whereas repression of Esr2b expression therein by 11KT masculinizes and defeminizes these behaviors

male-typical behaviors, probably because the masculinizing effect of 11KT/AR signaling is suppressed by the demasculinizing effect of E2/Esr2b signaling. It seems that mutual antagonism between Esr2b-mediated E2 signaling and AR-mediated 11KT signaling in adult life serves to establish and actively maintain sex-typical mating behaviors (Fig. 7.6).

Another study in medaka unveiled that *esr2b* is expressed exclusively in females due to positive autoregulation by E2 in several brain regions, including the ventral telencephalic (Vs/Vp) and magnocellular preoptic (PMm/PMg) nuclei (Hiraki et al. 2012), which have been implicated, by classic lesion and stimulation studies, in mating behaviors (Demski et al. 1975; Kyle and Peter 1982; Koyama et al. 1984; Satou et al. 1984). Vs/Vp and PMm/PMg are part of the evolutionarily conserved "social behavior network" and considered homologous to the bed nucleus of the stria terminalis/subpallial amygdala and the paraventricular nucleus in the rodent brain, respectively (Newman 1999; Goodson 2005; O'Connell and Hofmann 2011, 2012; Goodson and Kingsbury 2013; Herget et al. 2014), which are also relevant to mating behaviors (Veenema and Neumann 2008; Yao et al. 2017; Bayless et al. 2019). It is reasonable to assume that activation of E2/Esr2b signaling in Vs/Vp and PMm/PMg is responsible for the expression of female-typical mating behaviors, and that males do not display female-typical behaviors because E2/Esr2b signaling is not activated in these brain nuclei. Crucially, 11KT has an inhibitory effect on *esr2b* expression in Vs/Vp and PMm/PMg; high levels of 11KT secretion from the male testis may contribute to the defeminization of male mating behaviors by inhibiting E2/Esr2b signaling in these brain nuclei (Hiraki et al. 2012; Nishiike et al. 2021). This defeminizing effect of 11KT, along with its ability to promote male-typical (masculinize) behaviors, presumably induces and maintains the male pattern of mating behaviors.

It is noteworthy that the sexually dimorphic *esr2b* expression in Vs/Vp and PMm/PMg of medaka can be reversed between the sexes by altering the adult sex steroid milieu; the E2-dominated adult steroid milieu stimulates, whereas the 11KT-dominated adult steroid milieu inhibits, *esr2b* expression therein, regardless of sex (Hiraki et al. 2012; Nishiike et al. 2021). Sex-differentiated expression of steroid receptors and effects of the adult steroid milieu on it have also been reported in the brain of mammals and birds (Okubo et al. 2019). However, the degree of both sex differences and steroid effects is modest and much less extensive than *esr2b* expression in medaka. These differences between medaka and mammals/birds highlight the profound sexual dimorphism and adult steroid-dependent lability in the activation of E2/Esr2b signaling in behaviorally relevant nuclei of the teleost brain. These particular modes of E2/Esr2b signaling are most likely responsible for the adult sexual lability in teleost mating behaviors.

Evidence has accumulated in rodents that the effects of gonadal sex steroids early in life on brain and behavior are mediated, in part, by DNA methylation of steroid receptor genes, although there are some discrepancies between studies (Westberry et al. 2010; Kurian et al. 2010; Schwarz et al. 2010). This epigenetic modification limits the flexibility of receptor gene expression in relevant brain nuclei later in life in a sex-specific manner. Considering the lability of steroid receptor expression in adults, the teleost brain is unlikely to undergo such a developmental process. This suggests again that steroids early in life have little effect on sexual differentiation of brain and behavior in teleosts.

7.8 Neuropeptide B Is a Direct Mediator of E2/Esr2b Signaling and Required for Female Sexual Receptivity

Recent studies, particularly in mice employing optogenetic and chemogenetic manipulations, have identified sex steroid-responsive neural circuits that underlie sex-specific mating behaviors, as well as the effector genes downstream of sex steroids that mediate these behaviors (Bayless and Shah 2016; Chen and Hong 2018; Jennings and de Lecea 2020). Nonetheless, very few genes have been identified in any species that are sex-specific direct transcriptional targets of sex steroids for mediating sex-specific mating behaviors (Yang and Shah 2014). Given that the social behavior network comprises subsets of peptidergic neurons, neuropeptides and their receptors seem to be good candidates for direct targets of sex steroids. Work in the past few years in medaka has revealed that the gene encoding a neuropeptide, neuropeptide B (NPB), is an immediate downstream effector of E2/Esr2b signaling that mediates female receptivity to male courtship, as detailed below.

NPB, together with its close relative, neuropeptide W (NPW), was originally identified as a ligand for the orphan receptors GPR7 and GPR8 (now designated NPBWR1 and NPBWR2) (Fujii et al. 2002; Brezillon et al. 2003; Tanaka et al. 2003). In mammals, NPB and NPW have been implicated in diverse physiological processes, including energy homeostasis, food intake, inflammatory pain response, social interaction, and pituitary hormone secretion (Sakurai 2013; Watanabe and Yamamoto 2015). Teleost fish have two NPB genes (*npba* and *npbb*) and one NPBWR2 gene, while lacking the NPW and NPBWR1 genes. A search for genes differentially expressed between the sexes in the medaka brain identified *npba* as being much more highly expressed in females (Hiraki et al. 2014). It was subsequently found that *npba* is expressed exclusively in females in Vs/Vp and PMm/PMg, specifically in *esr2b*-positive neurons, as a result of direct transcriptional activation by E2/Esr2b signaling. Similarly, expression of *npbb* in these brain nuclei is nearly confined to females and is E2-dependent (Hiraki et al. 2014; Hiraki-Kajiyama et al. 2019; Nishiike et al. 2021). Behavioral testing revealed that both *npba*- and *npbwr2*-deficient female medaka require more time to accept males after receiving courtship stimulation. In addition, *npbwr2*-deficient females and females that are simultaneously deficient in both *npba* and *npbb* tend to accept males without being courted (Hiraki-Kajiyama et al. 2019). These findings suggest that NPB signaling in teleosts plays a significant role in female mate choice, possibly by facilitating the acceptance of males performing courtship display and the refusal of males exhibiting no courtship. Notably, the sexually dimorphic expression of *npba* and *npbb* in Vs/Vp and PMm/PMg can be reversed, to some extent, between female and male patterns in response to changes in the adult steroid milieu, as with *esr2b*. Furthermore, the morphological, transcriptional, and electrophysiological phenotypes of *npba*/*npbb*-expressing neurons that indicate cellular activation (e.g., large euchromatic nuclei with abundant cytoplasm, high levels of RNA polymerase II activity and histone marks of active transcription, and relatively high spontaneous firing rates) are all critically dependent on the adult E2 milieu (Kikuchi et al. 2019). Altogether, E2-dependent, and thus female-specific, NPB signaling is likely to be a crucial element of the neural circuitry that underlies sexual dimorphism and lability of teleost mating behaviors.

However, it is of note that the behavioral defects in females deficient in NPB signaling are much less severe than those in *esr2b*-deficient females. A possible explanation for this is that E2/Esr2b signaling regulates most, if not all, aspects of female mating behaviors by simultaneously affecting multiple behavior-related genes, including *npba*/*npbb*, and that each of these genes handles one or a few aspects separately (Fig. 7.6). This speculation is consistent with the view that sexually dimorphic social behaviors are regulated in a modular fashion by multiple sexually dimorphic genes that function downstream of sex steroid signaling (Xu et al. 2012).

7.9 Conclusions and Future Directions

Considering the evidence presented above, it is highly probable that, in teleost fish, neither early gonadal steroids nor the sex chromosome complement contributes much to the development of sex-typical mating behaviors; instead, steroids in adulthood serve to both differentiate the neural substrates and activate them to facilitate the expression of these behaviors in a reversible and transient manner. Recent evidence from studies in medaka further suggests that mutual antagonism between Esr2b-mediated E2 signaling and AR-mediated 11KT signaling in adult life underlies the induction, maintenance, and reversal of sex-typical mating behaviors. Individual adult fish can display a spectrum of sex-typical behaviors ranging from exclusively masculine to exclusively feminine, probably as a result of this mutual antagonism. These studies also suggest that the striking sexual dimorphism and steroid-dependent lability in Esr2b expression in the telencephalic and preoptic nuclei are the primary molecular basis for sexual differentiation and lability of teleost mating behaviors. The consequent sex-differentiated but reversible activation of downstream effectors, including Npba/Npbb, in response to the adult steroid milieu may allow for the transition of mating behaviors from one end of the sexual spectrum to the other.

Besides Npba/Npbb, several other neuropeptides, including gonadotropin-releasing hormones (Gnrh2 and Gnrh3) (Yamamoto et al. 1997; Ogawa et al. 2006; Okuyama et al. 2014; Marvel et al. 2021), nonapeptides (vasotocin, the teleost ortholog of mammalian vasopressin, and isotocin, the teleost ortholog of mammalian oxytocin) (Yokoi et al. 2015, 2020), and secretogranins (Scg2a and Scg2b) (Mitchell et al. 2020), have been shown to be involved in mating behaviors in teleosts. Sex-differential expression of these neuropeptides has been reported in the brains of many teleost species (e.g., Grober et al. 1994, 2002; Elofsson et al. 1997, 1999; Foran and Bass 1998; Ishizaki et al. 2004; Black et al. 2004; Maruska et al. 2007; Maruska 2009; Kuramochi et al. 2011; Kawabata et al. 2012). The expression of these peptides has also been reported to be significantly altered with sex change in gobies (*Trimma okinawae*) and wrasses (*Thalassoma bifasciatum*) (Grober and Sunobe 1996; Godwin et al. 2000; Todd et al. 2019). However, only limited information is available on whether the behaviorally relevant expression of these neuropeptides is under the control of sex steroids, and if so, through what signaling pathways (Yamashita et al. 2017; Narita et al. 2018). This issue certainly warrants future research to determine the signaling pathways initiated by sex steroids in the brain to mediate sex-typical behaviors.

Another important issue to be addressed in the future is the functional connection between E2/Esr2b signaling and PGF2α in regulating female mating behaviors. Evidence from several teleost species suggests that gonadal steroids, including E2, are not required for the execution of female mating behaviors, provided that sufficient PGF2α is available (Munakata and Kobayashi 2010). In goldfish, for example, ovariectomized and immature females, and even males, are sexually receptive to courting males when administered with PGF2α but not when administered with E2

(Stacey and Kyle 1983; Kobayashi and Stacey 1993). Similarly, in African cichlid, administration of PGF2α to females in a nonreproductive state elicits female-typical mating behaviors (Kidd et al. 2013) and, moreover, genetic disruption of the PGF2α receptor abolishes female receptivity (Juntti et al. 2016). These observations seem to be incompatible with the findings in medaka that defined the relevance of E2/Esr2b signaling to female mating behaviors (Nishiike et al. 2021). The reason for this incompatibility is currently unknown, but it could be due to species differences in the impact and mode of action of E2 and PGF2α. Alternatively, and more likely, E2/Esr2b signaling for female mating behaviors may involve E2 produced locally in the brain, rather than, or in addition to, circulating E2 derived from the ovary, and acts upstream of PGF2α signaling. This is in accordance with the fact stated above that the adult teleost brain has more than 100-fold greater aromatase activity than the adult mammalian and avian brains and thus produces greater amounts of E2 (Pasmanik and Callard 1985). Future work is needed to clarify the behavioral role of E2 produced in large amounts in the teleost brain.

Finally, future work is also necessary to ascertain whether the findings and implications presented in this review can be generalized across species in teleost fish. As the most diverse group of vertebrates (comprising half of all vertebrate species), teleosts exhibit a remarkable diversity in reproductive strategies. However, sex steroid regulation of sex-typical behaviors has been studied only in a few species. One should be cautious in assuming that findings from these species will apply to other species.

Acknowledgments This work was supported by the Ministry of Education, Culture, Sports, Science, and Technology (MEXT), Japan, and the Japan Society for the Promotion of Science (JSPS) (MEXT/JSPS grant numbers 17H06429 and 19H03044 (to KO)).

References

Adkins-Regan E (2009) Hormones and sexual differentiation of avian social behavior. Dev Neurosci 31:342–350

Agate RJ, Grisham W, Wade J, Mann S, Wingfield J, Schanen C, Palotie A, Arnold AP (2003) Neural, not gonadal, origin of brain sex differences in a gynandromorphic finch. Proc Natl Acad Sci U S A 100:4873–4878

Alward BA, Hilliard AT, York RA, Fernald RD (2019) Hormonal regulation of social ascent and temporal patterns of behavior in an African cichlid. Horm Behav 107:83–95

Alward BA, Laud VA, Skalnik CJ, York RA, Juntti SA, Fernald RD (2020) Modular genetic control of social status in a cichlid fish. Proc Natl Acad Sci U S A 117:28167–28174

Arnold AP (2017) A general theory of sexual differentiation. J Neurosci Res 95:291–300

Arnold AP (2020) Sexual differentiation of brain and other tissues: five questions for the next 50 years. Horm Behav 120:104691

Ball GF, Balthazart J, McCarthy MM (2014) Is it useful to view the brain as a secondary sexual characteristic? Neurosci Biobehav Rev 46:628–638

Balthazart J (2019) New concepts in the study of the sexual differentiation and activation of reproductive behavior, a personal view. Front Neuroendocrinol 55:100785

Balthazart J, Charlier TD, Barker JM, Yamamura T, Ball GF (2010) Sex steroid-induced neuroplasticity and behavioral activation in birds. Eur J Neurosci 32:2116–2132

Bao AM, Swaab DF (2011) Sexual differentiation of the human brain: relation to gender identity, sexual orientation and neuropsychiatric disorders. Front Neuroendocrinol 32:214–226

Bayless DW, Shah NM (2016) Genetic dissection of neural circuits underlying sexually dimorphic social behaviours. Philos Trans R Soc Lond Ser B Biol Sci 371:20150109

Bayless DW, Yang T, Mason MM, Susanto AAT, Lobdell A, Shah NM (2019) Limbic neurons shape sex recognition and social behavior in sexually naive males. Cell 176:1190–1205

Beyer C, Wozniak A, Hutchison JB (1993) Sex-specific aromatization of testosterone in mouse hypothalamic neurons. Neuroendocrinology 58:673–681

Black MP, Reavis RH, Grober MS (2004) Socially induced sex change regulates forebrain isotocin in *Lythrypnus dalli*. Neuroreport 15:185–189

Borg B (1994) Androgens in teleost fishes. Comp Biochem Physiol 109:219–245

Brezillon S, Lannoy V, Franssen JD, Le Poul E, Dupriez V, Lucchetti J, Detheux M, Parmentier M (2003) Identification of natural ligands for the orphan G protein-coupled receptors GPR7 and GPR8. J Biol Chem 278:776–783

Capel B (2017) Vertebrate sex determination: evolutionary plasticity of a fundamental switch. Nat Rev Genet 18:675–689

Chen P, Hong W (2018) Neural circuit mechanisms of social behavior. Neuron 98:16–30

Chen S, Zhang G, Shao C, Huang Q, Liu G, Zhang P, Song W, An N, Chalopin D, Volff JN, Hong Y, Li Q, Sha Z, Zhou H, Xie M, Yu Q, Liu Y, Xiang H, Wang N, Wu K, Yang C, Zhou Q, Liao X, Yang L, Hu Q, Zhang J, Meng L, Jin L, Tian Y, Lian J, Yang J, Miao G, Liu S, Liang Z, Yan F, Li Y, Sun B, Zhang H, Zhang J, Zhu Y, Du M, Zhao Y, Schartl M, Tang Q, Wang J (2014) Whole-genome sequence of a flatfish provides insights into ZW sex chromosome evolution and adaptation to a benthic lifestyle. Nat Genet 46:253–260

Court L, Vandries L, Balthazart J, Cornil CA (2020) Key role of estrogen receptor β in the organization of brain and behavior of the Japanese quail. Horm Behav 125:104827

Demski LS, Bauer DH, Gerald JW (1975) Sperm release evoked by electrical stimulation of the fish brain: a functional-anatomical study. J Exp Zool 191:215–232

Devlin RH, Nagahama Y (2002) Sex determination and sex differentiation in fish: an overview of genetic, physiological, and environmental influences. Aquaculture 208:191–364

Douard V, Brunet F, Boussau B, Ahrens-Fath I, Vlaeminck-Guillem V, Haendler B, Laudet V, Guiguen Y (2008) The fate of the duplicated androgen receptor in fishes: a late neofunctionalization event? BMC Evol Biol 8:336

Edwards DA, Burge KG (1971) Early androgen treatment and male and female sexual behavior in mice. Horm Behav 2:49–58

Elofsson U, Winberg S, Francis RC (1997) Number of preoptic GnRH-immunoreactive cells correlates with sexual phase in a protandrously hermaphroditic fish, the dusky anemonefish (*Amphiprion melanopus*). J Comp Physiol A 181:484–492

Elofsson UO, Winberg S, Nilsson GE (1999) Relationships between sex and the size and number of forebrain gonadotropin-releasing hormone-immunoreactive neurones in the ballan wrasse (*Labrus berggylta*), a protogynous hermaphrodite. J Comp Neurol 410:158–170

Foran CM, Bass AH (1998) Preoptic AVT immunoreactive neurons of a teleost fish with alternative reproductive tactics. Gen Comp Endocrinol 111:271–282

Force A, Lynch M, Pickett FB, Amores A, Yan YL, Postlethwait J (1999) Preservation of duplicate genes by complementary, degenerative mutations. Genetics 151:1531–1545

Fujii R, Yoshida H, Fukusumi S, Habata Y, Hosoya M, Kawamata Y, Yano T, Hinuma S, Kitada C, Asami T, Mori M, Fujisawa Y, Fujino M (2002) Identification of a neuropeptide modified with bromine as an endogenous ligand for GPR7. J Biol Chem 277:34010–33401

Gammerdinger WJ, Kocher TD (2018) Unusual diversity of sex chromosomes in African cichlid fishes. Genes 9:480

Ghosal R, Sorensen PW (2016) Male-typical courtship, spawning behavior, and olfactory sensitivity are induced to different extents by androgens in the goldfish suggesting they are controlled by different neuroendocrine mechanisms. Gen Comp Endocrinol 232:160–173

Godwin J (2010) Neuroendocrinology of sexual plasticity in teleost fishes. Front Neuroendocrinol 31:203–216

Godwin J, Sawby R, Warner RR, Crews D, Grober MS (2000) Hypothalamic arginine vasotocin mRNA abundance variation across sexes and with sex change in a coral reef fish. Brain Behav Evol 55:77–84

Goodson JL (2005) The vertebrate social behavior network: evolutionary themes and variations. Horm Behav 48:11–22

Goodson JL, Kingsbury MA (2013) What's in a name? Considerations of homologies and nomenclature for vertebrate social behavior networks. Horm Behav 64:103–112

Göppert C, Harris RM, Theis A, Boila A, Hohl S, Rüegg A, Hofmann HA, Salzburger W, Böhne A (2016) Inhibition of aromatase induces partial sex change in a cichlid fish: distinct functions for sex steroids in brains and gonads. Sex Dev 10:97–110

Grober MS, Sunobe T (1996) Serial adult sex change involves rapid and reversible changes in forebrain neurochemistry. Neuroreport 7:2945–2949

Grober MS, Fox SH, Laughlin C, Bass AH (1994) GnRH cell size and number in a teleost fish with two male reproductive morphs: sexual maturation, final sexual status and body size allometry. Brain Behav Evol 43:61–78

Grober MS, George AA, Watkins KK, Carneiro LA, Oliveira RF (2002) Forebrain AVT and courtship in a fish with male alternative reproductive tactics. Brain Res Bull 57:423–425

Hallgren SL, Linderoth M, Olsén KH (2006) Inhibition of cytochrome p450 brain aromatase reduces two male-specific sexual behaviours in the male Endler guppy (*Poecilia reticulata*). Gen Comp Endocrinol 147:323–328

Handa RJ, Weiser MJ, Zuloaga DG (2009) A role for the androgen metabolite, 5α-androstane-3β,17β-diol, in modulating oestrogen receptor β-mediated regulation of hormonal stress reactivity. J Neuroendocrinol 21:351–358

Hawkins MB, Thornton JW, Crews D, Skipper JK, Dotte A, Thomas P (2000) Identification of a third distinct estrogen receptor and reclassification of estrogen receptors in teleosts. Proc Natl Acad Sci U S A 97:10751–10756

Herget U, Wolf A, Wullimann MF, Ryu S (2014) Molecular neuroanatomy and chemoarchitecture of the neurosecretory preoptic-hypothalamic area in zebrafish larvae. J Comp Neurol 522:1542–1564

Hiraki T, Takeuchi A, Tsumaki T, Zempo B, Kanda S, Oka Y, Nagahama Y, Okubo K (2012) Female-specific target sites for both oestrogen and androgen in the teleost brain. Proc R Soc B 279:5014–5023

Hiraki T, Nakasone K, Hosono K, Kawabata Y, Nagahama Y, Okubo K (2014) Neuropeptide B is female-specifically expressed in the telencephalic and preoptic nuclei of the medaka brain. Endocrinology 155:1021–1032

Hiraki-Kajiyama T, Yamashita J, Yokoyama K, Kikuchi Y, Nakajo M, Miyazoe D, Nishiike Y, Ishikawa K, Hosono K, Kawabata-Sakata Y, Ansai S, Kinoshita M, Nagahama Y, Okubo K (2019) Neuropeptide B mediates female sexual receptivity in medaka fish, acting in a female-specific but reversible manner. Elife 8:e39495

Huffman LS, O'Connell LA, Hofmann HA (2013) Aromatase regulates aggression in the African cichlid fish *Astatotilapia burtoni*. Physiol Behav 112–113:77–83

Ikeuchi T, Todo T, Kobayashi T, Nagahama Y (1999) cDNA cloning of a novel androgen receptor subtype. J Biol Chem 274:25205–25209

Ishizaki M, Iigo M, Yamamoto N, Oka Y (2004) Different modes of gonadotropin-releasing hormone (GnRH) release from multiple GnRH systems as revealed by radioimmunoassay using brain slices of a teleost, the dwarf gourami (*Colisa lalia*). Endocrinology 145:2092–2103

Iwamatsu T, Kobayashi H, Hamaguchi S, Sagegami R, Shuo T (2005) Estradiol-17β content in developing eggs and induced sex reversal of the medaka (*Oryzias latipes*). J Exp Zool A Comp Exp Biol 303:161–167

Iwamatsu T, Kobayashi H, Sagegami R, Shuo T (2006) Testosterone content of developing eggs and sex reversal in the medaka (*Oryzias latipes*). Gen Comp Endocrinol 145:67–74

Jennings KJ, de Lecea L (2020) Neural and hormonal control of sexual behavior. Endocrinology 161:bqaa150

Juntti SA, Tollkuhn J, Wu MV, Fraser EJ, Soderborg T, Tan S, Honda S, Harada N, Shah NM (2010) The androgen receptor governs the execution, but not programming, of male sexual and territorial behaviors. Neuron 66:260–272

Juntti SA, Hilliard AT, Kent KR, Kumar A, Nguyen A, Jimenez MA, Loveland JL, Mourrain P, Fernald RD (2016) A neural basis for control of cichlid female reproductive behavior by prostaglandin F2α. Curr Biol 26:943–949

Kamiya T, Kai W, Tasumi S, Oka A, Matsunaga T, Mizuno N, Fujita M, Suetake H, Suzuki S, Hosoya S, Tohari S, Brenner S, Miyadai T, Venkatesh B, Suzuki Y, Kikuchi K (2012) A trans-species missense SNP in Amhr2 is associated with sex determination in the tiger pufferfish, *Takifugu rubripes* (fugu). PLoS Genet 8:e1002798

Katz Y, Eckstein B (1974) Changes in steroid concentration in blood of female *Tilapia aurea* (teleostei, cichlidae) during initiation of spawning. Endocrinology 95:963–967

Kawabata Y, Hiraki T, Takeuchi A, Okubo K (2012) Sex differences in the expression of vasotocin/isotocin, gonadotropin-releasing hormone, and tyrosine and tryptophan hydroxylase family genes in the medaka brain. Neuroscience 218:65–77

Kayo D, Zempo B, Tomihara S, Oka Y, Kanda S (2019) Gene knockout analysis reveals essentiality of estrogen receptor β1 (Esr2a) for female reproduction in medaka. Sci Rep 9:8868

Kidd MR, Dijkstra PD, Alcott C, Lavee D, Ma J, O'Connell LA, Hofmann HA (2013) Prostaglandin F2α facilitates female mating behavior based on male performance. Behav Ecol Sociobiol 67:1307–1315

Kikuchi K, Hamaguchi S (2013) Novel sex-determining genes in fish and sex chromosome evolution. Dev Dyn 242:339–353

Kikuchi Y, Hiraki-Kajiyama T, Nakajo M, Umatani C, Kanda S, Oka Y, Matsumoto K, Ozawa H, Okubo K (2019) Sexually dimorphic neuropeptide B neurons in medaka exhibit activated cellular phenotypes dependent on estrogen. Endocrinology 160:827–839

Kobayashi M, Stacey N (1993) Prostaglandin-induced female spawning behavior in goldfish (*Carassius auratus*) appears independent of ovarian influence. Horm Behav 27:38–55

Koyama Y, Satou M, Oka Y, Ueda K (1984) Involvement of the telencephalic hemispheres and the preoptic area in sexual behavior of the male goldfish, *Carassius auratus*: a brain-lesion study. Behav Neural Biol 40:70–86

Koyama T, Nakamoto M, Morishima K, Yamashita R, Yamashita T, Sasaki K, Kuruma Y, Mizuno N, Suzuki M, Okada Y, Ieda R, Uchino T, Tasumi S, Hosoya S, Uno S, Koyama J, Toyoda A, Kikuchi K, Sakamoto T (2019) A SNP in a steroidogenic enzyme is associated with phenotypic sex in *Seriola* fishes. Curr Biol 29:1901–1909

Kudwa AE, Bodo C, Gustafsson JA, Rissman EF (2005) A previously uncharacterized role for estrogen receptor β: defeminization of male brain and behavior. Proc Natl Acad Sci U S A 102: 4608–4612

Kuramochi A, Tsutiya A, Kaneko T, Ohtani-Kaneko R (2011) Sexual dimorphism of gonadotropin-releasing hormone type-III (GnRH3) neurons and hormonal sex reversal of male reproductive behavior in Mozambique tilapia. Zool Sci 28:733–739

Kurian JR, Olesen KM, Auger AP (2010) Sex differences in epigenetic regulation of the estrogen receptor-α promoter within the developing preoptic area. Endocrinology 151:2297–2305

Kyle AL, Peter RE (1982) Effects of forebrain lesions on spawning behaviour in the male goldfish. Physiol Behav 28:1103–1109

Le Page Y, Diotel N, Vaillant C, Pellegrini E, Anglade I, Mérot Y, Kah O (2010) Aromatase, brain sexualization and plasticity: the fish paradigm. Eur J Neurosci 32:2105–2115

Lephart ED (1996) A review of brain aromatase cytochrome P450. Brain Res Rev 22:1–26
Liu H, Todd EV, Lokman PM, Lamm MS, Godwin JR, Gemmell NJ (2017) Sexual plasticity: a fishy tale. Mol Reprod Dev 84:171–194
Lynch M, Force A (2000) The probability of duplicate gene preservation by subfunctionalization. Genetics 154:459–473
Maekawa F, Tsukahara S, Kawashima T, Nohara K, Ohki-Hamazaki H (2014) The mechanisms underlying sexual differentiation of behavior and physiology in mammals and birds: relative contributions of sex steroids and sex chromosomes. Front Neurosci 8:242
Margiotta-Casaluci L, Courant F, Antignac JP, Le Bizec B, Sumpter JP (2013) Identification and quantification of 5α-dihydrotestosterone in the teleost fathead minnow (*Pimephales promelas*) by gas chromatography-tandem mass spectrometry. Gen Comp Endocrinol 191:202–209
Marshall Graves JA, Peichel CL (2010) Are homologies in vertebrate sex determination due to shared ancestry or to limited options? Genome Biol 11:205
Martyniuk CJ, Bissegger S, Langlois VS (2013) Current perspectives on the androgen 5 alpha-dihydrotestosterone (DHT) and 5 alpha-reductases in teleost fishes and amphibians. Gen Comp Endocrinol 194:264–274
Maruska KP (2009) Sex and temporal variations of the vasotocin neuronal system in the damselfish brain. Gen Comp Endocrinol 160:194–204
Maruska KP, Mizobe MH, Tricas TC (2007) Sex and seasonal co-variation of arginine vasotocin (AVT) and gonadotropin-releasing hormone (GnRH) neurons in the brain of the halfspotted goby. Comp Biochem Physiol A Mol Integr Physiol 147:129–144
Marvel M, Levavi-Sivan B, Wong TT, Zmora N, Zohar Y (2021) Gnrh2 maintains reproduction in fasting zebrafish through dynamic neuronal projection changes and regulation of gonadotropin synthesis, oogenesis, and reproductive behaviors. Sci Rep 11:6657
McCarthy MM (2020) A new view of sexual differentiation of mammalian brain. J Comp Physiol A 206:369–378
McCarthy MM, Arnold AP (2011) Reframing sexual differentiation of the brain. Nat Neurosci 14: 677–683
McCarthy MM, Nugent BM, Lenz KM (2017) Neuroimmunology and neuroepigenetics in the establishment of sex differences in the brain. Nat Rev Neurosci 18:471–484
Mitchell K, Zhang WS, Lu C, Tao B, Chen L, Hu W, Trudeau VL (2020) Targeted mutation of secretogranin-2 disrupts sexual behavior and reproduction in zebrafish. Proc Natl Acad Sci U S A 117:12772–12783
Munakata A, Kobayashi M (2010) Endocrine control of sexual behavior in teleost fish. Gen Comp Endocrinol 165:456–468
Nagahama Y, Chakraborty T, Paul-Prasanth B, Ohta K, Nakamura M (2021) Sex determination, gonadal sex differentiation and plasticity in vertebrate species. Physiol Rev 101:1237–1308
Narita Y, Tsutiya A, Nakano Y, Ashitomi M, Sato K, Hosono K, Kaneko T, Chen RD, Lee JR, Tseng YC, Hwang PP, Ohtani-Kaneko R (2018) Androgen induced cellular proliferation, neurogenesis, and generation of GnRH3 neurons in the brain of mature female Mozambique tilapia. Sci Rep 8:16855
Newman SW (1999) The medial extended amygdala in male reproductive behavior. A node in the mammalian social behavior network. Ann N Y Acad Sci 877:242–257
Nishiike Y, Miyazoe D, Togawa R, Yokoyama K, Nakasone K, Miyata M, Kikuchi Y, Kamei Y, Todo T, Ishikawa-Fujiwara T, Ohno K, Usami T, Nagahama Y, Okubo K (2021) Estrogen receptor 2b is the major determinant of sex-typical mating behavior and sexual preference in medaka. Curr Biol 31:1699–1710
O'Connell LA, Hofmann HA (2011) The vertebrate mesolimbic reward system and social behavior network: a comparative synthesis. J Comp Neurol 519:3599–3639
O'Connell LA, Hofmann HA (2012) Evolution of a vertebrate social decision-making network. Science 336:1154–1157
Ogawa S, Taylor JA, Lubahn DB, Korach KS, Pfaff DW (1996) Reversal of sex roles in genetic female mice by disruption of estrogen receptor gene. Neuroendocrinology 64:467–470

Ogawa S, Lubahn DB, Korach KS, Pfaff DW (1997) Behavioral effects of estrogen receptor gene disruption in male mice. Proc Natl Acad Sci U S A 94:1476–1481

Ogawa S, Eng V, Taylor J, Lubahn DB, Korach KS, Pfaff DW (1998) Roles of estrogen receptor-α gene expression in reproduction-related behaviors in female mice. Endocrinology 139:5070–5081

Ogawa S, Chan J, Chester AE, Gustafsson JA, Korach KS, Pfaff DW (1999) Survival of reproductive behaviors in estrogen receptor β gene-deficient (βERKO) male and female mice. Proc Natl Acad Sci U S A 96:12887–12892

Ogawa S, Chester AE, Hewitt SC, Walker VR, Gustafsson JA, Smithies O, Korach KS, Pfaff DW (2000) Abolition of male sexual behaviors in mice lacking estrogen receptors α and β (αβERKO). Proc Natl Acad Sci U S A 97:14737–14741

Ogawa S, Akiyama G, Kato S, Soga T, Sakuma Y, Parhar IS (2006) Immunoneutralization of gonadotropin-releasing hormone type-III suppresses male reproductive behavior of cichlids. Neurosci Lett 403:201–205

Ogino Y, Tohyama S, Kohno S, Toyota K, Yamada G, Yatsu R, Kobayashi T, Tatarazako N, Sato T, Matsubara H, Lange A, Tyler CR, Katsu Y, Iguchi T, Miyagawa S (2018) Functional distinctions associated with the diversity of sex steroid hormone receptors ESR and AR. J Steroid Biochem Mol Biol 184:38–46

Okubo K, Takeuchi A, Chaube R, Paul-Prasanth B, Kanda S, Oka Y, Nagahama Y (2011) Sex differences in aromatase gene expression in the medaka brain. J Neuroendocrinol 23:412–423

Okubo K, Miyazoe D, Nishiike Y (2019) A conceptual framework for understanding sexual differentiation of the teleost brain. Gen Comp Endocrinol 284:113129

Okuyama T, Yokoi S, Abe H, Isoe Y, Suehiro Y, Imada H, Tanaka M, Kawasaki T, Yuba S, Taniguchi Y, Kamei Y, Okubo K, Shimada A, Naruse K, Takeda H, Oka Y, Kubo T, Takeuchi H (2014) A neural mechanism underlying mating preferences for familiar individuals in medaka fish. Science 343:91–94

Pasmanik M, Callard GV (1985) Aromatase and 5α-reductase in the teleost brain, spinal cord, and pituitary gland. Gen Comp Endocrinol 60:244–251

Paul-Prasanth B, Bhandari RK, Kobayashi T, Horiguchi R, Kobayashi Y, Nakamoto M, Shibata Y, Sakai F, Nakamura M, Nagahama Y (2013) Estrogen oversees the maintenance of the female genetic program in terminally differentiated gonochorists. Sci Rep 3:2862

Postlethwait J, Amores A, Cresko W, Singer A, Yan YL (2004) Subfunction partitioning, the teleost radiation and the annotation of the human genome. Trends Genet 20:481–490

Rissman EF, Early AH, Taylor JA, Korach KS, Lubahn DB (1997) Estrogen receptors are essential for female sexual receptivity. Endocrinology 138:507–510

Sakurai T (2013) NPBWR1 and NPBWR2: implications in energy homeostasis, pain, and emotion. Front Endocrinol 4:23

Satou M, Oka Y, Kusunoki M, Matsushima T, Kato M, Fujita I, Ueda K (1984) Telencephalic and preoptic areas integrate sexual behavior in hime salmon (landlocked red salmon, *Oncorhynchus nerka*): results of electrical brain stimulation experiments. Physiol Behav 33:441–447

Schwarz JM, Nugent BM, McCarthy MM (2010) Developmental and hormone-induced epigenetic changes to estrogen and progesterone receptor genes in brain are dynamic across the life span. Endocrinology 151:4871–4881

Scott AP, Sumpter JP, Stacey N (2010) The role of the maturation-inducing steroid, 17,20-β-dihydroxypregn-4-en-3-one, in male fishes: a review. J Fish Biol 76:183–224

Shu T, Zhai G, Pradhan A, Olsson PE, Yin Z (2020) Zebrafish *cyp17a1* knockout reveals that androgen-mediated signaling is important for male brain sex differentiation. Gen Comp Endocrinol 295:113490

Södersten P (1972) Mounting behavior in the female rat during the estrous cycle, after ovariectomy, and after estrogen or testosterone administration. Horm Behav 3:307–320

Stacey N, Kobayashi M (1996) Androgen induction of male sexual behaviors in female goldfish. Horm Behav 30:434–445

Stacey NE, Kyle AL (1983) Effects of olfactory tract lesions on sexual and feeding behavior in the goldfish. Physiol Behav 30:621–628

Sun LN, Jiang XL, Xie QP, Yuan J, Huang BF, Tao WJ, Zhou LY, Nagahama Y, Wang DS (2014) Transdifferentiation of differentiated ovary into functional testis by long-term treatment of aromatase inhibitor in Nile tilapia. Endocrinology 155:1476–1488

Takatsu K, Miyaoku K, Roy SR, Murono Y, Sago T, Itagaki H, Nakamura M, Tokumoto T (2013) Induction of female-to-male sex change in adult zebrafish by aromatase inhibitor treatment. Sci Rep 3:3400

Tanaka H, Yoshida T, Miyamoto N, Motoike T, Kurosu H, Shibata K, Yamanaka A, Williams SC, Richardson JA, Tsujino N, Garry MG, Lerner MR, King DS, O'Dowd BF, Sakurai T, Yanagisawa M (2003) Characterization of a family of endogenous neuropeptide ligands for the G protein-coupled receptors GPR7 and GPR8. Proc Natl Acad Sci U S A 100:6251–6256

Tobet SA, Baum MJ, Tang HB, Shim JH, Canick JA (1985) Aromatase activity in the perinatal rat forebrain: effects of age, sex and intrauterine position. Brain Res 355:171–178

Todd EV, Ortega-Recalde O, Liu H, Lamm MS, Rutherford KM, Cross H, Black MA, Kardailsky O, Marshall Graves JA, Hore TA, Godwin JR, Gemmell NJ (2019) Stress, novel sex genes, and epigenetic reprogramming orchestrate socially controlled sex change. Sci Adv 5: eaaw7006

Tohyama S, Ogino Y, Lange A, Myosho T, Kobayashi T, Hirano Y, Yamada G, Sato T, Tatarazako N, Tyler CR, Iguchi T, Miyagawa S (2017) Establishment of estrogen receptor 1 (ESR1)-knockout medaka: ESR1 is dispensable for sexual development and reproduction in medaka, *Oryzias latipes*. Develop Growth Differ 59:552–561

van Breukelen NA (2013) Androgen receptor antagonist impairs courtship but not aggressive behavior in the monogamous cichlid, *Amatitlania nigrofasciata*. Horm Behav 63:527–532

Veenema AH, Neumann ID (2008) Central vasopressin and oxytocin release: regulation of complex social behaviours. Prog Brain Res 170:261–276

Volff JN, Nanda I, Schmid M, Schartl M (2007) Governing sex determination in fish: regulatory putsches and ephemeral dictators. Sex Dev 1:85–99

Wai EH, Hoar WS (1963) The secondary sex characters and reproductive behavior of gonadectomized sticklebacks treated with methyl testosterone. Can J Zool 41:611–628

Watanabe N, Yamamoto M (2015) Neural mechanisms of social dominance. Front Neurosci 9:154

Westberry JM, Trout AL, Wilson ME (2010) Epigenetic regulation of estrogen receptor α gene expression in the mouse cortex during early postnatal development. Endocrinology 151:731–740

Xu X, Coats JK, Yang CF, Wang A, Ahmed OM, Alvarado M, Izumi T, Shah NM (2012) Modular genetic control of sexually dimorphic behaviors. Cell 148:596–607

Yamamoto N, Oka Y, Kawashima S (1997) Lesions of gonadotropin-releasing hormone-immunoreactive terminal nerve cells: effects on the reproductive behavior of male dwarf gouramis. Neuroendocrinology 65:403–412

Yamashita J, Kawabata Y, Okubo K (2017) Expression of isotocin is male-specifically up-regulated by gonadal androgen in the medaka brain. J Neuroendocrinol 29:e12545

Yang CF, Shah NM (2014) Representing sex in the brain, one module at a time. Neuron 82:261–278

Yang T, Shah NM (2016) Molecular and neural control of sexually dimorphic social behaviors. Curr Opin Neurobiol 38:89–95

Yao S, Bergan J, Lanjuin A, Dulac C (2017) Oxytocin signaling in the medial amygdala is required for sex discrimination of social cues. elife 6:e31373

Yazawa T, Inaba H, Imamichi Y, Sekiguchi T, Uwada J, Islam MS, Orisaka M, Mikami D, Ida T, Sato T, Miyashiro Y, Takahashi S, Khan MRI, Suzuki N, Umezawa A, Kitano T (2021) Profiles of 5α-reduced androgens in humans and eels: 5α-dihydrotestosterone and 11-ketodihydrotestosterone are active androgens produced in eel gonads. Front Endocrinol 12:657360

Yokoi S, Okuyama T, Kamei Y, Naruse K, Taniguchi Y, Ansai S, Kinoshita M, Young LJ, Takemori N, Kubo T, Takeuchi H (2015) An essential role of the arginine vasotocin system in mate-guarding behaviors in triadic relationships of medaka fish (*Oryzias latipes*). PLoS Genet 11:e1005009

Yokoi S, Naruse K, Kamei Y, Ansai S, Kinoshita M, Mito M, Iwasaki S, Inoue S, Okuyama T, Nakagawa S, Young LJ, Takeuchi H (2020) Sexually dimorphic role of oxytocin in medaka mate choice. Proc Natl Acad Sci U S A 117:4802–4808

Yong L, Thet Z, Zhu Y (2017) Genetic editing of the androgen receptor contributes to impaired male courtship behavior in zebrafish. J Exp Biol 220:3017–3021

Zhai G, Shu T, Xia Y, Lu Y, Shang G, Jin X, He J, Nie P, Yin Z (2018) Characterization of sexual trait development in *cyp17a1*-deficient zebrafish. Endocrinology 159:3549–3562

Chapter 8
Comparative Perspectives on the Function of Oxytocin in Fish and Mammals

Saori Yokoi, Larry J. Young, and Hideaki Takeuchi

Abstract The central oxytocin (OT) system regulates various behavioral processes, including social recognition, social bonding, and mate preference. In human studies, intranasal administration of OT facilitates the salience of social information mediated by visual information and motivational states in both sexes. Some rodent animal models, such as mice and prairie voles, have been used to investigate the neural/molecular mechanisms underlying central OT function. Elucidating how the OT system modulates visual processing is difficult using rodent models, however, because rodents mainly use chemical communication and olfactory investigation for social interactions. In this review, we propose medaka fish as an advantageous comparative model for investigating this issue. Medaka fish have the ability to visually recognize conspecifics and select the appropriate mate based on familiarity recognition. Medaka fish can use faces (head parts) for familiarity recognition, and may have a special neural system to recognize faces. OT receptors are distributed in homologous areas of the primary visual center in medaka fish and primates (optic tectum and superior colliculus, respectively). In addition, mutations of genes for OT or OT receptors (OTR) in medaka influence mate choice in a sex-specific manner. As human OT administration also has opposite effects depending on the sex, comparative studies between medaka and humans could help to elucidate the evolutionary

S. Yokoi
Faculty of Pharmaceutical Sciences, Hokkaido University, Hokkaido, Japan
e-mail: yokois@pharm.hokudai.ac.jp

L. J. Young
Center for Translational Social Neuroscience, Department of Psychiatry and Behavioral Sciences, Yerkes National Primate Research Center, Emory University, Atlanta, GA, USA

Center for Social Neural Networks, Faculty of Human Sciences, University of Tsukuba, Tsukuba, Japan
e-mail: lyoun03@emory.edu

H. Takeuchi (✉)
Graduate School of Natural Science and Technology, Okayama University, Okayama, Japan

Graduate School of Life Sciences, Tohoku University, Miyagi, Japan
e-mail: hideaki.takeuchi.a8@tohoku.ac.jp

M. Tanaka, M. Tachibana (eds.), *Spectrum of Sex*,
https://doi.org/10.1007/978-981-19-5359-0_8

roots and basic neural mechanisms underlying social salience in face detection, as well as how the OT system contributes to the sexual spectrum and diversity in mating strategies.

Keywords Social vision · Oxytocin · Social recognition · Superior colliculus

8.1 Introduction

Oxytocin (OXT; hereafter OT) is a nine amino-acid neuropeptide that is widely conserved among vertebrates. Previously, orthologous neuropeptides had different names based on small protein-coding differences, such as isotocin (fish) and mesotocin (birds). A universal nomenclature (oxytocin) was recently proposed on the basis of evolutionary relationships because orthologous genes coding peptide hormones stem from a common ancestral gene (Theofanopoulou et al. 2021). From fish to mammals, OT is mainly synthesized in the hypothalamus and stored in the posterior pituitary gland. OT is released to the bloodstream from the posterior pituitary gland and transported through axons as a neurohormone to a broad range of extrahypothalamic brain areas. Oxytocinergic neurons release OT from axonal terminals as well as dendrites/axons via nonsynaptic (volume) transmission (Ludwig and Leng 2006), which enables OT to reach broad neuronal targets (Oti et al. 2021). Central OT regulates various behavioral and physiologic processes, including stress, analgesia, social recognition, and mate preference. In the field of social neuroscience, many studies have focused on the effects of OT to facilitate affiliative behaviors toward familiar individuals. For example, studies in rodents indicate that OT signaling in the brain is essential for the onset of maternal nurturing behavior (Rilling and Young 2014; Numan and Young 2016; Froemke and Young 2021). In monogamous rodents (prairie voles), OT facilitates pair-bond formation in both males and females, and administration of an OT receptor antagonist into the brain prevents mating-induced partner preferences in both sexes (Young et al. 2001; Johnson et al. 2016; Walum and Young 2018). Therefore, OT is often considered by some to function as the "love hormone" in mammals.

8.2 Familiarity Recognition in Fish Species

A number of fish species have the ability to discriminate between familiar and nonfamiliar conspecifics and to make behavioral choices based on familiarity (i.e., prior social interaction between individuals) (Ward et al. 2020). In fish, forming shoals with familiar conspecifics can increase fitness by decreasing aggression within groups, facilitating cooperative foraging and escaping, and providing mating opportunities (Seppä et al. 2001; Nadler et al. 2021; Utne-Palm and Hart 2000; Doran et al. 2019). Shoaling preferences for familiar conspecifics are observed in a broad range of fish species, such as guppies (Magurran et al. 1994), sticklebacks

Table 8.1 List of fish species that exhibit shoaling preference based on familiarity recognition

Class/order	Species name	Reference
Actinopterygii		
Cypriniformes	European minnows (*Phoxinus phoxinu*)	Griffiths et al. (2007)
	Texas shiner (*Notropis amabilis*)	Farmer et al. (2004)
	Blacktail shiner (*Cyprinella venusta*)	Farmer et al. (2004)
	Golden shiner (*Notemigonus crysoleucas*)	Barber and Wright (2001)
	Zebrafish (*Danio rerio*)	Bruzzone et al. (2020)
Salmoniformes	Sea trout (*Salmo trutta*)	Höjesjö et al. (1998)
	Rainbow trout (*Oncorhynchus mykiss*)	Johnsson (1997)
Gasterosteiformes	Three-spined stickleback (*Gasterosteus aculeatus*)	Barber and Ruxton (2000)
Clupeiformes	Pacific herring, (*Clupea pallasii*)	Hay and McKinnell (2002)
Perciformes	Paradise fish (*Macropodus opercularis*)	Miklósi et al. (1992)
	Damselfish (*Chromis viridis*)	Nadler et al. (2021)
Cyprinodontiformes	Banded killifish (*Fundulus diaphanus*)	Lee-Jenkins and Godin (2010)
	Guppy (*Poecilia reticulata*)	Magurran et al. (1994)
Chondrichthyes		
Carcharhiniformes	Lemon sharks (*Negaprion brevirostris*)	Keller et al. (2017)

(Barber and Ruxton 2000), minnows (Griffiths et al. 2007), shiners (Farmer et al. 2004), cichlids (Lee-Jenkins and Godin 2010), zebrafish (Ribeiro et al. 2020), and sharks (Keller et al. 2017), among others (Krause et al. 2000; Griffiths 2003; Ward et al. 2020) (please see Table 8.1). Mate choice based on familiarity is also observed in some fish species. Guppy males more frequently exhibit courtship displays toward unfamiliar females compared with familiar females, which might enhance the spread of the male's genes and the generation of more offspring with divergent genetic backgrounds (Hughes et al. 1999; Kelley et al. 1999). In contrast, medaka females prefer to mate with visually familiarized males (Okuyama et al. 2014; Wang and Takeuchi 2017), which could influence a female's choice for dominant males in mate-guarding competitions (Yokoi et al. 2015, 2016). Given that the ability to recognize familiarity spans among fish species from chondrichthyes to teleosts, the molecular/neural mechanisms underlying familiarity recognition show remarkable convergence.

8.3 OT Function in Shoaling Preference Based on Familiarity Recognition in Zebrafish

Zebrafish have a strong shoaling preference toward conspecifics (Oliveira 2013) as well as the ability to discriminate familiar individuals (Gerlach and Lysiak 2006; Hinz et al. 2013; Bruzzone et al. 2020). Zebrafish can use visual information for social interaction and clearly respond to visual images and biologic motion conspecifics (Saverino and Gerlai 2008; Larsch and Baier 2018). In zebrafish, both pharmacologic and genetic approaches have been applied to investigate OT function in shoaling preference. Administration of the specific OT receptor (OTR) antagonist L-368,899 decreases shoaling behaviors in adult and larval zebrafish (Landin et al. 2020). In contrast, *otra*-mutant zebrafish display normal shoaling behavior, which might be due to genetic compensation of the *otra* mutation. Shoaling preference based on familiarity recognition, however, is defective in *otra* mutants. Zebrafish are able to discriminate between familiar and novel mates, and prefer to approach novel mates (novelty preference) (Bruzzone et al. 2020), whereas *otra* mutants are not able to discriminate between familiar and novel mates (Ribeiro et al. 2020). In the same behavioral paradigm, zebrafish tend to approach novel objects and *otra*-mutant fish exhibit impaired object recognition. Therefore, in zebrafish, OT is assumed to be required for memory recognition of familiar versus novel entities, not only in a social context (Ribeiro et al. 2020). In zebrafish, however, there are no reports of OT signaling in familiarity recognition in heterosexual interactions such as mate preference.

8.4 Familiarity Recognition and Mate Preference in Medaka Fish

Medaka fish (*Oryzias latipes*) are an attractive model for investigating the molecular/neural basis underlying mate preference based on familiarity recognition (Okuyama et al. 2017). Both medaka and zebrafish are model organisms for molecular genetics with similar size embryos, larvae, and adults. Like in zebrafish, state-of-the-art molecular-genetic methods, such as CRISPR/Cas9 (Ansai and Kinoshita 2014, 2017; Watakabe et al. 2018) and the tetracycline (Tet)-ON system (Hosoya et al. 2021), are available for application in medaka fish. Medaka fish diverged from zebrafish 314–332 million years ago (Kasahara et al. 2007) and have unique reproductive behaviors. While zebrafish females have an ovarian cycle of 5 days (Hisaoka and Firlit 1962), medaka females have an ovarian cycle of only 24 h and are ready to spawn eggs every morning. Thus, mating behaviors can be analyzed every morning using the same females, enabling investigation of social behaviors in a social context, including heterosexual interactions. This is a strong advantage of using medaka fish compared with zebrafish in the field of social neuroscience. In addition, female mate preference toward males can be easily judged on the basis of female behavior. Medaka mating behaviors consist of stereotypical behavioral components (Ono and Uematsu 1957). First, the male approaches the female and then the male exhibits a courtship display by swimming rapidly in front of the female. After the male wraps a receptive female with the anal and dorsal fins (crossing), the pair simultaneously release their gametes (spawning; see our video: https://www.youtube.com/watch?v=7HB5UaNrTSQ). If the female is not receptive, it tends to exhibit rejection behavior (Walter and Hamilton 1970). Importantly, medaka females exhibit mate preference on the basis of familiarity, i.e., medaka females visually discriminate between familiar and unfamiliar mates, and prefer to mate with familiar males (Okuyama et al. 2014). "Latency to mate" is the measure used to evaluate female mate preference and refers to the interval between the first male courtship and mating. This index negatively correlates with the degree of female acceptance for a focal male. Visual familiarization procedures, allowing a pair (male and female) to see each other through a transparent wall before a mating test, significantly decrease "the latency to mate" with the familiar male (Okuyama et al. 2014; Wang and Takeuchi 2017; Yokoi et al. 2020).

8.5 Sexual Dimorphic OT Function in Mate Preference in Medaka

In medaka fish, females prefer to choose familiar mates, whereas males court indiscriminately and irrespective of familiarity. A recent behavioral analysis of *ot*/*otr* mutants revealed that these mutations influence mate preference in a sex-specific manner (Fig. 8.1). The OT signaling (OT and OTR1; hereafter OTRa) is essential for

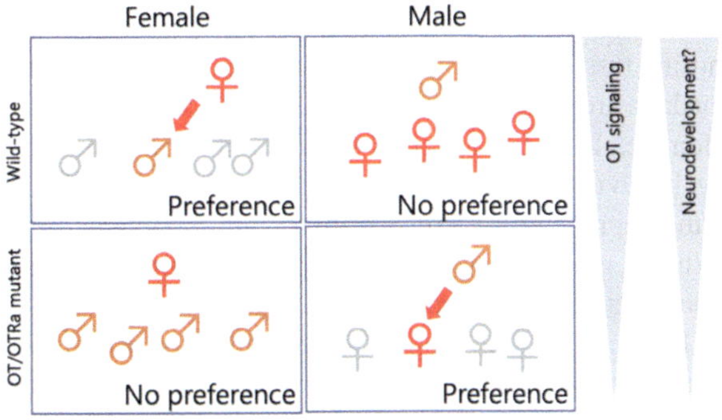

Fig. 8.1 Sexual dimorphism of OT signaling function in mate preference for familiar mates. Wild-type medaka females (left upper panel), but not males (right upper panel), tend to choose familiar mates as their mating partners. This tendency is reversed in *ot* and *otra* mutants. Females show no preference for familiar males (left lower panel) and males gain a mate preference for familiar females (right lower panel). The results of the transcriptome analysis imply that neurodevelopmental defects due to abnormal OT signaling cause these mutant phenotypes. Red arrows represent mate preference for familiar mates

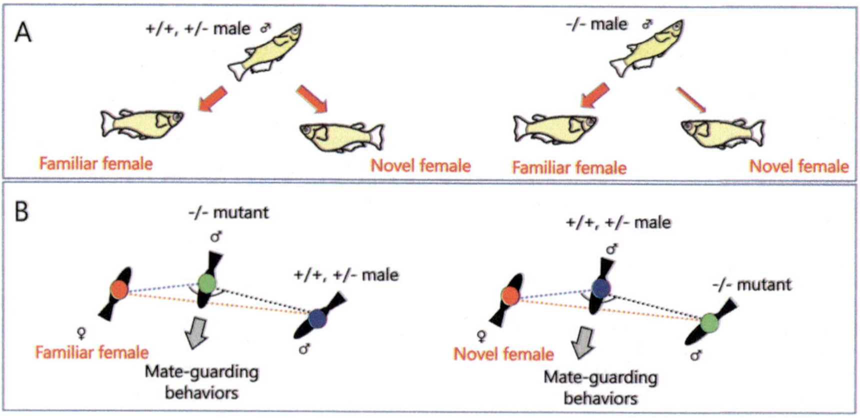

Fig. 8.2 Mate preference for familiar mates in *ot/otra*-mutant males. (**a**) Wild-type (+/+) and *ot/otra* heterozygous mutant (+/−) males exhibit courtship behavior toward novel females as well as toward familiar females. On the other hand, *ot/otra* homozygous mutant (−/−) males less frequently exhibit courtship behavior toward novel females than toward familiar females. (**b**) *ot/otra* homozygous mutant (−/−) males tend to be dominant against wild-type (+/+) or *ot/otra* heterozygous mutant (+/−) males in mate-guarding behaviors when familiar females are present (left). In contrast, *ot/otra* homozygous mutant males do not show mate-guarding behaviors toward novel females and tend to be subordinate against wild-type/heterozygous mutant males (right)

eliciting female mate preference for familiar males, while it is also required for the unrestricted and indiscriminate male mate choice. Visual familiarization does not influence the "latency to mate" in either *ot-* or *otra*-mutant females. Thus, *ot-* or *otra*-mutant females lose their mate preference for familiar males (Fig. 8.1). Either *ot-* or *otra*-mutant males more frequently exhibit courtship displays toward familiar females than novel females (Fig. 8.2a). The *ot-* or *otra*-mutant male-mating preference for familiar mates is also demonstrated in another behavioral paradigm, the male mate-guarding test (Yokoi et al. 2020). When two male and one female (triadic relationship) medaka are placed in a small tank, the two males compete against each other for the female. In the male-male competition, the dominant male exhibits "mate-guarding behavior": maintaining a dominant position near the female that interferes with the approach of the rival male toward the female (Yokoi et al. 2015, 2016, 2020). When a novel female is used as a target partner in the triadic relationship, *ot/otra* homozygous mutant males tend to be subordinate against the heterozygous mutant (Fig. 8.2b), which might be due to a loss of sexual motivation toward the novel female in *ot/otra*-mutant males. In contrast, *ot/otra* homozygous mutants have excessive mate preference for familiar females. In the triadic relationship, *ot/otra* homozygous mutant males tend to win against heterozygous mutant/wild-type males (Fig. 8.2b). Studies of the effects of OT on mate preference in mammals using monogamous rodents (prairie voles) revealed that OT signaling facilitates mate preference for familiar individuals in both males and females (Young et al. 2001; Johnson et al. 2016; Walum and Young 2018). In monogamous species, sexual dimorphism of the mating strategy is thought to be reduced because monogamous pairs share resource protection, parenting, and social support. Medaka studies showed that the conserved OT function can influence mate preference based on familiarity recognition from fish to mammals. It might be that OT function has evolved sex-specific functions at both neural and behavioral levels according to sexual dimorphism in the mating strategy.

8.6 Candidate Neuronal Targets of OT Signaling for Social Recognition in Mammals

The OTR distribution is closely related to the neural circuits underlying species-specific social behaviors. For example, OTR-expressing brain regions in the rodent brain (rats, mice, and voles) are prominently found in the olfactory pathway (Freeman and Young 2016), where OT signaling enhances the signal-to-noise response to social information, enhancing signal processing (Ford and Young 2021; Froemke and Young 2021). Rodents exhibit social interactions mainly based on chemical communication and olfactory investigation. Conspecific recognition as well as familiarity recognition is mediated by olfactory processing of major urinary proteins in the rodent (Brennan 2004). OTRs are also expressed in brain regions involved in the social decision-making network (O'Connell and Hofmann 2012), such as the amygdala and bed nucleus of the stria terminalis (Freeman and Young 2016). In contrast, in the primate brain, OTRs are concentrated in brain regions involved in visual processing. In humans, face recognition is a critically important ability for individual recognition, which enables the long-lasting formation of social relationships with multiple members (Calder et al. 2011). Visual information from individual faces provides large amounts of social information regarding sex, age, and emotional expression (Emery 2000). In nonhuman primates, eye-to-eye contact also functions as a social signal (Shepherd 2010; Harrod et al. 2020). Although there are some notable species-specific differences (Rogers Flattery et al. 2021), brain areas expressing OTRs in the primate brain are found in the visual pathway. Especially, OTR expression in the superior colliculus (SC) is conserved among four species of primates (human, titi monkey, rhesus monkey, and common marmoset) (Freeman and Young 2016). Interestingly, studies in cortically blind patients with damage to the visual cortex suggest the possible involvement of the SC in human face detection, such as another's gaze direction and emotional expressions (Morris et al. 2001; Burra et al. 2013). Furthermore, nonhuman primate studies provide neurophysiologic evidence that the SC functions as an innate rapid face-detection system in primates (Le et al. 2020). The SC neurons immediately respond (within 50 ms) to face-like patterns, but not to nonface patterns. Furthermore, SC neurons innervated by the central part of the receptive fields are selective for face-like patterns (Le et al. 2020). In humans, administration of OT acutely increases gaze to the eye region of human faces in populations with autistic syndrome (Andari et al. 2010; Auyeung et al. 2015). In studies of primate face recognition, most attention has been focused on the function of face-specific cortical regions such as the fusiform face area (Kanwisher et al. 1997). To our knowledge, however, there is no evidence demonstrating the involvement of OT signaling via SC neurons in social (face) recognition in primates.

8.7 Candidate Neuronal Targets of OT Signaling for Social Recognition in Medaka Fish

In fish species, OTRs are widely expressed throughout the forebrain and midbrain (Huffman et al. 2012; Yokoi et al. 2020; Schuppe et al. 2022), including the social decision-making network (O'Connell and Hofmann 2012). In the medaka brain, OTRs are expressed in brain regions related to social behaviors, such as the telencephalon (Dm and Vp), preoptic area, habenula, optic tectum, and hypothalamus (Fig. 8.3) (Yokoi et al. 2020). The Dm and Vp/Vv are homologous to the amygdala and lateral septum, respectively, in mammals (O'Connell and Hofmann 2012). The OTR distribution pattern does not differ between males and females. Importantly, in the medaka brain, OTRs are expressed in the optic tectum, whose mammalian equivalent is the SC (Basso et al. 2021). In zebrafish, the optic tectum is the primary processing center for visual information and functions to filter visual information in response to small objects (Del Bene et al. 2010). The optic tectum superficial layers receive input from retinal axons and the deeper layers convey the

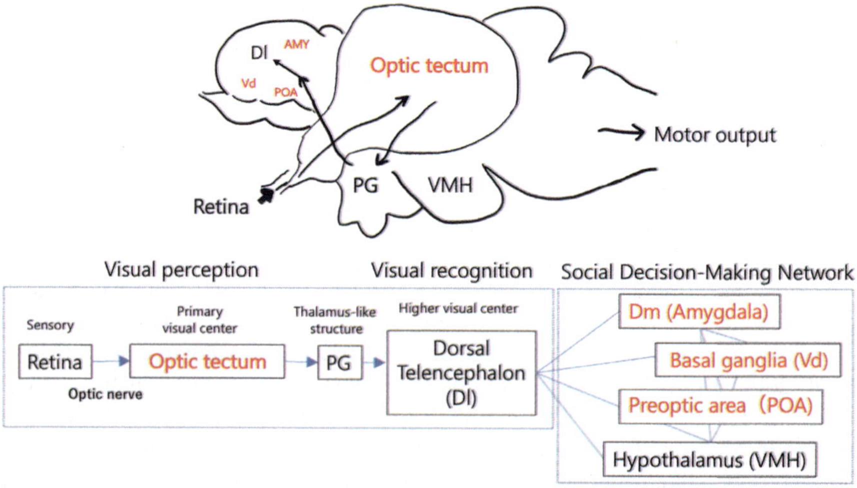

Fig. 8.3 Possible neural processing underlying mate preference on the basis of familiarity recognition in the medaka brain. OTR-expressing regions in the medaka brain are highlighted in red font. OTRs are not expressed in olfactory pathways such as the olfactory bulb. OT signaling might play a role in filtering social cues in visual pathways as well as modulating social decision-making networks

processed information to specific motor outputs (Sato et al. 2007), as well as to the secondary visual processing center (PG, preglomerular complex) linking it to forebrain structures (Bloch et al. 2020; Basso et al. 2021). The PG receives visual information from the optic tectum and projects to the lateral part of the dorsal telencephalic area, which is believed to be homologous to mammalian hippocampus (O'Connell and Hofmann 2012) and/or visual cortex (Yamamoto and Ito 2005; Mueller 2012) (Fig. 8.3). In zebrafish larvae, the optic tectum is innervated by oxytocinergic cells (Herget et al. 2017) and OTR mutants exhibit defects in detecting biologic motion, which mimics body motion and attracts real fish (Nunes et al. 2020), suggesting that OT could modulate primary perceptual mechanisms underlying the detection of social cues. Further medaka studies will clarify whether the optic tectum could function as a filter for visual information in response to social clues from conspecifics and whether the oxytocinergic system modulates the filtering of social information.

Importantly, medaka use visual cues from the face (head part) for familiarity recognition. A conditioning test (electric shock 2-alternative forced-choice [TAFC] design) revealed that medaka females also use face parts for individual recognition (Wang and Takeuchi 2017). Females can discriminate between two unfamiliar male faces in this behavioral paradigm. Furthermore, mating tests and TAFC revealed that medaka fish exhibit a "face inversion effect", in which recognizing an inverted (upside-down) face is more difficult compared with an upright face. It is more difficult for medaka females to recognize males when inverted faces are presented using a prism (Wang and Takeuchi 2017). In the primate brain, face-specific cortical regions such as the fusiform face area are involved in the configural processing for face recognition (Kanwisher et al. 1997) and this system is believed to be involved in the perception of an "upright face" but not an "inverted face." The configural recognition process is specific to faces and does not apply to stimuli from objects (Kanwisher et al. 1997). Considering that the face inversion effect also occurs in medaka fish, it is possible that medaka fish have a brain region specialized for social cognition based on visual information.

8.8 What Are the Molecular Targets of OT Signaling?

What are the molecular targets of OT signaling in the medaka fish brain? Transcriptome analysis of the medaka brain with *ot*/*otra* mutations suggests that the OT signaling pathway could enhance complement component C1q (Fig. 8.4). Mutations of *ot* or *otra* decrease the expression of three genes for components of the C1q complex (C1qa, C1qb, and C1qc) in the brains of both sexes (Yokoi et al. 2020). C1q was originally identified as a component of the innate immune system that directly promotes the engulfment of apoptotic cells by macrophages (Ogden et al. 2001; Nauta et al. 2002). Recent studies demonstrated that the C1q complex has an important role in normal synaptic pruning and brain development. During normal brain development, immature neurons overproduce synaptic formations and

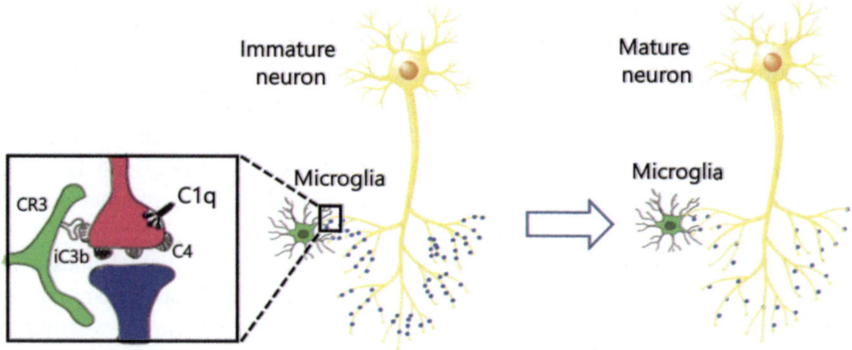

Fig. 8.4 Microglia synapse elimination mediated by C1q components. During postnatal development, excessive synapses are eliminated by microglia. Low-activity synapses are tagged by complement components (C1q, C4, and iC3b). Microglial cells recognize tagged synapses by binding iC3b through CR3 receptors and selectively eliminate low-activity synapses. High-activity synapses are protected from synaptic pruning. In medaka fish brain, either *ot* or *otra* mutations significantly decrease the expression level of three C1q component genes (C1qa, C1qb, and C1qc) irrespective of sex, implying that OT signaling positively regulates synaptic pruning mediated by this pathway. Neuronal images are from *TogoTV (©2016 DBCLS TogoTV)*

subsequent synaptic pruning provides for a lower density but more efficient synaptic connections (Fig. 8.4). C1q is localized to developing CNS synapses during periods of active synapse elimination and is required for normal brain wiring (Stephan et al. 2012). Medaka-mutant analyses imply that C1q is a promising candidate molecular target of OT signaling during normal neural development via synapse elimination. Therefore, defects in OT signaling may impair normal neurodevelopment via synapse elimination, which could lead to an inappropriate mate choice based on familiarity in medaka fish (Fig. 8.1). Consistently, perturbation of OT neurons during early zebrafish development leads to altered functional connectivity in some brain-regions associated with social behaviors and an impaired response to social stimuli (Nunes et al. 2021).

8.9 Future Prospects in Medaka Studies

The medaka optic tectum could be an important model for comparing the molecular/neural basis underlying how OT modulates face detection between the fish optic tectum and primate SC. The face inversion effect in medaka fish implies that medaka have specific visual processing pathways for face detection (Wang and Takeuchi 2017). State-of-the-art molecular genetic tools available for medaka fish will allow us to investigate this issue. The doxycycline-inducible Tet-ON system, a genetic tool for inducible and reversible transgene expression was recently reported to be applicable to medaka fish (Hosoya et al. 2021). In addition, efficient methods for targeted gene knock-in using the CRISPR/Cas9 system are available for medaka fish

(Watakabe et al. 2018), and allow for a reporter gene of interest, such as a fluorescent protein and/or optogenetic tools (Rost et al. 2017), to mimic the expression of endogenous target genes. The combination of a targeted gene knock-in system and Tet-ON system can visualize and modulate a subset of neurons expressing a specific marker gene in the medaka optic tectum. Comparative studies among vertebrates could help to elucidate the evolutionary roots and basic neural mechanisms underlying face detection.

Medaka fish with *ot/otra* mutations could be a promising model for studies of neurodevelopmental diseases such as autism spectrum disorders (ASD). Recent studies in mammals revealed an essential role of OT in early life for normal postnatal neurodevelopment. An early defect in the OT signaling could impair maturation, which might cause short-term and long-term pathologic consequences. For a long time, neurodevelopment disorders have been believed to result from the disruption of this early life synapse remodeling (Zoghbi 2003). ASD mouse models have a similar impairment in synapse remodeling, which derives from an enhanced turnover of excitatory synapses (Isshiki et al. 2014). Therefore, defects in synaptic pruning in the developing brain are considered a potential mechanism of ASD. Although complement component C1q is involved in fundamental neurodevelopmental pathways for the maintenance and elimination of dendrites and synapses, there are no reports of OT-dependent synaptic pruning mediated by C1q signaling (Magdalon et al. 2020). Further studies on OT function using medaka fish might shed light on the possible involvement of this novel pathway in human neurodevelopmental disease.

Recent human studies suggest that OT has sexually dimorphic effects on face recognition (Gao et al. 2016). Human intranasal administration of OT enhances social salience and the motivational state in both sexes (Bartz et al. 2011; McCall and Singer 2012), it could have opposite effects depending on the sex (Fischer-Shofty et al. 2013; Ditzen et al. 2013; Scheele et al. 2014). Investigation of the molecular mechanisms underlying sex-specific functions of OT signaling using genetic tools in primates, however, is difficult. The Medaka fish is an important comparative model to approach this issue. The single-cell transcriptome technique might be useful for comparing gene expression profiles of OTR-expressing neurons between sexes at the single cell level. This comparison might help to identify sex-specific subsets of OTR-expressing neurons in the medaka brain, although there is no significant difference in OTR distributions between sexes (Yokoi et al. 2020). Considering that OT signaling has essential roles in neurodevelopment, medaka studies will be valuable for clarifying how neurodevelopment mediated by the OT/C1q systems could contribute to the human sexual spectrum and diversity in mating strategy.

Acknowledgments This work was supported by the National Institute for Basic Biology Priority Collaborative Research Project 10-104 (to H.T.), 19-347 (to H.T.), and 21-335 (to H.T.); Joint Research Grant 01111904; Japan Society for the Promotion of Science (JSPS) KAKENHI Grants 21H04773, 20H04925 (to H.T.), 18H02479 (to H.T.), 19K16247 (to S.Y.), 19H04889 (to S.Y.), 21H0570801 (to S.Y.), Astellas Foundation for Research on Metabolic Disorders (to S.Y.), Takeda Science Foundation (to H.T. and S.Y.), and the natural science grant of the Mitsubishi Foundation.

LJY's contributions were supported by National Institute of Health Grants P50MH100023 and P51OD011132.

References

Andari E, Duhamel JR, Zalla T, Herbrecht E, Leboyer M, Sirigu A (2010) Promoting social behavior with oxytocin in high-functioning autism spectrum disorders. Proc Natl Acad Sci U S A 107:4389–4394

Ansai S, Kinoshita M (2014) Targeted mutagenesis using CRISPR/Cas system in medaka. Biol Open 3:362–371

Ansai S, Kinoshita M (2017) Genome editing of medaka. Methods Mol Biol 1630:175–188

Auyeung B, Lombardo MV, Heinrichs M, Chakrabarti B, Sule A, Deakin JB, Bethlehem RA, Dickens L, Mooney N, Sipple JA, Thiemann P, Baron-Cohen S (2015) Oxytocin increases eye contact during a real-time, naturalistic social interaction in males with and without autism. Transl Psychiatry 5:e507. https://doi.org/10.1038/tp.2014.146

Barber L, Ruxton GD (2000) The importance of stable schooling: do familiar sticklebacks stick together? Proc Biol Sci 267:151–155

Barber I, Wright HA (2001) How strong are familiarity preferences in shoaling fish? Anim Behav 61:975–979

Bartz JA, Zaki J, Bolger N, Ochsner KN (2011) Social effects of oxytocin in humans: context and person matter. Trends Cogn Sci 15:301–309

Basso MA, Bickford ME, Cang J (2021) Unraveling circuits of visual perception and cognition through the superior colliculus. Neuron 109:918–937

Bloch S, Hagio H, Thomas M, Heuzé A, Hermel JM, Lasserre E, Colin I, Saka K, Affaticati P, Jenett A, Kawakami K, Yamamoto N, Yamamoto K (2020) Non-thalamic origin of zebrafish sensory nuclei implies convergent evolution of visual pathways in amniotes and teleosts. Elife 9: e54945. https://doi.org/10.7554/eLife.54945

Brennan PA (2004) The nose knows who's who: chemosensory individuality and mate recognition in mice. Horm Behav 46:231–240

Bruzzone M, Gatto E, Lucon Xiccato T, Dalla Valle L, Fontana CM, Meneghetti G, Bisazza A (2020) Measuring recognition memory in zebrafish larvae: issues and limitations. PeerJ 8: e8890. https://doi.org/10.7717/peerj.8890

Burra N, Hervais-Adelman A, Kerzel D, Tamietto M, de Gelder B, Pegna AJ (2013) Amygdala activation for eye contact despite complete cortical blindness. J Neurosci 33:10483–10489

Calder A, Rhodes G, Johnson HM, Haxby J (2011) Oxford handbook of face perception. Oxford University Press, Oxford

Del Bene F, Wyart C, Robles E, Tran A, Looger L, Scott EK, Isacoff EY, Baier H (2010) Filtering of visual information in the tectum by an identified neural circuit. Science 330:669–673

Ditzen B, Nater UM, Schaer M, La Marca R, Bodenmann G, Ehlert U, Heinrichs M (2013) Sex-specific effects of intranasal oxytocin on autonomic nervous system and emotional responses to couple conflict. Soc Cogn Affect Neurosci 8:897–902

Doran C, Bierbach D, Laskowski KL (2019) Familiarity increases aggressiveness among clonal fish. Anim Behav 148:153–159

Emery NJ (2000) The eyes have it: the neuroethology, function and evolution of social gaze. Neurosci Biobehav Rev 24:581–604

Farmer NA, Ribble DO, Miller DG III (2004) Influence of familiarity on shoaling behaviour in Texas and blacktail shiners. J Fish Biol 64:776–782

Fischer-Shofty M, Levkovitz Y, Shamay-Tsoory SG (2013) Oxytocin facilitates accurate perception of competition in men and kinship in women. Soc Cogn Affect Neurosci 8:313–317

Ford CL, Young LJ (2021) Translational opportunities for circuit-based social neuroscience: advancing 21st century psychiatry. Curr Opin Neurobiol 68:1–8

Freeman SM, Young LJ (2016) Comparative perspectives on oxytocin and vasopressin receptor research in rodents and primates: translational implications. J Neuroendocrinol 28. https://doi.org/10.1111/jne.12382

Froemke RC, Young LJ (2021) Oxytocin, neural plasticity, and social behavior. Annu Rev Neurosci 44:359–381

Gao S, Becker B, Luo L, Geng Y, Zhao W, Yin Y, Hu J, Gao Z, Gong Q, Hurlemann R, Yao D, Kendrick KM (2016) Oxytocin, the peptide that bonds the sexes also divides them. Proc Natl Acad Sci U S A 113:7650–7654

Gerlach G, Lysiak N (2006) Kin recognition and inbreeding avoidance in zebrafish, *Danio rerio*, is based on phenotype matching. Anim Behav 71:1371–1377

Griffiths SW (2003) Learned recognition of conspecifics by fishes. Fish Fish 4:256–268

Griffiths SW, Ojanguren AF, Orpwood JE, Magurran AE, Armstrong JD (2007) Familiarity-biased patterns of association shift with time among European minnows. J Fish Biol 71:1602–1612

Harrod EG, Coe CL, Niedenthal PM (2020) Social structure predicts eye contact tolerance in nonhuman primates: evidence from a crowd-sourcing approach. Sci Rep 10:6971

Hay DE, McKinnell SM (2002) Tagging along: association among individual Pacific herring (*Clupea pallasi*) revealed by tagging. Can J Fish Aquat Sci 59:1960–1968

Herget U, Gutierrez-Triana JA, Salazar Thula O, Knerr B, Ryu S (2017) Single-cell reconstruction of oxytocinergic neurons reveals separate hypophysiotropic and encephalotropic subtypes in larval zebrafish. eNeuro 4. https://doi.org/10.1523/ENEURO.0278-16.2016

Hinz C, Kobbenbring S, Kress S, Sigman L, Müller A, Gerlach G (2013) Kin recognition in zebrafish, *Danio rerio*, is based on imprinting on olfactory and visual stimuli. Anim Behav 85: 925–930

Hisaoka KK, Firlit C (1962) Ovarian cycle and egg production in the zebrafish, *Brachydanio rerio*. Copeia 1962:788

Höjesjö J, Johnsson JI, Petersson E, Järvi T (1998) The importance of being familiar: individual recognition and social behavior in sea trout (*Salmo trutta*). Behav Ecol 9:445–451

Hosoya O, Chung M, Ansai S, Takeuchi H, Miyaji M (2021) A modified Tet-ON system minimizing leaky expression for cell-type specific gene induction in medaka fish. Develop Growth Differ 63:397–405

Huffman LS, O'Connell LA, Kenkel CD, Kline RJ, Khan IA, Hofmann HA (2012) Distribution of nonapeptide systems in the forebrain of an African cichlid fish, *Astatotilapia burtoni*. J Chem Neuroanat 44:86–97

Hughes KA, Du L, Rodd FH, Reznick DN (1999) Familiarity leads to female mate preference for novel males in the guppy, *Poecilia reticulata*. Anim Behav 58:907–916

Isshiki M, Tanaka S, Kuriu T, Tabuchi K, Takumi T, Okabe S (2014) Enhanced synapse remodelling as a common phenotype in mouse models of autism. Nat Commun 5:4742

Johnson ZV, Walum H, Jamal YA, Xiao Y, Keebaugh AC, Inoue K, Young LJ (2016) Central oxytocin receptors mediate mating-induced partner preferences and enhance correlated activation across forebrain nuclei in male prairie voles. Horm Behav 79:8–17

Johnsson JI (1997) Individual recognition affects aggression and dominance relations in rainbow trout, *Oncorhynchus mykiss*. Ethology 103:267–282

Kanwisher N, McDermott J, Chun MM (1997) The fusiform face area: a module in human extrastriate cortex specialized for face perception. J Neurosci 17:4302–4311

Kasahara M, Naruse K, Sasaki S, Nakatani Y, Qu W, Ahsan B, Yamada T, Nagayasu Y, Doi K, Kasai Y, Jindo T, Kobayashi D, Shimada A, Toyoda A, Kuroki Y, Fujiyama A, Sasaki T, Shimizu A, Asakawa S, Shimizu N, Hashimoto S, Yang J, Lee Y, Matsushima K, Sugano S, Sakaizumi M, Narita T, Ohishi K, Haga S, Ohta F, Nomoto H, Nogata K, Morishita T, Endo T, Shin IT, Takeda H, Morishita S, Kohara Y (2007) The medaka draft genome and insights into vertebrate genome evolution. Nature 447:714–719

Keller BA, Finger J-S, Gruber SH, Abel DC, Guttridge TL (2017) The effects of familiarity on the social interactions of juvenile lemon sharks, *Negaprion brevirostris*. J Exp Mar Biol Ecol 489: 24–31

Kelley JL, Graves JA, Magurran AE (1999) Familiarity breeds contempt in guppies. Nature 401: 661–662

Krause J, Butlin RK, Peuhkuri N, Pritchard VL (2000) The social organization of fish shoals: a test of the predictive power of laboratory experiments for the field. Biol Rev Camb Philos Soc 75: 477–501

Landin J, Hovey D, Xu B, Lagman D, Zettergren A, Larhammar D, Kettunen P, Westberg L (2020) Oxytocin receptors regulate social preference in zebrafish. Sci Rep 10:5435. https://doi.org/10.1038/s41598-020-61073-4

Larsch J, Baier H (2018) Biological motion as an innate perceptual mechanism driving social affiliation. Curr Biol 28:3523–3532

Le QV, Le QV, Nishimaru H, Matsumoto J, Takamura Y, Hori E, Maior RS, Tomaz C, Ono T, Nishijo H (2020) A prototypical template for rapid face detection is embedded in the monkey superior colliculus. Front Syst Neurosci 14. https://doi.org/10.3389/fnsys.2020.00005

Lee-Jenkins SS, Godin JG (2010) Social familiarity and shoal formation in juvenile fishes. J Fish Biol 76:580–590

Ludwig M, Leng G (2006) Dendritic peptide release and peptide-dependent behaviours. Nat Rev Neurosci 7:126–136

Magdalon J, Mansur F, Teles e Silva AL, de Goes VA, Reiner O, Sertié AL (2020) Complement system in brain architecture and neurodevelopmental disorders. Front Neurosci 14:23. https://doi.org/10.3389/fnins.2020.00023

Magurran AE, Seghers BH, Shaw PW, Carvalho GR (1994) Schooling preferences for familiar fish in the guppy, Poecilia reticulata. J Fish Biol 45:401–406

McCall C, Singer T (2012) The animal and human neuroendocrinology of social cognition, motivation and behavior. Nat Neurosci 15:681–688

Miklósi Á, Haller J, Csányi V (1992) Different duration of memory for conspecific and heterospecific fish in the paradise fish (*Macropodus opercularis* L.). Ethology 90:29–36

Morris JS, DeGelder B, Weiskrantz L, Dolan RJ (2001) Differential extrageniculostriate and amygdala responses to presentation of emotional faces in a cortically blind field. Brain 124: 1241–1252

Mueller T (2012) What is the thalamus in zebrafish? Front Neurosci 6:64–64

Nadler LE, McCormick MI, Johansen JL, Domenici P (2021) Social familiarity improves fast-start escape performance in schooling fish. Commun Biol 4:897

Nauta AJ, Trouw LA, Daha MR, Tijsma O, Nieuwland R, Schwaeble WJ, Gingras AR, Mantovani A, Hack EC, Roos A (2002) Direct binding of C1q to apoptotic cells and cell blebs induces complement activation. Eur J Immunol 32:1726–1736

Numan M, Young LJ (2016) Neural mechanisms of mother-infant bonding and pair bonding: similarities, differences, and broader implications. Horm Behav 77:98–112

Nunes AR, Carreira L, Anbalagan S, Blechman J, Levkowitz G, Oliveira RF (2020) Perceptual mechanisms of social affiliation in zebrafish. Sci Rep 10:3642

Nunes AR, Gliksberg M, Varela SAM, Teles M, Wircer E, Blechman J, Petri G, Levkowitz G, Oliveira RF (2021) Developmental effects of oxytocin neurons on social affiliation and processing of social information. J Neurosci 41:8742–8760

O'Connell LA, Hofmann HA (2012) Evolution of a vertebrate social decision-making network. Science 336:1154–1157

Ogden CA, deCathelineau A, Hoffmann PR, Bratton D, Ghebrehiwet B, Fadok VA, Henson PM (2001) C1q and mannose-binding lectin engagement of cell surface calreticulin and CD91 initiates macropinocytosis and uptake of apoptotic cells. J Exp Med 194:781–795

Okuyama T, Yokoi S, Abe H, Isoe Y, Suehiro Y, Imada H, Tanaka M, Kawasaki T, Yuba S, Taniguchi Y, Kamei Y, Okubo K, Shimada A, Naruse K, Takeda H, Oka Y, Kubo T, Takeuchi

H (2014) A neural mechanism underlying mating preferences for familiar individuals in medaka fish. Science 343:91–94

Okuyama T, Yokoi S, Takeuchi H (2017) Molecular basis of social competence in medaka fish. Develop Growth Differ 59:211–218

Oliveira R (2013) Mind the fish: zebrafish as a model in cognitive social neuroscience. Front Neural Circuits 7:131. https://doi.org/10.3389/fncir.2013.00131

Ono Y, Uematsu T (1957) Mating ethogram in *Oryzias latipes* (with 1 text-figure). J Fac Sci Hokkaido Univ Ser VI Zool 13:197–202

Oti T, Satoh K, Uta D, Nagafuchi J, Tateishi S, Ueda R, Takanami K, Young LJ, Galione A, Morris JF, Sakamoto T, Sakamoto H (2021) Oxytocin influences male sexual activity via non-synaptic axonal release in the spinal cord. Curr Biol 31:103–114

Ribeiro D, Nunes AR, Gliksberg M, Anbalagan S, Levkowitz G, Oliveira RF (2020) Oxytocin receptor signalling modulates novelty recognition but not social preference in zebrafish. J Neuroendocrinol 32:e12834

Rilling JK, Young LJ (2014) The biology of mammalian parenting and its effect on offspring social development. Science 345(6198):771–776

Rogers Flattery CN, Coppeto DJ, Inoue K, Rilling JK, Preuss TM, Young LJ (2021) Distribution of brain oxytocin and vasopressin V1a receptors in chimpanzees (*Pan troglodytes*): comparison with humans and other primate species. Brain Struct Funct 227:1907. https://doi.org/10.1007/s00429-021-02369-7

Rost BR, Schneider-Warme F, Schmitz D, Hegemann P (2017) Optogenetic tools for subcellular applications in neuroscience. Neuron 96:572–603

Sato T, Hamaoka T, Aizawa H, Hosoya T, Okamoto H (2007) Genetic single-cell mosaic analysis implicates ephrinB2 reverse signaling in projections from the posterior tectum to the hindbrain in zebrafish. J Neurosci 27:5271–5279

Saverino C, Gerlai R (2008) The social zebrafish: behavioral responses to conspecific, heterospecific, and computer-animated fish. Behav Brain Res 191:77–87

Scheele D, Striepens N, Kendrick KM, Schwering C, Noelle J, Wille A, Schläpfer TE, Maier W, Hurlemann R (2014) Opposing effects of oxytocin on moral judgment in males and females. Hum Brain Mapp 35:6067–6076

Schuppe ER, Zhang MD, Perelmuter JT, Marchaterre MA, Bass AH (2022) Oxytocin-like receptor expression in evolutionarily conserved nodes of a vocal network associated with male courtship in a teleost fish. J Comp Neurol 530(6):903–922

Seppä T, Laurila A, Peuhkuri N, Piironen J, Lower N (2001) Early familiarity has fitness consequences for Arctic char (*Salvelinus alpinus*) juveniles. Can J Fish Aquat Sci 58:1380–1385

Shepherd S (2010) Following gaze: gaze-following behavior as a window into social cognition. Front Integr Neurosci 4:5. https://doi.org/10.3389/fnint.2010.00005

Stephan AH, Barres BA, Stevens B (2012) The complement system: an unexpected role in synaptic pruning during development and disease. Annu Rev Neurosci 35:369–389

Theofanopoulou C, Gedman G, Cahill JA, Boeckx C, Jarvis ED (2021) Universal nomenclature for oxytocin-vasotocin ligand and receptor families. Nature 592:747–755

Utne-Palm AC, Hart PJB (2000) The effects of familiarity on competitive interactions between three-spined sticklebacks. Oikos 91:225–232

Walter RO, Hamilton JB (1970) Head-up movements as an indicator of sexual unreceptivity in female medaka, *Oryzias latipes*. Anim Behav 18:125–127

Walum H, Young LJ (2018) The neural mechanisms and circuitry of the pair bond. Nat Rev Neurosci 19:643–654

Wang MY, Takeuchi H (2017) Individual recognition and the 'face inversion effect' in medaka fish (*Oryzias latipes*). Elife 6:e24728. https://doi.org/10.7554/eLife.24728

Ward AJW, Kent MIA, Webster MM (2020) Social recognition and social attraction in group-living fishes. Fronti Ecol Evol 8:15. https://doi.org/10.3389/fevo.2020.00015

Watakabe I, Hashimoto H, Kimura Y, Yokoi S, Naruse K, Higashijima S (2018) Highly efficient generation of knock-in transgenic medaka by CRISPR/Cas9-mediated genome engineering. Zool Lett 4:3. https://doi.org/10.1186/s40851-017-0086-3

Yamamoto N, Ito H (2005) Fiber connections of the anterior preglomerular nucleus in cyprinids with notes on telencephalic connections of the preglomerular complex. J Comp Neurol 491: 212–233

Yokoi S, Okuyama T, Kamei Y, Naruse K, Taniguchi Y, Ansai S, Kinoshita M, Young LJ, Takemori N, Kubo T, Takeuchi H (2015) An essential role of the arginine vasotocin system in mate-guarding behaviors in triadic relationships of medaka fish (*Oryzias latipes*). PLoS Genet 11:e1005009

Yokoi S, Ansai S, Kinoshita M, Naruse K, Kamei Y, Young LJ, Okuyama T, Takeuchi H (2016) Mate-guarding behavior enhances male reproductive success via familiarization with mating partners in medaka fish. Front Zool 13:21. https://doi.org/10.1186/s12983-016-0152-2

Yokoi S, Naruse K, Kamei Y, Ansai S, Kinoshita M, Mito M, Iwasaki S, Inoue S, Okuyama T, Nakagawa S, Young LJ, Takeuchi H (2020) Sexually dimorphic role of oxytocin in medaka mate choice. Proc Natl Acad Sci U S A 117:4802–4808

Young LJ, Lim MM, Gingrich B, Insel TR (2001) Cellular mechanisms of social attachment. Horm Behav 40:133–138

Zoghbi HY (2003) Postnatal neurodevelopmental disorders: meeting at the synapse? Science 302: 826–830

Chapter 9
Orchestration of the Synthesis of Sex Hormones and their Roles in Establishing Sex Differences in Mammals

Takashi Baba

Abstract This chapter presents classical knowledge as well as recent findings on sex hormones. First, the synthesis of hormones is reviewed, and a tight linkage with core energy metabolism is introduced. The reader can thus realize the presence of a sophisticated regulation system governed by a single transcription factor, Ad4BP/SF-1. Several examples of the important roles of sex hormones in reproduction and the establishment of sex differences in multiple organs/cell types are introduced. Furthermore, the latest knowledge on sex differences in skeletal muscle and the adrenal cortex is presented. Finally, the involvement of sex hormones in establishing the sex spectrum is discussed.

Keywords Sex hormones · Sex differences · Skeletal muscle · Adrenal cortex

9.1 Sex Hormones

9.1.1 Discovery of Sex Hormones

Sex hormones are key molecules in endocrine regulation that induce sex differences. The concept of a "sex hormone" was first provided by Berthold et al. in 1849 (& NA, 1996). They observed the disappearance of male-specific larger crests and male-specific behavior from roosters on the removal of their testes. These male-specific characteristics were recovered by the re-implantation of the testes in the castrated roosters. These observations revealed the presence of humoral factors secreted from the testis that were necessary for maintaining male-specific characteristics. Currently, androgens and estrogens are known to induce male and female characteristics.

T. Baba (✉)
Department of Molecular Biology, Graduate School of Medical Sciences, Kyushu University, Fukuoka, Japan
e-mail: baba.takashi.547@m.kyushu-u.ac.jp

M. Tanaka, M. Tachibana (eds.), *Spectrum of Sex*,
https://doi.org/10.1007/978-981-19-5359-0_9

9.1.2 Regulation of Sex Hormone Synthesis

Sex hormones are primarily synthesized in the testis and ovary using cholesterol as a starting material. As CYP11A1 (P450 side-chain cleavage), an enzyme that mediates the first step of steroidogenesis, localizes in the inner mitochondrial membrane, cholesterol is transported to the specific location to be utilized for the synthesis. The transportation of cholesterol from the outer to the inner mitochondrial membrane has been shown to be mediated by steroid acute regulatory protein (STAR). The delivery of cholesterol from the endoplasmic reticulum, where cholesterol is synthesized de novo to the outer mitochondrial membrane, was recently shown to be mediated by MGARP (also known as OSAP or HUMMR) (Jinn et al. 2015). A potent androgen, testosterone, is synthesized in testicular Leydig cells from cholesterol through four enzymatic reactions, mediated by CYP11A1, HSD3B (3β-Hydroxysteroid dehydrogenase), CYP17A1 (17α-hydroxylase/17,20-lyase P450), and HSD17B3 (17-β-Hydroxysteroid dehydrogenase type III) (Miller 1988, 2013; Waterman and Bischof 1997; Morohashi et al. 2013) (Fig. 9.1). Synthesized testosterone is secreted into the blood to be delivered to the entire body; in target cells, testosterone is converted to a more potent androgen, 5α-DHT, by SRD5A1/2. In the ovary, androstane-3,17-dione produced in theca cells is transferred to granulosa cells, where CYP19A1 (aromatase P450) mediates the conversion of androstane-3,17-dione to estrone. Thereafter, HSD17B1 (17β-Hydroxysteroid dehydrogenase type I) mediates the conversion of estrone to a potent estrogen, 17β-estradiol.

9.1.3 Orchestration of Sex Hormone Synthesis by Ad4BP/SF-1

Interestingly, all genes encoding the above enzymes (*Cyp11a1*, *Hsd3b1*, *Cyp17a1*, *Hsd17b1/3*, and *Cyp19a1*) and *Star* genes have been reported as the targets of a transcription factor, Ad4BP/SF-1 (Ad4-binding protein/steroidogenic factor-1, NR5A1) (Morohashi et al. 1992; Val et al. 2003; Hammer et al. 2005; Hoivik et al. 2010). Thus, it is considered that Ad4BP/SF-1 is a master regulator of sex hormone synthesis in gonads, which governs the transcription of all genes involved. It should be noted that *Mgarp* genes are also directly regulated by Ad4BP/SF-1 (Baba et al. 2018).

Cholesterol is delivered into peripheral cells, including the testis and ovary, through blood flow. In addition, cholesterol can be synthesized de novo using acetyl-CoA as a starting material. Twenty enzymatic reactions are necessary to produce cholesterol from acetyl-CoA. The transcription factor SREBP-2 is known to transactivate all genes involved in cholesterogenesis in response to lowered cholesterol content in the cytoplasm (Freeman and Ascoli 1982; Brown and Goldstein 1999; Horton et al. 2002). In addition, we recently found that *Ad4BP/SF-1* knockdown resulted in a decrease in the expression of all 20 cholesterogenic genes

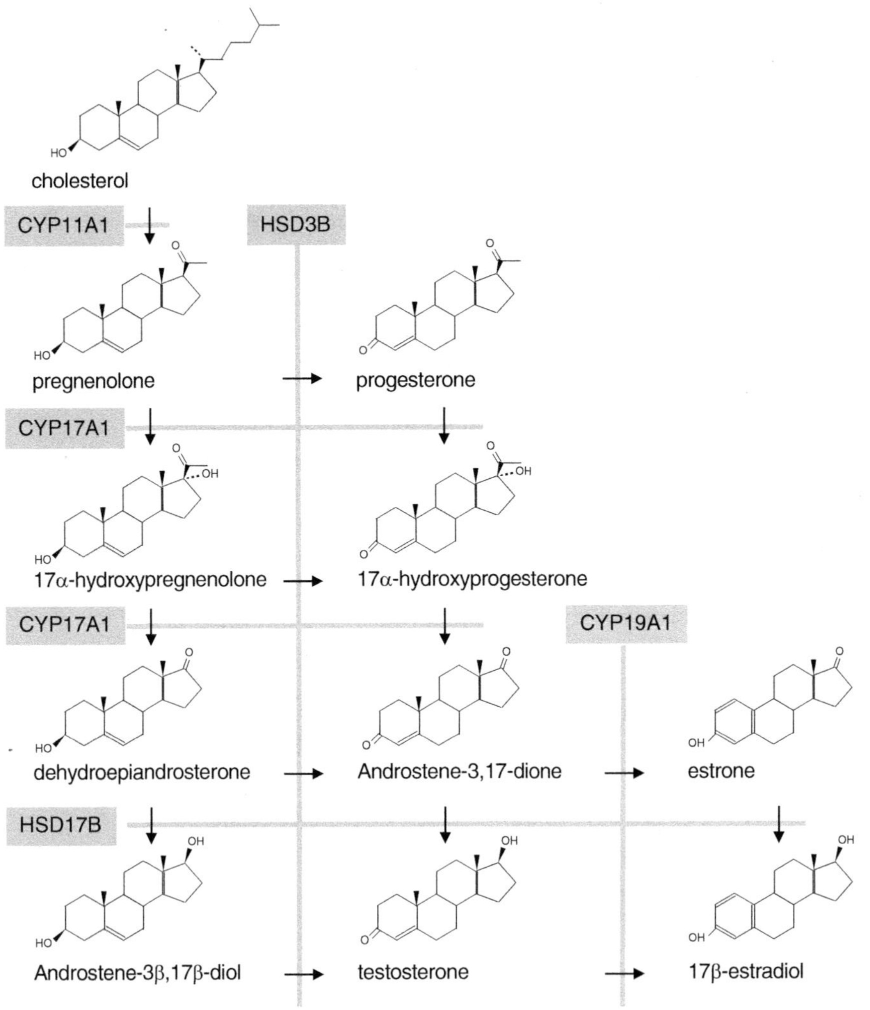

Fig. 9.1 Synthetic pathway of sex hormones from cholesterol

(Baba et al. 2018). Moreover, ChIP-sequencing revealed that Ad4BP/SF-1 bound to many cholesterogenic gene loci. As expected, Ad4BP/SF-1 was observed to physically interact with SREBP-2, and this interaction was shown to be necessary for the high-level expression of cholesterogenic genes. Together, these results indicate that Ad4BP/SF-1 is involved in the regulation of cholesterogenic genes in cooperation with SREBP-2, at least in steroidogenic cells.

How is acetyl-CoA supplied for cholesterogenesis? In addition to fatty acid β-oxidation, glycolysis is considered a major acetyl-CoA-supplying pathway. The final product of glycolysis, pyruvate, is converted to acetyl-CoA by the pyruvate dehydrogenase complex in the mitochondrial matrix; thereafter, it is used as a carbon

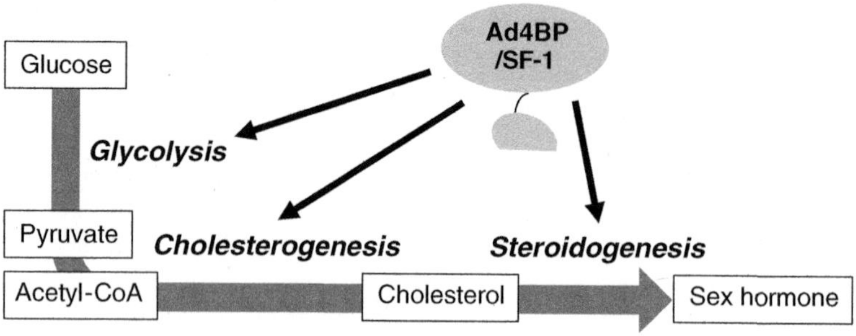

Fig. 9.2 Orchestration of multiple metabolic pathways by Ad4BP/SF-1. Ad4BP/SF-1 regulates expression of all genes involved in glycolysis, cholesterogenesis, and steroidogenesis, thereby governs the synthesis of sex hormones from glucose

source in the TCA cycle. One of the intermediates of the TCA cycle, citrate, can be translocated to the cytosol, where it is converted back into acetyl-CoA by the function of ACLY (Chypre et al. 2012). We demonstrated that *Acly* is directly regulated by Ad4BP/SF-1. We also examined the involvement of Ad4BP/SF-1 in glycolysis. Consequently, transcriptome analysis and ChIP-sequencing showed that Ad4BP/SF-1 regulates all glycolytic genes (Baba et al. 2014). Furthermore, *Ad4BP/SF-1* knockdown lowered glycolytic activity. Taken together, these findings suggest that Ad4BP/SF-1 governs sex hormone synthesis by orchestrating multiple metabolic pathways from glucose to sex hormones (Fig. 9.2). Notably, the synthesis of sex hormones is tightly linked to core energy metabolism.

9.2 Sex Hormone Receptors

How do sex hormones exert their functions? Unlike other hormones for cell membrane receptors, sex hormones are lipophilic and can transverse the cell membrane to reach the cytoplasm. In the cytoplasm, sex hormones bind to their specific receptors, androgen receptor (AR) and estrogen receptors (ERα and ERβ), to transduce signaling. These sex hormone receptors belong to the nuclear receptor superfamily (Olefsky 2001). Nuclear receptors have a common structure with five functional domains: an N-terminal regulatory domain, a DNA-binding domain, a hinge region, a ligand-binding domain (LBD), and a C-terminal regulatory domain. Upon the binding of ligands, including sex hormones, to the LBD, the receptors undergo conformational changes to become active. The activated receptors translocate into the nucleus where they bind to their cognate DNA sequences and induce the transcription of adjacent genes. The cDNA of the estrogen receptor was cloned by Pierre Chambon in 1985, and the gene product is presently known as ERα (Walter et al. 1985). Another gene encoding β type of the receptor (ERβ) was cloned from rat prostate, human testis, and mouse ovary in 1996–1997 (Kuiper et al. 1996; Mosselman et al. 1996; Tremblay et al. 1997). The cDNA of AR was cloned in

1988 (Faber et al. 1991). cDNA cloning and determination of the structural organization of the genes have provided valuable tools to elucidate the functions of sex hormone receptors in the androgen and estrogen signaling pathways. Indeed, a number of knockout mice have been generated to elucidate the functions of sex hormones in multiple organs/cells.

9.3 Roles of Sex Hormones in Reproduction

9.3.1 *Role of Testosterone in Male Secondary Reproductive Organs*

Testosterone is produced by testicular Leydig cells in adult males. In contrast, in fetal males, androstane-3,17-dione produced by Leydig cells is transferred to Sertoli cells and converted to testosterone (O'Shaughnessy et al. 2000; Shima et al. 2013). Thus, both Leydig cells and Sertoli cells are necessary for testosterone production during the fetal stage. In mice, Sertoli and Leydig cells differentiate at embryonic day 11.5 (E11.5) and E12.5, respectively. The produced testosterone plays a role in the development of secondary reproductive organs of the male type, such as the epididymis, vas deferens, and seminal vesicles. All these organs are derived from the Wolffian ducts. Originally, both fetal males and females have two ducts: the Wolffian duct and Mullerian duct. When Sertoli and Leydig cells differentiate, the anti-Mullerian hormone (AMH) secreted from Sertoli cells induces the regression of Mullerian ducts, and testosterone helps the development of Wolffian ducts into secondary reproductive tissues. Consequently, Wolffian duct-derived organs develop in males, and Mullerian duct-derived organs, such as the uterus, develop in females. Thus, testosterone plays an important role in the development of sexually dimorphic secondary reproductive organs.

Testosterone is also necessary for sex determination of external genitalia, that is, a penis. The external genitalia of both sexes are derived from a common undifferentiated primordium, the genital tubercle (GT). In mice, GT develops from the cloacal region at E10.5. The conspicuous morphological sexual differences in GT appear at E16.5. The masculinization of GT is regulated by androgen. Indeed, when male fetuses were treated with antiandrogenic chemicals such as flutamide at E15.5–E17.5, their GTs would be demasculinized, and their morphologies would be similar to those of female GTs (Genitalia et al. 2009; Welsh et al. 2010; Matsushita et al. 2018). Considering the ability of androgen to cause the development of male-type GT, what happens when androgens are ectopically administered into female mice during the fetal stage? When female fetuses were treated with androgen at E15.5–E16.5, their GTs would masculinize. Based on these observations, it has been established that androgen plays a crucial role in the normal development of male external genitalia and that the critical time window of androgen-induced GT masculinization is E15.5–E16.5.

9.3.2 *Role of Testosterone in Spermatogenesis*

In addition to its indispensable role in the sex determination of somatic cells, androgen plays an important role in germ cells, namely spermatogenesis. Spermatogenesis occurs in the seminiferous tubules and involves three major cell types: peritubular myoid cells (PTM), Sertoli cells, and germ cells. To investigate the role of androgen in each cell type, cell-specific *Ar*-knockout mice were generated and their phenotypes are described briefly below. PTM-specific KO mice showed nearly no elongated spermatids and were infertile, indicating that AR in PTM is necessary for sperm maturation (Welsh et al. 2009). Sertoli cell-specific KO mice showed reduced testicular weight and were infertile. The number of postmeiotic spermatids were significantly reduced or absent by the elimination of AR in Sertoli cells, indicating that AR is necessary for germ cell meiosis (Chang et al. 2004; De Gendt et al. 2004; Holdcraft and Braun 2004). Germ cell-specific KO mice revealed normal testes, spermatogenesis, and fertility, revealing that AR expression in germ cells is dispensable (Zhou et al. 2002). This observation is consistent with the finding that *Ar* is not expressed or has low expression in germ cells. The aforementioned and other knockout mouse models for *Ar* revealed the following roles of AR in spermatogenesis: (1) maintenance of the blood–testis barrier; (2) regulation of meiosis; (3) maintenance of Sertoli cell-spermatid adhesion; and (4) release of sperm. Through the regulation of these multiple events, AR-mediated androgen signaling plays an indispensable role in achieving normal spermatogenesis and male fertility (Smith and Walker 2014).

9.4 Sex Hormone-Induced Sex Differences

9.4.1 *Sex Determination of the Brain*

In addition to the sex determination of reproductive tissues, sex hormones are known to function in the sex determination of the brain. The female sex hormone 17-β-estradiol is produced by the conversion of testosterone, which is mediated by aromatase P450 (CYP19A1). The mouse *Cyp19a1* gene has a unique structure, where the gene has multiple untranslated first exons driven by multiple promoters. The promoters are activated in a tissue-specific manner such that the gene is expressed in multiple tissues, including the gonads, adipose tissues, chondrocytes, bone, brain, skin, fetal liver, and placenta (Simpson et al. 2000). The expression of *Cyp19a1* in multiple tissues suggests that estrogen can be locally produced, and that local estrogen plays a role in the tissues where it is produced. For example, during the perinatal period, testosterone produced by testicular Leydig cells in male mice is transferred to the brain and converted to estradiol by CYP19A1. Estradiol produced in the brain is known to be essential for the development of the brain to the male type. Indeed, it has been reported that the sexual behavior of *Cyp19a1*-knockout mice is severely impaired (Honda et al. 1998; Ogawa et al. 1999; Robertson et al. 2001; Toda et al. 2001).

9.4.2 Role of Estrogen in Bone

The role of the female sex hormone, estrogen, in bone cells has been well understood by observing postmenopausal females. As osteoporosis is one of the most evident symptoms in postmenopausal women, it has been accepted that estrogen has osteoprotective effects. Bone tissue is composed of three types of cells: osteoblasts, osteoclasts, and osteocytes. Osteoblasts and osteocytes are involved in the formation and mineralization of bone tissue, while osteoclasts are involved in tissue absorption. Estrogen exerts osteoprotective effects through direct action on bone tissue as well as through indirect actions, such as the modulation of cytokine production by immune cells and the increased induction of pituitary gland hormones. The direct action of estrogen on each type of bone cell was examined in bone cell type-specific *Esr1* (ERα)-knockout mice. ERα deletion in cells of the osteoblast lineage suggests that ERα plays an important role in the maintenance of bone metabolism (Almeida et al. 2013; Määttä et al. 2013; Melville et al. 2014). Osteoclastic ERα-null mice showed that osteoclastic ERα shortens the lifespan of osteoclasts by promoting apoptosis (Martin-Millan et al. 2010). In osteocyte ERα-knockout mice, ERα in osteocytes was found to play a significant role in maintaining bone mass by regulating osteoblastic bone formation only in females (Kato et al. 2009; Kondoh et al. 2014).

9.4.3 Roles of Sex Hormones in Establishing Sex Differences in Skeletal Muscles

Skeletal muscle size is defined by its weight and cross-sectional area (CSA). It has been shown that both muscle weight and CSA are sexually dimorphic: skeletal muscles are heavier and larger in males and lighter and smaller in females (Haizlip et al. 2015). What establishes this sex difference in muscles? A clear answer was provided by conventional experiments with gonadectomized mice, followed by hormone replenishment/replacement. Male muscles became lighter and smaller on the removal of the testes, whereas female muscles were not significantly affected by the removal of ovaries. The sizes of gonadectomized male muscles are recovered by the replenishment of testosterone, while female muscles become heavier and larger by the replacement of estradiol by testosterone. These observations indicate that testosterone has the ability to make skeletal muscles larger and that sex hormones are one of the major determinants of skeletal muscle size (Aguet et al. 2020). Currently, skeletal muscles are widely recognized to be one of the most sensitive target organs for sex hormones. Notably, because testosterone has anabolic effects on skeletal

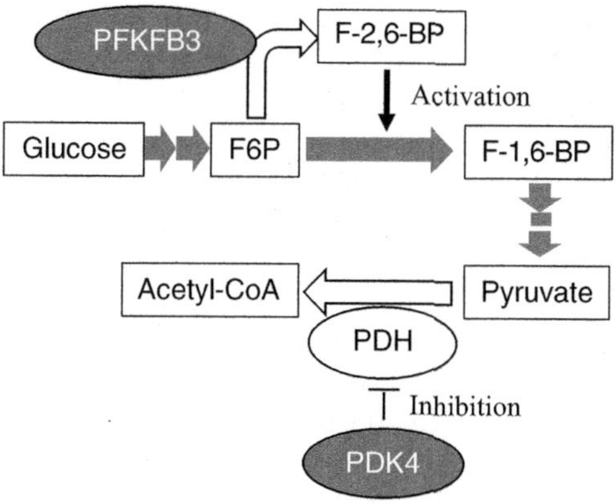

Fig. 9.3 Sexually dimorphic metabolic modulation underlying male-predominant glycolysis and female-predominant fatty acid utilization

muscles as described above, the use of sex hormones and their derivatives has been strictly prohibited in athletes.

In addition to muscle size, muscle performance is known to be different between males and females: muscles in males can generate more force, while muscles in female demonstrate greater endurance. What causes this sex difference? This difference may be due to the sexually dimorphic content of different types of muscle fibers. Skeletal muscles comprise multiple types of muscle fibers. In mice, there are four types of fibers: type I, type IIa, type IIx, and type IIb (Schiaffino et al. 1989; Pette and Staront 1997). Only type I fibers are classified as slow-twitch fibers, where energy production is highly dependent on mitochondrial oxidative phosphorylation. Type IIa, IIx, and IIb fibers are classified as fast-twitch fibers, where energy production primarily depends on glycolysis, although the extent of dependency varies among the three types of fibers. Because fast-twitch fibers are good at generating force and slow-twitch fibers are good at endurance, sexually dimorphic muscle performance is supposed to be due to the difference in the ratio of the fast-twitch and slow-twitch fibers. However, it is not known whether there are any intrinsic differences in the metabolism in the same type of fibers between males and females (i.e., male type IIB fibers vs. female type IIB fibers).

Our study on isolated type IIb fibers recently provided an answer to this question. We attempted to identify male-biased and female-biased genes using transcriptome analysis and found 68 male-biased and 60 female-biased genes, with >1.5-fold higher expression than that in the opposite sex (Christianto et al. 2021). Among the identified sex-biased genes, we focused on *Pfkfb3*, whose expression in type IIb fibers was three-fold higher in males than in females. PFKFB3 plays a crucial role in glycolytic regulation by producing fructose-2,6-bisphosphate, which robustly activates PFKM (a type of phosphofructokinase-1 found in muscles) (Obach et al. 2004; Ma et al. 2020) (Fig. 9.3). Therefore, the male-enriched expression of *Pfkfb3* suggested that the glycolytic activity in type IIb fibers in the quadriceps would be higher in males than in females. As expected, the glycolytic activity was indeed higher in males than in females, and *Pfkfb3* knockdown in male type IIb fibers decreased the activity of glycolysis to a level similar to that in female type IIb fibers.

In addition to *Pfkfb3*, we identified *Pdk4* as a potentially interesting sex-biased gene, whose expression was higher in females than in males in the presence of the female sex hormone, 17β-estradiol. PDK4 phosphorylates pyruvate dehydrogenase to inhibit its activity. As pyruvate dehydrogenase converts pyruvate to acetyl-CoA, its inhibition limits the supply of acetyl-CoA from glycolysis to the TCA cycle (Pettersen et al. 2019) (Fig. 9.3). In response to the inhibition of pyruvate dehydrogenase, fatty-acid β-oxidation is activated to provide acetyl-CoA to the TCA cycle as well as NADH to mitochondrial oxidative phosphorylation. Indeed, *Pdk4* knockdown in female type IIb fibers attenuated the dependency of oxidative phosphorylation on fatty acid β-oxidation. These results showed a clear difference in the metabolism between male and female type IIb fibers: glycolysis was preferred in male type IIb fibers and β-oxidation was preferred in female type IIb fibers.

The basis of the sex difference in metabolism in muscle fibers is the sexually dimorphic expression of two key metabolic genes, *Pfkfb3* and *Pdk4*. *Pdk4* expression is regulated by estrogen. In contrast, *Pfkfb3* expression is not controlled by androgen, but by some yet unknown testis-derived factors. It is necessary to identify these testis-derived factors to further understand the molecular mechanism underlying sexually dimorphic metabolism in the same types of muscle fibers. Further, the investigation of whether the difference in metabolism can be found in other types of fibers, especially in slow-twitch fibers, would provide valuable information.

9.4.4 Roles of Androgen in Establishing Sex Differences in the Adrenal Cortex

Sexually dimorphic organ size can also be observed in the mouse adrenal gland. The female adrenal gland is larger than the male adrenal gland, and these size differences are observed from 3 weeks after birth. As the secretion of a sufficient amount of testosterone from the testis to the circulation is initiated approximately 3 weeks after birth, the involvement of sex hormones has been suggested. Indeed, castration increases the adrenal size, whereas testosterone replenishment in females reduces the adrenal size, indicating that the male sex hormone, testosterone, is responsible for the sexually dimorphic size of the adrenal gland.

The adrenal gland is composed of the medulla and the cortex. The adrenal cortex consists of three layers: the zone glomerulosa (zG), zona fasciculata (zF), and zona reticularis. Which part of the adrenal gland is affected by testosterone and contributes to the difference in gland size? Although the testosterone receptor, AR, is expressed in all cell types in the adrenal gland, including in blood vessel cells, testosterone affects only the number of zF cells by regulating cell proliferation (Grabek et al. 2019). It should be noted that zF cells differentiate from zG cells. The rate of differentiation of zF cells from zG cells is reduced by testosterone, but the detailed mechanism underlying the action of testosterone is yet to be elucidated. To understand the mechanism of the action of testosterone in zF cells, genome-wide analysis, such as ChIP-seq using AR antibody, is necessary.

Cortisol is produced and secreted by zF cells. In rodents, the plasma cortisol concentration reportedly exhibits a sex difference and is higher in females than in males. In contrast, a similar difference in plasma cortisol levels has not been reported in humans. However, Cushing's disease, which accompanies an excess secretion of cortisol, is reported more frequently in women than in men (Levasseur et al. 2019). Cushing's disease is caused by the excess secretion of ACTH from ACTH-producing tumors in the pituitary gland. As sex differences in the frequency of tumor occurrence have not been reported, differences in the sensitivity of zF cells to ACTH might be the cause of the high frequency of Cushing's disease in women.

9.5 Sex Spectrum

Male individuals have XY sex chromosomes in all cells, while female individuals have XX chromosomes. This is one of the most fundamental differences between males and females; in other words, cells have their own sex. A well-known significance of having the Y chromosome is in determining the gonadal sex as male by the function of sex determining factor on Y (SRY) (Koopman et al. 1990; Kashimada and Koopman 2010). However, even after sex determination, sex chromosome composition is sexually dimorphic throughout life. One may wonder whether different compositions of sex chromosomes may result in sexually dimorphic outputs in each cell. From this point of view, it may be interesting to focus on X- or Y-linked histone-modifying enzyme genes. It is established that there are four such genes: *Utx* and *Smcx* on the X chromosome, and *Uty* and *Smcy* on the Y chromosome. Gene products of all four genes are histone lysine demethylases that target H3K27 (UTX/Y) or H3K4 (SMCX/Y). The methylation of H3K27 and H3K4 is closely related to gene suppression and activation, respectively. Therefore, it can be hypothesized that these sex chromosome-linked demethylases establish sexually dimorphic epigenome landscapes to induce sex-biased gene expression profiles.

Another important sexual dimorphism is the production and secretion of sex hormones. Previous studies have shown that sex hormones function in the induction and, in some cases, the modulation of sexual characteristics such as the size of skeletal muscle fibers. As described above, sex hormones exert their roles by binding and activating specific receptors that function as transcription factors. Whether transcription factors can bind to their target sites is highly dependent on the epigenetic status of the target locus. For example, a transcription factor cannot bind to a target genomic region if the locus is heterochromatinized by the tri-methylation of histone H3K27. Thus, it is possible to expect crosstalk between the epigenome established by sex chromosome-linked genes and the transcription factors activated by sex hormones to establish sex-biased gene expression. The hypothetical concept of the "sex spectrum" is that all cells possess their own sex that can be located at any intermediate position between a typical male and a typical female. It is important to consider that crosstalk between sex chromosomes and sex hormones provides an approximate position of sex on the spectrum, and if either is unexpectedly disturbed for some reason, this position can also be shifted.

References

NA (1996) Arnold Adolph Berthold and the transplantation of testes. Endocrinologist 6(3): 164–168. https://doi.org/10.1097/00019616-199605000-00002

Aguet F et al (2020) The impact of sex on gene expression across human tissues. Science 369(6509). https://doi.org/10.1126/SCIENCE.ABA3066

Almeida M et al (2013) Estrogen receptor-α signaling in osteoblast progenitors stimulates cortical bone accrual. J Clin Investig 123(1). https://doi.org/10.1172/JCI65910

Baba T et al (2014) Glycolytic genes are targets of the nuclear receptor Ad4BP/SF-1. Nat Commun 5:3634. https://doi.org/10.1038/ncomms4634

Baba T et al (2018) Ad4BP/SF-1 regulates cholesterol synthesis to boost the production of steroids. Commun Biol 1(1):18. https://doi.org/10.1038/s42003-018-0020-z

Brown MS, Goldstein JL (1999) A proteolytic pathway that controls the cholesterol content of membranes, cells, and blood. Proc Natl Acad Sci U S A 96(20):11041–11048. https://doi.org/10.1073/pnas.96.20.11041

Chang C et al (2004) Infertility with defective spermatogenesis and hypotestosteronemia in male mice lacking the androgen receptor in Sertoli cells. Proc Natl Acad Sci U S A 101(18). https://doi.org/10.1073/pnas.0307306101

Christianto A, Baba T, Takahashi F, Inui K, Inoue M, Suyama M, Ono Y, Ohkawa Y, Morohashi KI (2021) Sex differences in metabolic pathways are regulated by Pfkfb3 and Pdk4 expression in rodent muscle. Commun Biol 4(1):1264. https://doi.org/10.1038/s42003-021-02790-y

Chypre M, Zaidi N, Smans K (2012) ATP-citrate lyase: a mini-review. Biochem Biophys Res Commun 422(1):1–4. https://doi.org/10.1016/j.bbrc.2012.04.144

Faber PW et al (1991) The mouse androgen receptor. Functional analysis of the protein and characterization of the gene. Biochem J 278(1):269–278. https://doi.org/10.1042/bj2780269

Freeman DA, Ascoli M (1982) Studies on the source of cholesterol used for steroid biosynthesis in cultured Leydig tumor cells. J Biol Chem 257(23):14231–14238. https://doi.org/10.1016/s0021-9258(19)45370-2

De Gendt K et al (2004) A Sertoli cell-selective knockout of the androgen receptor causes spermatogenic arrest in meiosis. Proc Natl Acad Sci U S A 101(5):1327–1332. https://doi.org/10.1073/pnas.0308114100

Genitalia E et al (2009) Genetic interactions of the androgen and Wnt/β-catenin pathways for the masculinization of external genitalia. Mol Endocrinol 23(6):871–880. https://doi.org/10.1210/me.2008-0478

Grabek A et al (2019) The adult adrenal cortex undergoes rapid tissue renewal in a sex-specific manner. Cell Stem Cell 25(2):290–296.e2. https://doi.org/10.1016/j.stem.2019.04.012

Haizlip KM, Harrison BC, Leinwand LA (2015) Sex-based differences in skeletal muscle kinetics and fiber-type composition. Physiology 30(1):30–39. https://doi.org/10.1152/physiol.00024.2014

Hammer GD, Parker KL, Schimmer BP (2005) Minireview: transcriptional regulation of adrenocortical development. Endocrinology 146(3):1018–1024. https://doi.org/10.1210/en.2004-1385

Hoivik EA et al (2010) Molecular aspects of steroidogenic factor 1 (SF-1). Mol Cell Endocrinol 315(1–2):27–39. https://doi.org/10.1016/j.mce.2009.07.003

Holdcraft RW, Braun RE (2004) Androgen receptor function is required in Sertoli cells for the terminal differentiation of haploid spermatids. Development 131(2):459–467. https://doi.org/10.1242/dev.00957

Honda SI et al (1998) Disruption of sexual behavior in male aromatase-deficient mice lacking exons 1 and 2 of the cyp19 gene. Biochem Biophys Res Commun 252(2):445–449. https://doi.org/10.1006/bbrc.1998.9672

Horton JD, Goldstein JL, Brown MS (2002) SREBPs: activators of the complete program of cholesterol and fatty acid synthesis in the liver. J Clin Investig 109(9):1125–1131. https://doi.org/10.1172/JCI0215593

Jinn S et al (2015) SnoRNA U17 regulates cellular cholesterol trafficking. Cell Metab 21(6): 855–867. https://doi.org/10.1016/j.cmet.2015.04.010

Kashimada K, Koopman P (2010) Sry: the master switch in mammalian sex determination. Development 137(23):3921–3930. https://doi.org/10.1242/dev.048983

Kato S, Kondo S, Imai Y (2009) Sex steroid hormones exhibit osteoprotecitve effects through nuclear receptors expressing in bone cells. Bone 44. https://doi.org/10.1016/j.bone.2009.01.055

Kondoh S et al (2014) Estrogen receptor α in osteocytes regulates trabecular bone formation in female mice. Bone 60:68–77. https://doi.org/10.1016/j.bone.2013.12.005

Koopman P et al (1990) Expression of a candidate sex-determining gene during mouse testis differentiation. Nature 348(6300):450–452. https://doi.org/10.1038/348450a0

Kuiper GGJM et al (1996) Cloning of a novel estrogen receptor expressed in rat prostate and ovary. Proc Natl Acad Sci U S A 93(12):5925–5930. https://doi.org/10.1073/pnas.93.12.5925

Levasseur A, Dumontet T, Martinez A (2019) Sexual dimorphism in adrenal gland development and tumorigenesis. In: Current opinion in endocrine and metabolic research. Elsevier Ltd, Amsterdam, pp 60–65. https://doi.org/10.1016/j.coemr.2019.07.008

Ma H et al (2020) C-Src promotes tumorigenesis and tumor progression by activating PFKFB3. Cell Rep 30(12):4235–4249.e6. https://doi.org/10.1016/j.celrep.2020.03.005

Määttä JA et al (2013) Inactivation of estrogen receptor in bone-forming cells induces bone loss in female mice. FASEB J 27(2):478–488. https://doi.org/10.1096/fj.12-213587

Martin-Millan M et al (2010) The estrogen receptor-α in osteoclasts mediates the protective effects of estrogens on cancellous but not cortical bone. Mol Endocrinol 24(2):323–334. https://doi.org/10.1210/me.2009-0354

Matsushita S et al (2018) Regulation of masculinization: androgen signalling for external genitalia development. Nat Rev Urol 15(6):358–368. https://doi.org/10.1038/s41585-018-0008-y

Melville KM et al (2014) Female mice lacking estrogen receptor-alpha in osteoblasts have compromised bone mass and strength. J Bone Miner Res 29(2):370–379. https://doi.org/10.1002/jbmr.2082

Miller WL (1988) Molecular biology of steroid hormone synthesis. Endocr Rev 9(3):295–318. https://doi.org/10.1210/edrv-9-3-295

Miller WL (2013) Steroid hormone synthesis in mitochondria. Mol Cell Endocrinol 379(1–2): 62–73. https://doi.org/10.1016/j.mce.2013.04.014

Morohashi K, Baba T, Tanaka M (2013) Steroid hormones and the development of reproductive organs. Sex Dev 7(1–3):61–79. https://doi.org/10.1159/000342272

Morohashi KI et al (1992) A common trans-acting factor, Ad4-binding protein, to the promoters of steroidogenic P-450s. J Biol Chem 267(25):17913–17919. https://doi.org/10.1016/s0021-9258(19)37129-7

Mosselman S, Polman J, Dijkema R (1996) ERβ: identification and characterization of a novel human estrogen receptor. FEBS Lett 392(1):49–53. https://doi.org/10.1016/0014-5793(96)00782-X

O'Shaughnessy PJ et al (2000) Localization of 17β-hydroxysteroid dehydrogenase/17-ketosteroid reductase isoform expression in the developing mouse testis - androstenedione is the major androgen secreted by fetal/neonatal Leydig cells. Endocrinology 141(7):2631–2637. https://doi.org/10.1210/endo.141.7.7545

Obach M et al (2004) 6-Phosphofructo-2-kinase (pfkfb3) gene promoter contains hypoxia-inducible factor-1 binding sites necessary for transactivation in response to hypoxia. J Biol Chem 279(51): 53562–53570. https://doi.org/10.1074/jbc.M406096200

Ogawa S et al (1999) Survival of reproductive behaviors in estrogen receptor β gene-deficient (βERKO) male and female mice. Proc Natl Acad Sci U S A 96(22):12887–12892. https://doi.org/10.1073/pnas.96.22.12887
Olefsky JM (2001) Nuclear receptor minireview series. J Biol Chem 276(40):36863–36864. https://doi.org/10.1074/jbc.R100047200
Pette D, Staront RS (1997) Mammalian skeletal muscle fiber type transitions. Int Rev Cytol 170: 143–223. https://doi.org/10.1016/s0074-7696(08)61622-8
Pettersen IKN et al (2019) Upregulated PDK4 expression is a sensitive marker of increased fatty acid oxidation. Mitochondrion 49:97–110. https://doi.org/10.1016/j.mito.2019.07.009
Robertson KM et al (2001) Characterization of the fertility of male aromatase knockout mice. J Androl 22(5):825–830. https://doi.org/10.1002/j.1939-4640.2001.tb02587.x
Schiaffino S et al (1989) Three myosin heavy chain isoforms in type 2 skeletal muscle fibres. J Muscle Res Cell Motil 10(3):197–205. https://doi.org/10.1007/BF01739810
Shima Y et al (2013) Contribution of Leydig and Sertoli cells to testosterone production in mouse fetal testes. Mol Endocrinol 27(1):63–73. https://doi.org/10.1210/me.2012-1256
Simpson E et al (2000) The role of local estrogen biosynthesis in males and females. Trends Endocrinol Metab 11(5):184–188. https://doi.org/10.1016/S1043-2760(00)00254-X
Smith LB, Walker WH (2014) The regulation of spermatogenesis by androgens. Semin Cell Dev Biol 30:2–13. https://doi.org/10.1016/j.semcdb.2014.02.012
Toda K et al (2001) A loss of aggressive behaviour and its reinstatement by oestrogen in mice lacking the aromatase gene (Cyp19). J Endocrinol 168(2):217–220. https://doi.org/10.1677/joe.0.1680217
Tremblay GB et al (1997) Cloning, chromosomal localization, and functional analysis of the murine estrogen receptor β. Mol Endocrinol 11(3):353–365. https://doi.org/10.1210/mend.11.3.9902
Val P et al (2003) SF-1 a key player in the development and differentiation of steroidogenic tissues. Nucl Recept 1(1):–8. https://doi.org/10.1186/1478-1336-1-8
Walter P et al (1985) Cloning of the human estrogen receptor cDNA. Proc Natl Acad Sci U S A 82(23):7889–7893. https://doi.org/10.1073/pnas.82.23.7889
Waterman MR, Bischof LJ (1997) Diversity of ACTH (cAMP)-dependent transcription of bovine steroid hydroxylase genes. FASEB J 11(6). https://doi.org/10.1096/fasebj.11.6.9194522
Welsh M et al (2009) Androgen action via testicular peritubular myoid cells is essential for male fertility. FASEB J 23(12):4218–4230. https://doi.org/10.1096/fj.09-138347
Welsh M et al (2010) Critical androgen-sensitive periods of rat penis and clitoris development. Int J Androl 33(1):e144–e152. https://doi.org/10.1111/j.1365-2605.2009.00978.x
Zhou Q et al (2002) Localization of androgen and estrogen receptors in adult male mouse reproductive tract. J Androl 23(6):870–881. https://doi.org/10.1002/j.1939-4640.2002.tb02345.x

Chapter 10
The Role of Sex Spectrum Differences in Reproductive Strategies and the Endocrine Mechanisms Underlying It

Takefumi Kikusui

Abstract During the process of sexual evolution, mammals have acquired different reproductive and social roles for males and females. Differences in the cost of sperm and egg have given rise to differences in the reproductive strategies of males and females. For example, distinct sex differences are observed in mating forms and investment in fostering. It can be assumed that sexual spectrum was brought about because individuals were able to successfully win in their reproductive strategies. Low-ranking males can escape the attacks of monopolizing males by concealing their "maleness." And such a strategy can be described as a "spectrum." We focused on testosterone as the molecular mechanism of this sex spectrum. After genetic sex determination, this sex spectrum may be brought about by regulating testosterone secretion. In this article, I introduce the relationship between reproductive forms and sex differences, as well as the function of testosterone in this context, and reconsider the meaning of sexual plasticity from the view of evolution.

Keywords Mating system · Sex differences · Testosterone · Social structure · Mouse ultrasonic vocalization

10.1 Introduction

The protobiont arose approximately 4 billion years ago and has evolved into various organisms to adapt to the global environmental changes. The basis of evolution is natural selection and sexual selection is the driving force, as proposed by Charles Darwin. In other words, individuals better adapted to the environment obtain more chances to reproduce and, as a result, more genes expressing that favorable trait are

T. Kikusui (✉)
School of Veterinary Medicine, Azabu University, Sagamihara, Kanagawa, Japan

Center for Human and Animal Symbiosis Science, Azabu University, Sagamihara, Kanagawa, Japan
e-mail: kikusui@azabu-u.ac.jp

M. Tanaka, M. Tachibana (eds.), *Spectrum of Sex*,
https://doi.org/10.1007/978-981-19-5359-0_10

transmitted to the next generation. In "On the Origin of Species" (Darwin 1859), Charles Darwin theorized the evolution of sexual phenotype through sexual selection and in particular regarding the preference of females for males (Darwin 1859). Based on this theory, current molecular bioscience has contributed greatly to elucidating this sex difference in sexual phenotypes.

A characteristic function of mammals, including humans, is to nurture the fetus through the placenta and to increase the survival rate of offspring through parental behaviors, including lactation. Compared to oviparous species such as fish, which lay eggs, or ovoviviparous species such as reptiles, which incubate eggs in the female's body and then give birth to pups, mammals may have few pups. In addition, newborn pups are often immature in terms of thermoregulatory and motor functions, and their parents must devote a lot of resources to them, including nurturing. Ostensibly, this form of mammalian reproduction deviates from the primary objective of reproductive strategy, which is to efficiently transmit more genes to the next generation. However, studies on mammals have revealed that the environment in which they are raised modifies their epigenetic status and regulates gene expression; thus, many mammals give birth in an immature and high in plasticity, which means that the infant and juvenile can obtain various types of information from the environment during development and acquire functions that are appropriate for that individual (Goldberg et al. 2007). In other words, the evolutionary strategy of mammals is that each individual has acquired plastic functions that allow it to adapt to its environment, dramatically increasing its chances of survival. During the development of a pup from an egg, information on its growth strategy, via endocrine dynamics in utero, mother–child relationships, interactions with siblings, and ultimately interactions within the group, is relevant to reproductive strategy. This chapter describes the classical finding of the function of hormones to regulate the sex-specific behavior and the recent findings of the molecular mechanisms underlying the development of sex spectrum behavior in males.

10.2 Social Structure and Sexual Diversity

The type of herd that animals form depends on its breeding pattern. For example, lions that live in grasslands are unusual among felines in that they form pride, which they hunt in, and a few dominant males have the right to reproduce in the pride. In even-toed ungulates, such as deer, the reproductive opportunities for males in the herd are even more strictly controlled, and it is common for a single mature male to be present in a harem with as many as 50 females. The breeding pattern determines the members of a herd. The difference between male and female breeding strategies is because of the size of their gametes. That is, males have small and mobile sperm (spermatids), while females have large eggs (female gametes); as a result, the number of offspring produced in a lifetime is generally larger in males than in females. For example, Genghis Khan, who is believed to have left the largest number of offspring in humans, is said to have had more than 200 direct descendants

(Lkhagvasuren et al. 2016). This is tens of times more than the number of children a single woman can give birth to. However, males who become subordinates have little or no chance to reproduce. Thus, while many females have the chance to produce offspring, many males die without producing any offspring. However, for animals of monogamous species, the chances of males having their own offspring increases, but at the same time they relinquish the opportunity to produce many offspring for various females. Females of monogamous species are also less likely to mate with a variety of males. In these species, males and females have enthusiastic caregiving and nurturing activities. In other words, the shift to monogamy involves a tradeoff between genetic diversity and reproductive strategies that emphasize caregiving (Gubernick and Teferi 2000).

The smallest herd types are only parent–infant combinations, where males and females are monogamous and their offspring are together. This is the most common type of flock in birds, accounting for 40–50% of all flocks. In mammals, the proportion of herds of this type is overwhelmingly low, with approximately 3% of species being monogamous. Typical monogamous mammals include prairie voles, marmosets, and raccoon dogs (Kleiman 1977). Groups larger than single families are referred to as polygyny, comprising multiple mature females and a limited number of males, or polygamy, when multiple males and females are present. It is becoming evident that the composition of group members is generally related to the difficulty of obtaining food resources, especially reproductive and food resources for females, and the size of the living area to obtain them (Lukas and Clutton-Brock 2013).

As alluded to above, a polygyny group comprises of many mature females and either a solitary male or a small number of immature males. Lions, fur seals, seals, deer, and gorillas form harem-type herds. In this setting, male competition for females is common and males who win the competition may monopolize several females (West and Packer 2002). In such cases, individuals who lose the competition between males lose their chance to reproduce. Polygyny can also be divided into several patterns (Altmann et al. 1977), one of which is resource-protective polygyny, in which males have rich food resources and nests, and mate with females that fall within their range. The quality of the territory determines how many females can mate. Another type of polygyny encountered is the so-called harem-type polygyny. In the harem type, males fight with other males to prevent them from mating with females in their territory. It differs from resource-protective polygyny described above in that the female herself is the object of defense (Altmann et al. 1977). In lek-style polygyny, males gather in a small area that has little to do with resources (called an arena) and desperately appeal to females (Payne 1984). Males who participate in these appeals are called leks. In the end, the male with the best appeal obtains the chance to mate with the surrounding females. This type of polygyny is widely observed in birds and fish.

In mammalian species with polygyny or harem breeding patterns, there are many notable sex differences in the appearance, body size, and behavior of males and females (Fig. 10.1). This is most likely because of sexual selection by females. In fact, in monogamy, the body size ratio of males to females is about 1.05 to 1.1. In polygyny, the ratio is approximately 1.55 to 1.75 (Kenagy and Trombulak 1986). For example, elephant seals exhibit a harem type of polygyny and females of the

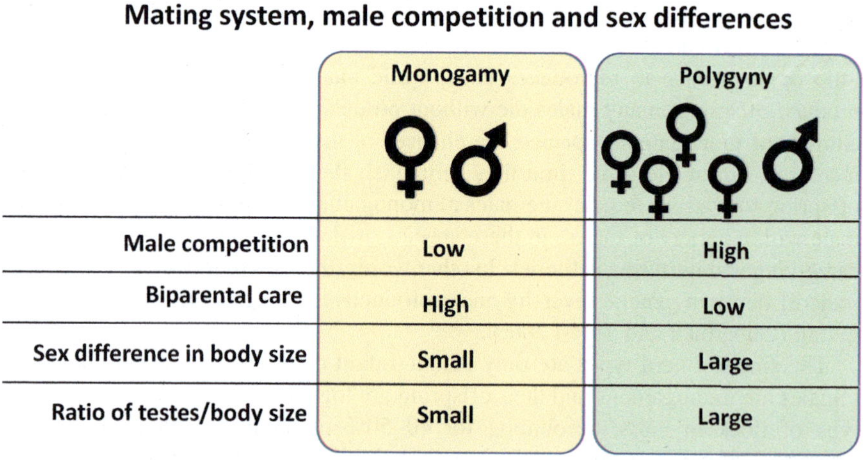

Fig. 10.1 Comparison of male competition, biparental care, the sex difference in body size, and the ratio of testes/body size in monogamous and polygynous mammalian species

species weigh approximately 500 kg while males weigh four times as much, about 2 tons. In monogamy, there is less competition between males and less mating during the breeding season, so the testes do not need to be as highly functioning. If the weight of the testes of a monogamous male is 1, the equivalent weight in polygamous male who engages in sperm competition is about twice as much, ranging from 1.8 to 2 (Kenagy and Trombulak 1986).

10.3 Androgens Create Sex Differences in Sexual Behavior

Many of the behaviors observed in the process of mating are unique to animals. The male kingfisher presents a fish to the female, while others, such as the peacock, spread their large feathers in front of the female as a sexual display. The recognition of each other's sex and the concomitant sexual motivation is driven and, in most cases, mating is the result of the female's acceptance of the male's courtship (Asaba et al. 2014a). The key stimulus for the initiation of these sexual behavior processes is the reception of sex signals in males from females. A series of behaviors are initiated when sexual motivation is driven by the acceptance of the partner's sexual signals. This sexual motivation results from the proper processing of sensory information from olfactory, visual, and auditory senses in the presence of sex hormones, such as androgens and estrogens (Asaba et al. 2014a). This series of mechanisms of sexual behavior can be observed clearly in the mating behavior of rodents in the laboratory.

The expression of male sexual behavior in rats is androgen dependent. Sexual behaviors of male rats upon encountering females start with display behaviors and vocalization of courtship ultrasonic vocalizations (USVs), followed by mounting,

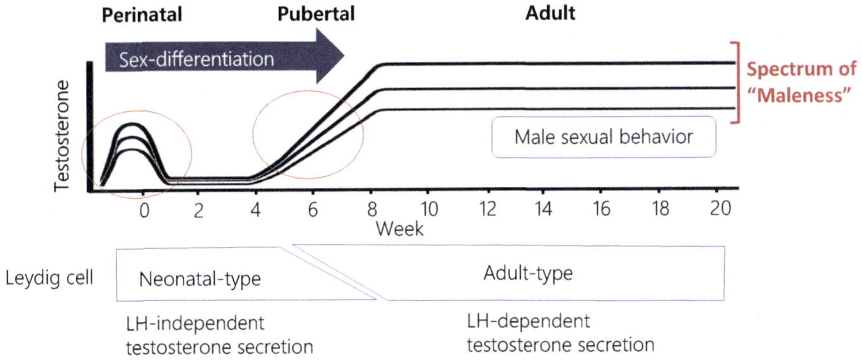

Fig. 10.2 The sexual differentiation of male rodents. Androgens such as testosterone are initially secreted by neonatal testes during the perinatal period. The Leydig cell in the testes can secrete testosterone independently of luting hormone (LH) in the circulating blood. During the pubertal period, adult-type Leydig cells are developed and can secrete testosterone dependently on LH. These testosterone secretions can differ by the individual; modulated by social and environmental factors. The differences in testosterone concentrations in these periods modulate the male sexual behavior in adulthood and form a "sex spectrum" in maleness

thrusting, introjection, and ejaculation when the female shows tolerance (Asaba et al. 2014a). The onset of sexual behavior in males requires the secretion of androgens from the testes following sexual maturation (Simerly 2002). Androgens act on androgen receptors in the brain or are converted to estrogen by P450 aromatase and act on estrogen receptors to induce male sexual behavior (Simerly 2002). Three points have been found to be important for the action of androgens on the brain (Fig. 10.2): (1) the androgen action in the early developmental period during the perinatal period (androgen shower, organizational effects), (2) the sexual maturation process from about 3 to 6 weeks of age in rats, and (3) the androgen action after growth (activational effects) (Berenbaum and Beltz 2011). In rats, androgens secreted from the testes during fetal life and shortly after birth enter the neurons in the brain, where they are aromatized by P450 aromatase and become estradiol (Cooke et al. 1998; Morris et al. 2004). Estrogen then acts on the limbic system and hypothalamus of the brain to form a male-type neural circuit via the activation of estrogen receptors. Estrogen of ovarian origin is also present in females, but it does not penetrate into the brain because alpha-fetoprotein binds to estrogen in the blood and cannot cross the blood-brain barrier (Cooke et al. 1998; Morris et al. 2004). A brain unaffected by androgens shifts to demasculinization and feminization. Therefore, the default sex of the rat brain is female, and when androgens act on it, it is derived from the male type. In primates, androgens have been shown to work directly on brain sex differentiation without being converted to estrogen (Berenbaum and Beltz 2011). Since humans are primates like monkeys, it is expected that the male brain is formed through androgen receptors in the same way (Berenbaum and Beltz 2011). The role of androgens in sex differentiation of the brain varies from species to species, but is thought to mirror their role in masculinization of the brain.

Thus, the male brain formed during the perinatal period is re-activated by androgens after maturation, resulting in the induction of male behavior. Male behavior is androgen-dependent, and if the testes of male rats are surgically removed, they no longer exhibit male sexual behavior (Cooke et al. 1998; Morris et al. 2004). However, when androgens are administered to castrated males, male sexual behavior is observed again. In contrast, no matter the amount of androgens administered to a sexually mature female, the female will not show male sexual behavior. This means that the formation of a male brain by androgen action during the perinatal period is a prerequisite for the functioning of androgens after sexual maturity. Interestingly, because of the sensitivity of the androgen receptors expressed in the brain, post-maturity androgens are not strongly affected by high or low blood concentrations as long as they are secreted in certain amounts (Wu and Gore 2010). This susceptibility depends on one's genetic background and the amount of sexual experience that one has had. For example, sexually experienced humans, dogs (Creed 1989), and rhesus monkeys have been reported to exhibit sexual behavior for several years, even after they have been castrated to prevent the production of male hormones (Sreenivasan and Weinberger 2016). In some countries, there is a history of castration for male sex offenders, but some men have been reported to still have erections and engage in sexual activity 10 years after the punishment was imposed (Sreenivasan and Weinberger 2016). Thus, the amount of androgens present in adulthood does not entirely determine the performance of sexual behavior.

10.4 Hormonal Controlling of the Sex Spectrum

As alluded to above, sex differences in mammals are brought by androgens produced by the androgen synthesis pathway originating from the Y chromosome; genetic information is converted into hormones that build and maintain sex differences. While the genetic information is one dimensional, the hormone concentration is on a spectrum of high to low concentrations and the male sex is promoted in a concentration-dependent manner. This spectrum of androgen-induced masculinization encompasses not only reproductive behavior but also neurophysiological functions such as cognition, emotion, and the stress response (McCarthy and Arnold 2011).

An example of the generation of this spectrum is the fetal implantation position. Mice, rats, and other animals are multiparous, producing 6–8 fetuses in a single pregnancy (Clark et al. 1992). In this setting, females sandwiched between two male fetuses are stimulated to become male like by testosterone secreted by the male fetuses. Similarly, males sandwiched between male fetuses become more male like. Thus, depending on the implantation position during the fetal period, a spectrum begins to emerge.

Even after birth, grooming by the dam is known to change the male characteristics (Champagne et al. 2007; Parent and Meaney 2008). Males whose mothers groomed them vigorously showed more male sexuality after growth, while those whose mothers did not groom them showed much less. This is probably due to the fact that mothers invest in their male offspring to become socially dominant and produce more offspring. To date, it is not known how mothers select particular male pups, but such social environments also produce a spectrum of male sexuality.

The relative contributions of androgens and estrogens to brain sex differentiation and their critical time horizons also vary by function (McCarthy and Arnold 2011). Administration of testosterone propionate to female mice in the neonatal period induces male sexual behavior, but not male-type vulnerability to social stress (Kikusui et al. 2013). Furthermore, the expression pattern of sex steroid receptors in the brain of newborn mice is sexually dimorphic depending on their location (Mogi et al. 2015). In perinatal mice, androgen receptor mRNA expression in the hypothalamus and prefrontal cortex (PFC) is higher in males than in females; ERβ mRNA expression in the PFC is higher in female mice immediately after birth. These results suggest that neonatal testosterone organizes the masculinization of brain function in a regionally and functionally specific manner.

It is possible that this androgen-induced male spectrum has emerged not only because of the sexual display to females through sexual selection by females, but also because of the development of aggression among males, and the suppression of infanticide by high testosterone (Brown 1986). This spectrum of phenotypes contributes to the final reproductive success, namely fitness. The variation in masculinity has evolved with the appropriate social status and environment. In other words, the efficiency of self-gene transmission to the next generation may be enhanced by controlling whether male sexuality should be elevated or suppressed under appropriate circumstances. The latter part is an example of male mice USVs, which is a spectrum phenotype.

10.5 Male and Female Mouse Communication Via Ultrasonic Vocalizations

Holy and Guo (2005) showed that male mice emit song-like vocalizations to female mice using calls in the high ultrasonic range, which is inaudible to humans (Holy and Guo 2005). We previously examined the structure of USVs in two strains of mice (C57BL6 and BALB/cA) and found that the patterns of syllable occurrence, as well as the frequency of occurrence of these syllables, differed significantly (Kikusui et al. 2011). Next, we tested the hypothesis that male pups can achieve vocal learning from the environment, from their fathers for example, which would mean that they would emit USVs similar to the song of their nurturing parents as seen in human language. The aforementioned two strains of mice were subjected to cross-fostering soon after birth, and the audible environment during development was reversed. However, even with cross-foster manipulation, the song characteristics of these two strains were maintained, and each sang the same song as its genetic parent. This indicates that complex USVs are genetically controlled in mice (Kikusui et al. 2011).

How diverse is the song? Koide et al. at the National Institute of Genetics examined the structure of the song in wild mice from around the world. They examined nine mouse strains, including KJR from Korea, MSM and JF-1 from Japan, and BFM/2 from France and found that the frequency of syllables differed greatly among the mouse strains (Sugimoto et al. 2011). The possibility that the frequency of syllables is genetically controlled has already been described, and it was expected that the similarity would be similar to that of genetic distance, that is, genetically close strains of mice would have similar syllables and genetically distant strains would have different syllables. However, there is no consistency between genetic distance and the structure of USVs, indicating that there is no strong selection pressure on the evolutionary process of song, even though USVs structure itself is genetically controlled. This means that evolution (the acquisition of diversity) occurred with little selection pressure. This is known as genetic drift and is probably responsible for the structure of USVs in mice (Sugimoto et al. 2011).

Subsequently, we examined the social functions of this genetically defined song, that is, whether the song affects the ability to attract female mice. When females of each strain were presented with two USVs, the C57BL6 females showed more of an approach to the BALB song, while the BALB females were more attracted to the C57BL6 song (Asaba et al. 2014b). This indicated that female mice are attracted to the USVs of males from genetically different strains to their own. In addition, experiments were conducted to change the auditory environment during development by cross-fostering immediately after the female mice gave birth. If the female mice's preference for the song was due to imprinting during the infant, the preference should be reversed in adulthood. It was found that C57BL6 mice raised by BALBs showed a preference for the USVs of C57BL6 mice, while BALBs raised by C57BL6 mice showed a preference for the USVs of BALBs (Asaba et al. 2014b). The reversal of song preference after cross-fostering manipulation indicates that this preference is imprinted in the female brain through sexual imprinting during the infant period.

Sexual imprinting is a phenomenon observed in more than 100 species of birds, as well as in fish and mammals, in which the preference of the opposite sex for a future mating partner is influenced by interactions with parents and other adult animals during early developmental periods (Ten Cate and Vos 1999). Some birds spend time with other animals soon after hatching from their eggs and show courtship toward those animals with whom they spend time.

10.6 Sex Spectrum of Ultrasonic Vocalization in Male Mice

Male sex signals to females usually function as competitive signals for males. For example, ESP1 in the tears of male mice enhances sexual behavior in females, but also induces aggression toward other males (Hattori et al. 2016). The chirping of birds has an attractive effect on females, but it is also used as a signal for territorial competition between males (Gil and Gahr 2002). In other words, in situations where males compete with each other, male signals have a variety of functions. If strong male signaling increases the frequency of attacks from surrounding males, then it is necessary to consider both the benefits and costs of male signaling. If a male cannot monopolize the females, it would be more beneficial to reduce the emission of male signals and hide away from the surrounding males. Thus, a spectrum of male signals is thought to have emerged.

A spectrum of USVs is also observed when a male mouse encounters a female. Some sexually mature male mice do not emit any sound when they encounter females, and even when paired with such males, females rarely produce pups (Asaba et al. 2017); this suggests that the maleness of USVs is strongly suppressed in these mice. In addition, when male mice are reared in groups and a hierarchy is formed, individuals at the top of the hierarchy often emit USVs, while those at the bottom of the hierarchy emit fewer USVs (Wang et al. 2014). This may be due to the fact that they hide their male characteristics to escape attacks from dominant individuals. In addition, as male mice age and their testosterone levels decrease, the number of times they emit USVs decreases (Kanno and Kikusui 2018). This suggests that the spectrum of USV emissions is caused by the strength of sexual motivation.

How is the development of USVs in male mice that show a sexual spectrum controlled? To investigate the effects of testosterone, we examined the USV emissions in females treated with testosterone (Kikusui et al. 2021a). Testosterone was administered in three stages: perinatally, at sexual maturation, and adulthood. Females treated after sexual maturation had the same USV emission as untreated females and did not become male type. When treated from the time of sexual maturity, a certain degree of USV emission was observed. In addition, when treated from the perinatal period, USV emission was almost equal to that of males (Kikusui et al. 2021a). This suggests that the neural circuits involved in USV emission are formed by testosterone during the perinatal period and that high and low testosterone concentrations during this period are probably the initial trigger for generating the spectrum characteristics of USVs.

Mice USVs are not monotonous but consist of complex syllables. Based on the results of the above experiments, we investigated whether perinatal testosterone treatment could also modulate the complex phoneme structure of syllables (Kikusui et al. 2021a). More complex syllables were observed in males than in females. This change was also observed in the testosterone-treated females. Unexpectedly, the testosterone-treated females had higher syllable frequencies than the males. This is inconsistent with the known decrease in vocal frequency caused by testosterone in general; the higher the testosterone level, the lower the frequency of vocalization. We further analyzed the transmission patterns of the syllables. Interestingly, the syllable transmission pattern of the testosterone-treated females remained female (Kikusui et al. 2021b). In other words, the syllable transmission between males and females was not altered by testosterone treatment, leaving open the possibility of genetic control. This is a notable finding relating to sex differences in mammalian behavior.

10.7 Neural Circuits Responsible for the Spectrum of USVs

Sex differences in neural circuits that control USV emission have also been reported. USV vocalization in males is caused by the reception of female pheromones, which are then transmitted to the reproductive center of the hypothalamus (Asaba et al. 2014a). We previously found that male sexual behaviors, such as mounting, are mediated by the vomeronasal neural circuit, whereas USV emissions induced by female pheromones are mainly mediated by the main olfactory system (Matsuo et al. 2015). Most of the female pheromone signals involved in USV emissions are transmitted from the main olfactory system to the medial nucleus of the amygdala (MeA), where a clear sex difference is observed in the response to urinary signals of the opposite sex (Fig. 10.3). In males, female pheromones strongly activate the posterior-dorsal part of the MeA, whereas in females, the posterior-ventral part of the MeA is activated. Conversely, in males, male pheromones activate the ventral caudal part of the meA, whereas in females, they activate the dorsal caudal part of it (Kikusui et al. 2018). Similar results were reported by Durac et al. (Bergan et al. 2014). This indicates that signaling in the MeA is switched between males and females and that this switching may control sexual behavior, sexual motivation, and USV emission in males and females. Since this switching is controlled by testosterone during the perinatal period, as mentioned above, we can assume that the switching itself becomes a spectrum, resulting in a spectrum of USV emissions.

Sex differences in the MeA at the molecular level have been investigated. The morphology of neurons has been compared, with the cell bodies of males found to be larger than those of females (Petrulis et al. 2017). There are also differences in

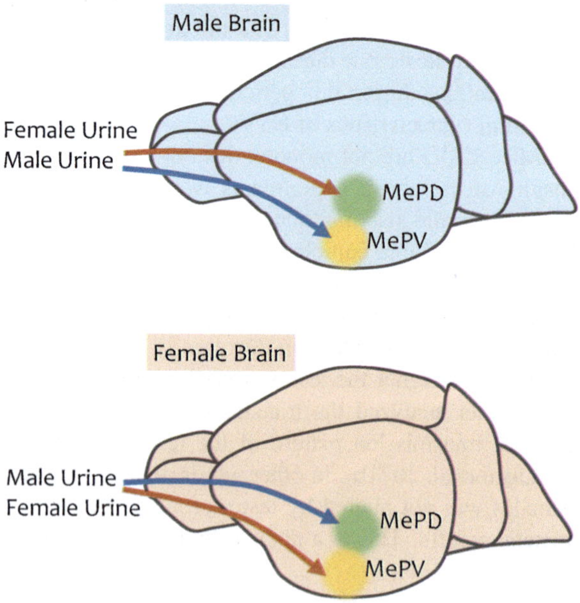

Fig. 10.3 Sex differences in the response to sexual signals (urine pheromone) in the medial amygdala (MeA) in males (upper) and females (lower). Female urine mainly activated the post-dorsal part of MeA (MePD) in males but the post-ventral part of MeA (MePV) in females. Contrary, male urine activated MePV in males and MePD in females. These results suggest that MeA has a switching function of sex-related information processing in the brain

dendrite orientation in both the posterior-dorsal and posterior-ventral parts of the MeA in male and female rats (Dall'Oglio et al. 2008). Most of these differences occur in neurons that express androgen and estrogen receptors, but sex differences in dendrites have also been found in the anterior and posterior parts of the MeA (Rasia-Filho et al. 2004). Neurochemically, male rats have higher concentrations of opioid receptors, substance P, and cholecystokinin in the posterior-dorsal part of the MeA than females (Petrulis et al. 2017). Following advances in molecular biology, it has been shown that the transcription factors Dbx1 and Foxp2 are expressed in the MeA during embryogenesis and determine the generation of a subclass of inhibitory neurons in the MeA (Lischinsky et al. 2017). In addition, there are sex differences between neurons expressing Dbx1 and those expressing Foxp2, and they respond differently to pheromones. There is also a marked difference in the spike patterns of action potentials between neuronal subgroups. This suggests that there is a heterogeneous interweaving of cells in the same nucleus and that it is important to look at the spectrum of sex differences at cellular levels within the same nucleus, rather than comparing internuclear levels.

Differences in the size of the MeA have been attributed to androgen- and estrogen receptor-mediated signaling in neurons. There is no expansion in volume or cell size in androgen-insensitive rats (Petrulis et al. 2017). The developmental mechanisms controlling sexual dimorphism in the MeA, such as cell size and chemical properties, may be due to perinatal hormones. However, the mechanisms that maintain sexual dimorphism during adulthood appear to be more complex. In adult rats, sex differences in volume and neuronal size of the posterior-dorsal part of the MeA are dependent on blood androgen concentration, but sex differences in total neuronal number are independent of blood androgen concentration (Morris et al. 2008). The neuropeptide cholecystokinin is observed in the posterior-dorsal part of the MeA and is more abundant in males, but castration reduces the immunoreactivity of cholecystokinin (Simerly and Swanson 1987).

Sex differences in the chemical properties of the MeA are associated with sex differences in social behavior, especially mating and aggression. Male mice have more P450 aromatase-positive cells in the MeA than female mice, and fibers from these neurons are denser in males than in females (Wu et al. 2009). These aromatase-containing neurons are also associated with aggression in mice, suggesting that they are incorporated into neural circuits involved in sex-specific behavior. It is likely that testosterone during the perinatal and sexual maturation periods acts on androgen receptors and is converted to estradiol by aromatase, which then acts on estrogen receptors resulting in the acquisition of sexual functions (Wu et al. 2009). In addition, mating experience and the presentation of olfactory stimuli of the opposite sex increase the expression of c-fos in androgen receptor (AR) immunoreactive cells of male Syrian and golden hamsters in the MeA (Wood and Newman 1993; Blake and Meredith 2011). However, there are discrete subpopulations of neurons in the MeA that respond to different components of sexual behavior (Kollack-Walker and Newman 1997; Bergan et al. 2014), and androgens and estrogens have different effects on neuronal function. It should be noted that androgens and estrogens seem to work differently in different subpopulations of the MeA. Interestingly, sex differences in the chemical structure of the MeA are also observed in eusocial, non-sexually mature animals such as naked mole rats; males have more AR-positive nuclei than females, even though there are no sex differences in region size or cell number (Holmes et al. 2008). The behavioral importance of these differences in AR expression in the MeA is still under investigation.

The effects of prenatal androgen exposure on human amygdala function have been examined using positron emission tomography imaging in women diagnosed with congenital adrenal hyperplasia (CAH), as women diagnosed with CAH have elevated androgens during fetal life (Merke and Bornstein 2005). The disease is usually diagnosed at birth, and treatment normalizes androgen levels. However, it is also known that girls with CAH often behave in a more masculine manner than their unaffected sisters (Berenbaum 1999). Although functional connectivity between the amygdala and hypothalamus differs between unaffected boys and girls, it does not differ between girls diagnosed with CAH and unaffected girls (Ciumas et al. 2009). Prenatal exposure to androgens may not affect brain function to the same extent in humans and rodents, but another possibility is that androgens may need to be present in adulthood in order to induce sex differences in connectivity.

10.8 Summary

In the process of sexual evolution, mammals have acquired differences in the reproductive and social roles of males and females. In particular, differences in the investment in nurturing have resulted in distinct sex differences. Through this process, males have become more competitive, and the reproductive success of dominant males has increased, but at the same time phenotypes have emerged in which males avoid competition with other males. In mammals, various male phenotypes are regulated by hormones. In social environments, social status, and encounters with the opposite sex, the secretion of testosterone is regulated to increase an individual's adaptability by displaying appropriate "masculinity." Such a strategy may be expressed as a "spectrum." We are beginning to understand the molecular mechanisms underlying the emergence of the endocrine sex spectrum. One of the main and crucial question to be addressed is why and how the mammals can obtain such a plasticity and spectrum of sexuality, which is encoded in the genes. Going forward, it is hoped that a deeper understanding of sexual selection, as proposed by Charles Darwin, will be achieved by understanding the underlying molecular mechanisms.

References

Altmann SA, Wagner SS, Lenington S (1977) Two models for the evolution of polygyny. Behav Ecol Sociobiol 2:397–410. https://doi.org/10.1007/bf00299508

Asaba A, Hattori T, Mogi K, Kikusui T (2014a) Sexual attractiveness of male chemicals and vocalizations in mice. Front Neurosci 8:231. https://doi.org/10.3389/fnins.2014.00231

Asaba A, Okabe S, Nagasawa M et al (2014b) Developmental social environment imprints female preference for male song in mice. PLoS One 9:e87186. https://doi.org/10.1371/journal.pone.0087186

Asaba A, Osakada T, Touhara K et al (2017) Male mice ultrasonic vocalizations enhance female sexual approach and hypothalamic kisspeptin neuron activity. Horm Behav 94:53–60

Berenbaum SA (1999) Effects of early androgens on sex-typed activities and interests in adolescents with congenital adrenal hyperplasia. Horm Behav 35(1):102–110

Berenbaum SA, Beltz AM (2011) Sexual differentiation of human behavior: effects of prenatal and pubertal organizational hormones. Front Neuroendocrinol 32:183–200. https://doi.org/10.1016/j.yfrne.2011.03.001

Bergan JF, Ben-Shaul Y, Dulac C (2014) Sex-specific processing of social cues in the medial amygdala. Elife 3:e02743. https://doi.org/10.7554/eLife.02743

Blake CB, Meredith M (2011) Change in number and activation of androgen receptor-immunoreactive cells in the medial amygdala in response to chemosensory input. Neuroscience 190:228–238. https://doi.org/10.1016/j.neuroscience.2011.05.056

Brown RE (1986) Social and hormonal factors influencing infanticide and its suppression in adult male long-Evans rats (Rattus norvegicus). J Comp Psychol 100:155–161

Champagne FA, Curley JP, Keverne EB, Bateson PPG (2007) Natural variations in postpartum maternal care in inbred and outbred mice. Physiol Behav 91:325–334. https://doi.org/10.1016/j.physbeh.2007.03.014

Ciumas C, Lindén Hirschberg A, Savic I (2009) High fetal testosterone and sexually dimorphic cerebral networks in females. Cereb Cortex 19:1167–1174. https://doi.org/10.1093/cercor/bhn160

Clark MM, vom Saal FS, Galef BG Jr (1992) Intrauterine positions and testosterone levels of adult male gerbils are correlated. Physiol Behav 51:957–960. https://doi.org/10.1016/0031-9384(92)90077-f

Cooke B, Hegstrom CD, Villeneuve LS, Breedlove SM (1998) Sexual differentiation of the vertebrate brain: principles and mechanisms. Front Neuroendocrinol 19:323–362. https://doi.org/10.1006/frne.1998.0171

Creed KE (1989) Effect of castration on penile erection in the dog. Neurourol Urodyn 8:607–614. https://doi.org/10.1002/nau.1930080609

Dall'Oglio A, Gehlen G, Achaval M, Rasia-Filho AA (2008) Dendritic branching features of posterodorsal medial amygdala neurons of adult male and female rats: further data based on the Golgi method. Neurosci Lett 430:151–156. https://doi.org/10.1016/j.neulet.2007.10.051

Darwin C (1859) On the origin of species by means of natural selection. John Murray, London

Gil D, Gahr M (2002) The honesty of bird song: multiple constraints for multiple traits. Trends Ecol Evol 17:133–141. https://doi.org/10.1016/S0169-5347(02)02410-2

Goldberg AD, Allis CD, Bernstein E (2007) Epigenetics: a landscape takes shape. Cell 128:635–638. https://doi.org/10.1016/j.cell.2007.02.006

Gubernick DJ, Teferi T (2000) Adaptive significance of male parental care in a monogamous mammal. Proc Biol Sci 267:147–150. https://doi.org/10.1098/rspb.2000.0979

Hattori T, Osakada T, Matsumoto A et al (2016) Self-exposure to the male pheromone ESP1 enhances male aggressiveness in mice. Curr Biol 26:1229–1234

Holmes MM, Goldman BD, Forger NG (2008) Social status and sex independently influence androgen receptor expression in the eusocial naked mole-rat brain. Horm Behav 54:278–285. https://doi.org/10.1016/j.yhbeh.2008.03.010

Holy TE, Guo Z (2005) Ultrasonic songs of male mice. PLoS Biol 3:e386

Kanno K, Kikusui T (2018) Effect of sociosexual experience and aging on number of courtship ultrasonic vocalizations in male mice. Zool Sci 35:208–214

Kenagy GJ, Trombulak SC (1986) Size and function of mammalian testes in relation to body size. J Mammal 67:1–22. https://doi.org/10.2307/1380997

Kikusui T, Kajita M, Otsuka N et al (2018) Sex differences in olfactory-induced neural activation of the amygdala. Behav Brain Res 346:96–104

Kikusui T, Nakanishi K, Nakagawa R et al (2011) Cross fostering experiments suggest that mice songs are innate. PLoS One 6:e17721. https://doi.org/10.1371/journal.pone.0017721

Kikusui T, Shima Y, Sonobe M et al (2021a) Testosterone regulates the emission of ultrasonic vocalizations and mounting behavior during different developmental periods in mice. Dev Psychobiol 63(4):725–733

Kikusui T, Shimokawa Y, Kanbara N et al (2013) Organizational effects of estrogen on male-type vulnerability to early weaning. Horm Behav 64:37–43. https://doi.org/10.1016/j.yhbeh.2013.04.007

Kikusui T, Sonobe M, Yoshida Y et al (2021b) Testosterone increases the emission of ultrasonic vocalizations with different acoustic characteristics in mice. Front Psychol 12:680176. https://doi.org/10.3389/fpsyg.2021.680176

Kleiman DG (1977) Monogamy in mammals. Q Rev Biol 52:39–69. https://doi.org/10.1086/409721

Kollack-Walker S, Newman SW (1997) Mating-induced expression of c-fos in the male Syrian hamster brain: role of experience, pheromones, and ejaculations. J Neurobiol 32:481–501. https://doi.org/10.1002/(sici)1097-4695(199705)32:5<481::aid-neu4>3.0.co;2-1

Lischinsky JE, Sokolowski K, Li P et al (2017) Embryonic transcription factor expression in mice predicts medial amygdala neuronal identity and sex-specific responses to innate behavioral cues. Elife 6:e21012. https://doi.org/10.7554/eLife.21012

Lkhagvasuren G, Shin H, Lee SE et al (2016) Molecular genealogy of a Mongol Queen's family and her possible kinship with Genghis khan. PLoS One 11:e0161622. https://doi.org/10.1371/journal.pone.0161622

Lukas D, Clutton-Brock TH (2013) The evolution of social monogamy in mammals. Science 341:526–530. https://doi.org/10.1126/science.1238677

Matsuo T, Hattori T, Asaba A et al (2015) Genetic dissection of pheromone processing reveals main olfactory system-mediated social behaviors in mice. Proc Natl Acad Sci U S A 112:311. https://doi.org/10.1073/pnas.1416723112

McCarthy MM, Arnold AP (2011) Reframing sexual differentiation of the brain. Nat Neurosci 14:677

Merke DP, Bornstein SR (2005) Congenital adrenal hyperplasia. Lancet 390(10108):2194–2210

Mogi K, Takanashi H, Nagasawa M, Kikusui T (2015) Sex differences in spatiotemporal expression of AR, ERα, and ERβ mRNA in the perinatal mouse brain. Neurosci Lett 584:88–92

Morris JA, Jordan CL, Breedlove SM (2004) Sexual differentiation of the vertebrate nervous system. Nat Neurosci 7:1034–1039. https://doi.org/10.1038/nn1325

Morris JA, Jordan CL, Breedlove SM (2008) Sexual dimorphism in neuronal number of the posterodorsal medial amygdala is independent of circulating androgens and regional volume in adult rats. J Comp Neurol 506:851–859. https://doi.org/10.1002/cne.21536

Parent CI, Meaney MJ (2008) The influence of natural variations in maternal care on play fighting in the rat. Dev Psychobiol 50:767–776. https://doi.org/10.1002/dev.20342

Payne RB (1984) Sexual selection, lek and arena behavior, and sexual size dimorphism in birds. Ornithol Monogr 33:1–52. https://doi.org/10.2307/40166729

Petrulis A, Fiber JM, Swann JM (2017) The medial amygdala, hormones, pheromones, social behavior network, and mating behavior. In: Pfaff DW, Joëls M (eds) Hormones, brain and behavior, vol 1, 3rd edn. Academic Press, pp 329–343. https://doi.org/10.1016/B978-0-12-803592-4.00011-0

Rasia-Filho AA, Fabian C, Rigoti KM, Achaval M (2004) Influence of sex, estrous cycle and motherhood on dendritic spine density in the rat medial amygdala revealed by the Golgi method. Neuroscience 126(4):839–847

Simerly RB (2002) Wired for reproduction: organization and development of sexually dimorphic circuits in the mammalian forebrain. Annu Rev Neurosci 25:507–536. https://doi.org/10.1146/annurev.neuro.25.112701.142745

Simerly RB, Swanson LW (1987) Castration reversibly alters levels of cholecystokinin immunoreactivity within cells of three interconnected sexually dimorphic forebrain nuclei in the rat. Proc Natl Acad Sci U S A 84:2087–2091. https://doi.org/10.1073/pnas.84.7.2087

Sreenivasan S, Weinberger LE (2016) Surgical castration and sexual recidivism risk. In: Phenix A, Hoberman HM (eds) Sexual offending: predisposing antecedents, assessments and management. Springer, New York, New York, NY, pp 769–777

Sugimoto H, Okabe S, Kato M et al (2011) A role for strain differences in waveforms of ultrasonic vocalizations during male-female interaction. PLoS One 6:e22093. https://doi.org/10.1371/journal.pone.0022093

Ten Cate C, Vos DR (1999) Sexual imprinting and evolutionary processes. Adv Stud Behav 28:1–31

Wang F, Kessels HW, Hu H (2014) The mouse that roared: neural mechanisms of social hierarchy. Trends Neurosci 37:674–682. https://doi.org/10.1016/j.tins.2014.07.005

West PM, Packer C (2002) Sexual selection, temperature, and the lion's mane. Science 297:1339–1343. https://doi.org/10.1126/science.1073257

Wood RI, Newman SW (1993) Mating activates androgen receptor-containing neurons in chemosensory pathways of the male Syrian hamster brain. Brain Res 614:65–77. https://doi.org/10.1016/0006-8993(93)91019-o

Wu D, Gore AC (2010) Changes in androgen receptor, estrogen receptor alpha, and sexual behavior with aging and testosterone in male rats. Horm Behav 58:306–316. https://doi.org/10.1016/j.yhbeh.2010.03.001

Wu MV, Manoli DS, Fraser EJ et al (2009) Estrogen masculinizes neural pathways and sex-specific behaviors. Cell 139:61–72. https://doi.org/10.1016/j.cell.2009.07.036

Part III
Environmental Regulation of the Sex Spectrum

Chapter 11
Symbiont-Induced Sexual and Reproductive Manipulation in Insects

Susumu Katsuma, Kanako Hirota, and Tomohiro Muro

Abstract Some insect symbionts are known to manipulate the reproduction and sexual development of their hosts for their own benefit. To accomplish this, these selfish sex ratio distorters must disturb the host's sex determination pathways. Recent progress in molecular biological and bioinformatic methods has enabled us to identify the symbiont effectors. In this chapter, we introduce the diversity of insect sex determination systems and dosage compensation systems, both of which are tightly associated with one another. Moreover, we describe highly tuned symbiont strategies for host sexual manipulation by that target the sex determination and/or dosage compensation machineries by taking advantage of their own factors.

Keywords Sexual manipulation · Symbiont · Sex determination · Dosage compensation · Insect

11.1 Sex Determination System in Insects

Insect sex determination systems have been extensively studied in a model insect, the fruit fly, *Drosophila melanogaster*. In this species, the primary signal for sex determination is the number of X chromosomes, which reflects the dose of X-linked signaling elements (XSE) (Hopkins and Kopp 2021) (Fig. 11.1). The amount of XSE proteins on the X chromosome determines the expression of RNA-binding protein gene *Sex-lethal* (*Sxl*). *Sxl* is expressed and translated into a functional SXL protein only in females, which triggers a multi-step regulation in female-specific splicing of sex-determining genes. SXL protein controls the sex determination cascade by directing pre-mRNA splicing of *transformer* (*tra*) transcripts to produce the RNA-binding protein TRA. TRA regulates sex-specific splicing of genes that encode transcription factors, including *doublesex* (*dsx*) (Hopkins and Kopp 2021).

S. Katsuma (✉) · K. Hirota · T. Muro
Department of Agricultural and Environmental Biology, Graduate School of Agricultural and Life Sciences, The University of Tokyo, Tokyo, Japan
e-mail: skatsuma@g.ecc.u-tokyo.ac.jp; hirota-kanako1102@g.ecc.u-tokyo.ac.jp; muro-tomohiro@g.ecc.u-tokyo.ac.jp

M. Tanaka, M. Tachibana (eds.), *Spectrum of Sex*,
https://doi.org/10.1007/978-981-19-5359-0_11

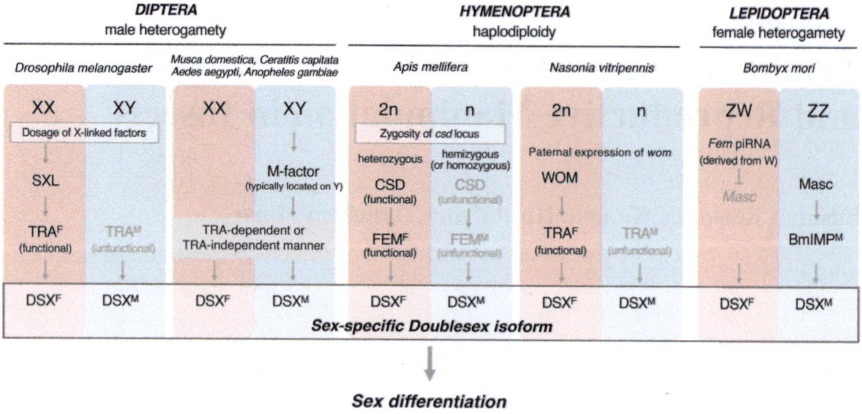

Fig. 11.1 Diversity in insect sex determination systems. Representative sex determination pathways in Diptera, Hymenoptera, and Lepidoptera are shown. For TRA, FEM, DSX, and BmIMP proteins, superscripts F and M indicate female- and male-type isoforms, respectively

On the other hand, owing to the absence of SXL protein, *tra* mRNA with a premature stop codon is transcribed in males, resulting in the absence of functional TRA protein. *dsx* is sex-specifically spliced and translated into sex-specific DSX proteins. DSX potentially binds to thousands of genome loci, leading to sex-specific gene expression and subsequent sexual differentiation (Clough et al. 2014).

Accumulating evidence has shown that primary factors in sex determination cascades in insects are surprisingly diverse, although to date there are only a small number of insects whose sex determination systems, especially primary sex determiners, have been characterized. Instead, *dsx* homologs have been shown to act commonly at the downstream end in the sex differentiation cascade of insects (Fig. 11.1). Several dipteran insects other than *D. melanogaster* possess different primary sex determiners, called M factors. Recent progress in molecular biological methods such as deep sequencing by next-generation sequencers (NGSs) and genome editing has identified the genes encoding these M factors and revealed that most of them are non-homologous. The M factor genes *Yob* and *Guy1* in the malaria mosquito *Anopheles gambiae* and *Anopheles stephensi*, respectively, both encode a 56-amino acid protein with a helix-loop-helix motif (Krzywinska et al. 2016; Criscione et al. 2016), whereas *Nix* in *Aedes aegypti* (Hall et al. 2015) and *Aedes albopictus* (Liu et al. 2020) encodes proteins with RNA recognition motifs. These mosquito M factors have been shown to induce male-specific *dsx* splicing, which is independent of *tra*. In the house fly *Musca domestica*, the M factor gene *Mdmd* encodes a paralog of a generic splicing factor (Sharma et al. 2017), whereas the M factor gene *MoY* of the medfly *Ceratitis capitata* encodes a small protein that does not bear any similarity to known proteins (Meccariello et al. 2019). MDMD and MOY both induce *tra* splicing in males, resulting in the absence of functional TRA and subsequent male-specific splicing of *dsx*.

Hymenopteran insects are haplodiploids, in which males are derived from haploid (unfertilized) eggs and females are derived from diploid (fertilized) eggs. Sex determination in this system has been deeply studied in the honeybee *Apis mellifera*. In this insect, sex is determined by heterozygosity at a single gene, *complementary sex determiner* (*csd*). Honeybees that are heterozygous at the *csd* gene develop into females, while homozygotes and hemizygotes at *csd* become males. The *csd* gene encodes an arginine/serine-rich protein that is structurally similar to TRA protein and contains a hypervariable region (Beye et al. 2003). There are over 50 different *csd* alleles among natural honeybee populations (Hasselmann and Beye 2004; Lechner et al. 2013). Further genome analysis identified a new gene, *feminizer* (*fem*), which is located 12 kilobases upstream of the *csd* locus (Hasselmann et al. 2008). Interestingly, *fem* encodes a protein that has an arginine/serine-rich domain with a high degree of sequence identity to the CSD protein, suggesting that these two paralogous genes are derived from an ancestrally conserved progenitor gene. RNA interference (RNAi) of the female-specific *fem* splice variant results in production of male progenies, and RNAi of female allelic *csd* transcripts results in expression of the male-specific *fem* variant. These results demonstrate that *fem* is essential for female development and conveys the signal from *csd* heterozygosity to the downstream sex determination cascade in the honeybee.

Most parasitoid wasps do not possess *csd* homologs (Whiting 1967). In the parasitoid wasp *Nasonia vitripennis*, a primary instructor gene was identified and named *wasp overruler of masculinization* (*wom*). *wom* is not maternally provided in unfertilized (haploid) eggs, whereas it is paternally provided in fertilized eggs at the early diploid embryonic stage (2 to 5 hours postoviposition [hpo]) (Zou et al. 2020). Parental RNAi of *wom* in early diploid embryos decreases both *wom* and *tra* expression levels, suggesting that WOM acts upstream of *tra* and activates zygotic *tra* transcription (from around 5 hpo), leading to female development. *Wom* encodes a 580-amino acid protein with a P53-like DNA-binding domain and a coiled-coil domain. Genome analysis revealed that this gene emerged via gene duplication and genomic rearrangements. Identification and functional characterization of *csd* and *wom* suggest that unique genes have evolved upstream of *tra*, which enables paternally mediated onset of female development in haplodiploid reproduction.

The sex determination system in lepidopteran insects has been primarily studied using the mulberry silkworm, *Bombyx mori*. Yoshimaro Tanaka discovered that *B. mori* females have a W chromosome (Tanaka 1916), and Haruo Hasimoto subsequently reported that one copy of the W chromosome is sufficient for determining femaleness, regardless of the copy number of Z chromosomes (Hasimoto 1933). These findings strongly suggest that the W chromosome encodes a putative feminizing factor, called *Feminizer* (*Fem*). In addition, transgenic and knockout approaches have shown that *B. mori dsx* homolog (*Bmdsx*) is required for sexual differentiation of *B. mori* (Suzuki et al. 2003, 2005; Xu et al. 2017a, b). In 2014, we identified a female-specific PIWI-interacting RNA (piRNA) as the feminizer of *B. mori* (Kiuchi et al. 2014). This piRNA is produced from the W-linked precursor RNA *Fem* (Fig. 11.2a), and *Fem* piRNA forms a complex with *B. mori* PIWI protein Siwi. The *Fem* piRNA–Siwi complex cleaves a protein-coding mRNA transcribed from the Z chromosome.

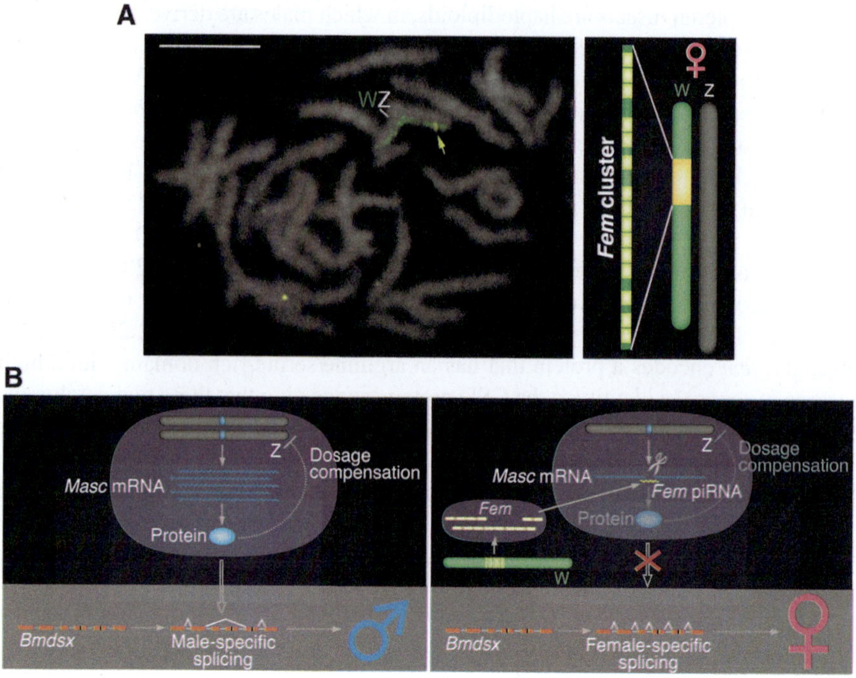

Fig. 11.2 Sex determination pathway in *Bombyx mori*. (**a**) Left: FISH identification of W chromosome (a green-signal painted chromatid) and *Fem* (yellow signals on the W: yellow arrow). Bar, 10 μm. Right: A drawing of the *Fem* cluster on the W chromosome. (**b**) The sex of *B. mori* is determined by the piRNA pathway. *Fem* RNAs are transcribed from the sex-determining region of the W chromosome and are cleaved by the maternally transmitted *Masc* piRNA–BmAgo3 complex. The *Fem* piRNA–Siwi complex then cleaves *Masc* mRNA, which encodes a protein required for both masculinization and dosage compensation. (Reprinted from Proc Jpn Acad, Ser B 94 (2018) with permission from The Japan Academy)

This Z-linked gene, which we named *Masculinizer* (*Masc*), encodes a protein required for both masculinization and dosage compensation (Fig. 11.2b). *Fem* and *Masc* mRNA both participate in the piRNA biogenesis pathway called the ping-pong cycle, where the piRNA amplification loop is maintained by two *B. mori* PIWI proteins, Siwi and BmAgo3 (Kawaoka et al. 2008). These findings revealed that the piRNA-mediated interaction between the two different sex chromosomes is the primary signal for the sex determination cascade in *B. mori*.

Although *Masc* homologs have been identified in several lepidopteran insects and their masculinizing activity has been verified in organisms and/or cultured cells (Fukui et al. 2015, 2018; Lee et al. 2015; Wang et al. 2019; Harvey-Samuel et al. 2020; Visser et al. 2021), piRNAs or other factors required for post-transcriptional regulation of *Masc* have not been reported in non-*Bombyx* species. Therefore, at present, it remains unclear whether the primary sex determiner is common in lepidopteran insects with a WZ/ZZ sex chromosome constitution. Lepidopteran species that are monosomic (Z0) in females and ZZ in males (e.g., *Samia* species) likely do not use sex-specific piRNAs because they do not have female-specific sex chromosomes.

11.2 Relationship Between Sex Determination and Dosage Compensation Cascades in Insects

Most insects have an XY sex chromosome system, where males have X and Y sex chromosomes and females have two X chromosomes. On the other hand, females are the heterogametic or hemigametic sex in lepidopteran insects. In this system, males have two Z sex chromosomes and females have Z and W sex chromosomes or one Z chromosome. Dosage compensation is a common mechanism that equilibrates X- or Z-linked gene expression with the autosomes. *D. melanogaster* uses an XY system, in which males compensate for reduced dosage of X-linked genes by hypertranscription of X-linked genes about twofold through global hyperacetylation of H4K16 on the hemizygous X chromosome (Lucchesi 1978; Ellison and Bachtrog 2019). In females, dosage compensation is repressed via inhibition of translation of *male-specific lethal 2* (*msl-2*) by SXL, a master switch for sex determination in *D. melanogaster*. MSL2 is a subunit of the chromatin remodeling dosage compensation complex (DCC); therefore, mutations in *msl-2* lead to male-specific lethality (Penalva and Sanchez 2003). Unlike SXL, other sex-determining factors, TRA and its downstream switch DSX, are not involved in X chromosome dosage compensation, suggesting that disruption or ectopic expression of these factors do not induce sex-specific lethality. Recently, a transgenic approach revealed that M factor *Guy1* regulates dosage compensation by increasing X chromosome gene expression and confers female-specific lethality in *An. stephensi* (Qi et al. 2019).

In *B. mori*, Z-linked gene *Masc* encodes a master regulator of masculinization (Kiuchi et al. 2014). Unexpectedly, we found that embryonic knockdown of *Masc* mRNA results in a failure of not only masculinization but also dosage compensation; depletion of *Masc* mRNA causes feminization and aberrant upregulation of Z-linked genes in male embryos (Fig. 11.2b). Moreover, failure of dosage compensation results in male-specific embryonic death in *B. mori* (Kiuchi et al. 2014). Our subsequent studies clearly showed that modification of *Masc* by transgenic and genome editing approaches enables us to manipulate sexual differentiation and sex-specific lethality in *B. mori* (Sakai et al. 2016; Xu et al. 2017a, b; Kiuchi et al. 2019). These findings suggest that *Masc* will be an ideal target to induce sex-specific death in lepidopteran insects. In conclusion, the sex determination cascade is tightly coupled with the dosage compensation system and a failure of dosage compensation results in sex-specific death in insects of different orders.

11.3 Symbiont-Induced Sexual and Reproductive Manipulation in Insects

All insects are considered to be infected by a large variety of microbes, including bacteria, fungi, and viruses, which range from parasitic to commensal or mutualistic. Among these, heritable endosymbionts are most tightly linked to the survival and reproduction of host insects. The endosymbiont resides within the cytoplasm and is maternally inherited, although intrasperm paternal inheritance of *Rickettsia* residing within the nuclei has been exceptionally reported in the leafhopper *Nephotettix cincticeps* (Watanabe et al. 2014). Maternal inheritance of the symbiont triggers alternation of the phenotypes of host insects, including sex ratio distortion and reproduction (Fig. 11.3).

The most extensively studied symbiont that induces sexual and reproductive manipulation in insects is *Wolbachia*. *Wolbachia* are a maternally inherited intracellular bacteria that infect over 66% of insect species (Hilgenboecker et al. 2008). *Wolbachia* infection is a well-known example in which host sex ratios and reproductive strategies are altered to increase the proportion of the infected females in the population. *Wolbachia*-induced phenotypes include parthenogenesis, feminization, cytoplasmic incompatibility (CI), and male killing, each of which is thought to be adaptive for *Wolbachia* by enhancing the production of infected females (Werren et al. 2008) (Fig. 11.3). Among them, the most common manipulation is CI, in which embryonic death occurs in crosses between infected males and uninfected females, whereas females infected with the same strain of *Wolbachia* rescue this lethality. In 2017, the *Wolbachia* genes underlying CI were identified (LePage et al. 2017;

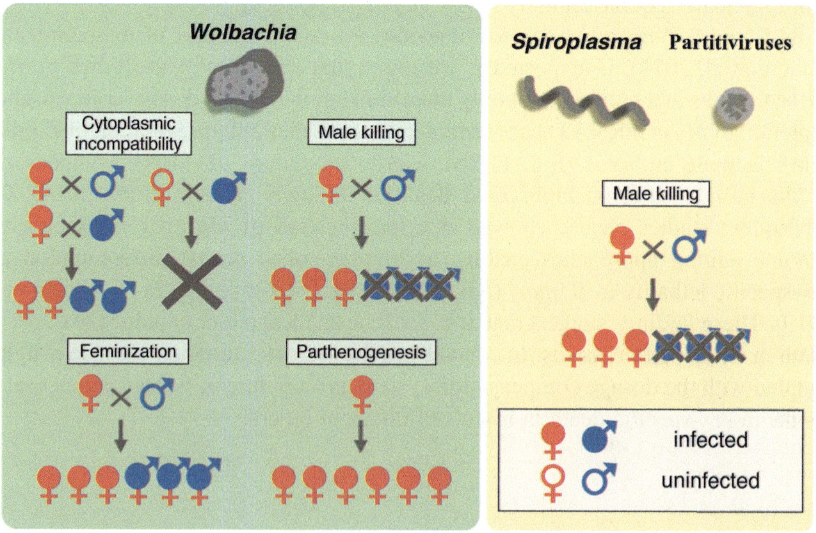

Fig. 11.3 Sexual and reproductive manipulation by *Wolbachia*, *Spiroplasma*, and partitiviruses

Beckmann et al. 2017). According to the hypothesis that the genes responsible for CI are present in all CI-inducing *Wolbachia* strains and absent or mutated in non-CI strains, and they are expressed in the gonads of infected insects, two *w*Mel genes, *WD0631* and *WD0632*, which are adjacent and coevolving within the eukaryotic association module of prophage WO, were selected. Both genes were expressed in the testes of *w*Mel-infected *D. melanogaster* males, and the expression levels were decreased significantly in aged males, which is consistent with the observation that CI strength decreases significantly in aged males (Reynolds and Hoffmann, 2002).

To examine whether these two genes are essential for CI induction, transgenic *D. melanogaster* strains expressing each gene were generated and CI phenotypes were assessed by measuring the percentage of embryos that hatched into larvae. Although CI was not observed in crossings between males expressing *WD0631* or *WD0632* and wild-type females, expression of both genes in the same males significantly reduced hatch rates compared with wild-type crosses with no effect on sex ratios. Moreover, this lethal phenotype was completely rescued by crossing *w*Mel-infected females. These results indicate that these two genes are the CI factors. They were named as *cifA* (*WD0631*) and *cifB* (*WD0632*), respectively (LePage et al. 2017).

The mechanism of CI is frequently explained by a modification–rescue or toxin–antidote system in which sperm is modified by *Wolbachia* toxin, and this modification can be rescued in the egg by a different *Wolbachia* factor. The method by which CI factors induce CI in host insects was revealed using *Wolbachia w*Pip proteins CidA and CidB (Beckmann et al. 2017). CidB is a *Wolbachia* deubiquitylating enzyme (DUB) that induces CI by cleaving ubiquitin from substrates. *cidB* is located in a two gene-containing operon, and the other gene *cidA* encodes a protein that binds CidB. CidA was previously identified as a protein (WPA0282) associated with *Wolbachia*-modified mosquito sperm. CI was observed in crossings between transgenic *D. melanogaster* males expressing the *cidA-cidB* operon and wild-type females. Interestingly, transgenic expression of CidB with a catalytically inactive DUB did not induce CI. CidB alone induced toxicity in yeast, and this effect depended on DUB activity. These results could explain the "modification–rescue" hypothesis for *Wolbachia*-induced CI.

Feminization of genetic males by *Wolbachia* infection is rarely observed but has been reported in several insect species, including two *Eurema* butterfly species. In Japanese populations of *E. hecabe* (Hiroki et al. 2002), two *Wolbachia* strains were identified, *w*CI (*w*HecCI2 or *w*HecFem1) and *w*Fem (*w*HecFem2), which induce CI and feminization, respectively (Hiroki et al. 2002, 2004). Treatment of *E. hecabe* larvae with tetracycline at different developmental stages suppresses the *w*HecFem2 titer and results in emergence of adult butterflies with sexually intermediate traits (Narita et al. 2007). This indicates that the feminizing action of *Wolbachia* is not complete at the embryonic stages and continues during larval development in male larvae. Quantitative PCR assay revealed that *w*Fem-infected *E. mandarina* females have only one Z chromosome and lose the W chromosome or possess a dysfunctional one, whereas the W chromosome is present in females without *w*Fem infection (Kern et al. 2015). Tetracycline treatment of *w*Fem-infected adult females resulted in production of both Z0 and ZZ eggs (Kageyama et al. 2017a). These findings demonstrate that *w*Fem infection disrupts maternal Z chromosome inheritance, resulting in no male progenies with ZZ. Combined with the observation that *w*Fem removal by tetracycline results in all-male progeny phenotype, *w*Fem possesses the factor that can feminize *E. mandarina* genetic males.

Parthenogenesis-inducing *Wolbachia* have been described in arthropods with haplodiploid sex determination, including wasps and thrips (Werren et al. 2008). In this host manipulation system, *Wolbachia* infection disrupts the cell cycle at the early embryonic stages, leading to diploid development in unfertilized eggs; infected females produce daughters from unfertilized eggs instead of sons. Many species of *Trichogramma* wasps are infected by *Wolbachia* and contain mixtures of sexual and asexual populations. Antibiotic treatment of parthenogenetic female wasps results in production of male progenies, indicating that asexual reproduction is induced by *Wolbachia* infection (Stouthamer et al. 1990, 1993). Transfer experiments of parthenogenesis-inducing *Wolbachia* from *Trichogramma dendrolimi* to *Trichogramma evanescens* did not produce a distinct phenotype of *Wolbachia* infection, probably due to the low *Wolbachia* density in *T. evanescens* (Watanabe et al. 2013), which suggests that parthenogenesis-inducing *Wolbachia* in *Trichogramma* wasps are well adapted to their hosts.

Wolbachia infection also restores the host oogenesis defects by interacting with the host RNA machinery. Some *Sxl* mutant strains with partial loss-of-function mutations are viable but possess severe defects in germline development in *D. melanogaster* females (Starr and Cline 2002). Through screening of *Wolbachia* genes that exhibit negative effects on the growth and survival of *D. melanogaster* cultured cells, Ote et al. (2016) identified a novel gene *toxic manipulator of oogenesis* (*TomO*) from the *w*Mel genome. Although *TomO* homologs were found in some *Wolbachia* groups, their sequences were quite different. For example, TomO homologs of *w*Ri and *w*Ana both belong to the same *Wolbachia* group (supergroup A) as *w*Mel and are split into two proteins, and a homolog of *w*Pip (supergroup B) possesses two ankyrin repeats that are absent in *w*Mel TomO. Interestingly, the germline-restricted overexpression of TomO in *Sxl* mutant females restores germ stem cells (GSCs). Further experiments revealed that TomO enhances GSC maintenance by increasing the Nanos (Nos) protein level through binding to *nos* mRNA (Ote et al. 2016). These functions were also observed in a TomO homolog of *w*Pip (Ote and Yamamoto 2018), suggesting a conserved strategy for host–*Wolbachia* interaction by TomO protein.

11.4 Bacterial Male Killing in Insects

Male killing has independently evolved in at least six bacterial taxa, including *Spiroplasma* and *Wolbachia* (Hurst and Frost 2015). Researchers have considered that male killing increases the fitness of infected females in the ways that reduce inbreeding (Dannowski et al. 2009), reduce brother-sister competition when the resources are limited (Koop et al. 2009), and/or provide nutrients to infected, surviving sisters when male killing occurs at embryonic stages and the sisters cannibalize dead brothers (Elnagdy et al. 2011). Male killing can also reduce the effective population size of the host (Berec et al. 2016) and alter sexual preference (Hornett et al. 2006).

11.5 *Spiroplasma*-Induced Male Killing in Insects

Spiroplasma are Gram-positive bacteria that are widely observed in plants and arthropods (Regassa and Gasparich 2006). Male killing induced by *Spiroplasma* has been reported in ladybird beetles, butterflies (e.g., *Danaus chrysippus*), and moths (e.g., *Ostrinia zaguliaevi*) (Hurst et al. 1999; Majerus et al. 1999; Jiggins et al. 2000; Tinsley and Majerus 2006; Tabata et al. 2011). Molecular phylogenetic analyses of bacterial 16S rRNA, *dnaA*, and *rpoB* genes have revealed that the male-killing *Spiroplasma* in *Drosophila* and those in ladybird beetles, butterflies, and moths belong to different clades, the *citri* clade and the *ixodetis* clade, respectively.

Spiroplasma poulsonii is a symbiotic bacterium that infects *D. melanogaster*. *S. poulsonii*-induced male killing was first discovered in *D. melanogaster* about 70 years ago (Williamson and Poulson 1979). This male killing was speculated to be caused by a secreted protein called "androcidin" (Oishi 1971). *S. poulsonii* does not kill *D. melanogaster* males that lack the male-specific lethal (MSL) components, but on the other hand, it induces female death in strains that ectopically express the MSL complex (Veneti et al. 2005; Cheng et al. 2016; Harumoto et al. 2016), suggesting that *S. poulsonii* targets the MSL complex to establish male killing. Male-specific embryonic death by *S. poulsonii* is induced through the host apoptotic pathway (Harumoto et al. 2014), and the transcriptional levels of the genes involved in DNA damage and apoptosis are enhanced only in infected male embryos (Harumoto et al. 2016). Cytological analysis demonstrated that DNA damage occurs on the male X chromosome associated with the MSL complex (Harumoto et al. 2016). These findings clearly indicate that the male-killing *Spiroplasma* targets the MSL components on the male X chromosome to induce male-specific apoptotic death in *D. melanogaster* embryos.

During the maintenance of *D. melanogaster* strains artificially infected by *S. poulsonii*, Harumoto and Lemaitre found a *S. poulsonii* mutant that does not kill the male progeny completely, irrespective of the genetic background of the *D. melanogaster* strain and bacterial titer (Harumoto and Lemaitre 2018). To identify the responsible gene in the *S. poulsonii* genome, they determined and compared the genome sequences of wild-type (complete male killing) and mutant (incomplete male killing) *S. poulsonii* strains and identified a candidate gene in which an 828 base-pair-long deletion was detected in the mutant genome. This gene, which they named *Spaid* (*S. poulsonii androcidin*), is located on a putative plasmid and encodes a 1065-amino acid protein with ankyrin (ANK) repeats and a DUB domain (Harumoto and Lemaitre 2018). Transgenic expression of a codon-optimized *Spaid* construct in *D. melanogaster* reproduced the phenotypes for male killing, which included all-female progeny and apoptotic cell death in embryos, indicating that this gene is a long-sought male-killing gene in *S. poulsonii*. Subsequent experiments revealed that Spaid expression triggers apoptosis even in female embryos when the MSL complex is expressed in females (Harumoto and Lemaitre 2018). Moreover, they observed that Spaid expression in male embryos resulted in DNA fragmentation and the formation of chromatin bridges, most of which overlapped with the signals of MSL1, a component of the MSL complex. These results clearly indicate that Spaid mediates its toxic effects via the dosage compensation machinery. Along with *Spiroplasma* Spaid, *Wolbachia w*Pip CI-inducing protein CidB also possesses a DUB domain (Beckmann et al. 2017), suggesting that insect ubiquitin pathways are commonly used in sexual and reproductive manipulation by insect symbiotic bacteria.

11.6 *Wolbachia*-Induced Male Killing in Insects

Wolbachia-induced male killing has been reported in three insect orders, Coleoptera, Diptera, and Lepidoptera, where male killing occurs during embryogenesis (early male killing) or larval stages (late male killing). In general, antibiotic (tetracycline) treatment of *Wolbachia*-infected insects results in the production of male progeny and restores the 1:1 sex ratio.

In 2019, the first discovery of a candidate gene for *Wolbachia*'s male killing was reported. This gene, *WO-mediated killing* (*wmk*), was identified from *w*Mel *Wolbachia* (Perlmutter et al. 2019). As with the discovery of CI genes *cifA* and *cifB* (LePage et al. 2017), *wmk* was isolated from the eukaryotic association module of *Wolbachia* prophage WO. This prophage module was shown to be enriched with many sequences predicted to have eukaryotic functions (Bordenstein and Bordenstein 2016). *wmk* is expressed in *D. melanogaster* embryos infected with *Wolbachia* and encodes a putative transcriptional regulator containing two helix-turn-helix (HTH) XRE family DNA-binding domains. *Wmk* homologs are commonly found in the genomes of phage WO-containing *Wolbachia*, including known male-killing strains, *w*Bol from the butterfly *Hypolimnas bolina* and *w*CauB from the moth *Cadra cautella* (Sasaki et al. 2002). Transgenic *D. melanogaster* expressing codon-optimized *wmk* exhibit a significant female bias (male/female = 0.65), but not a complete male-killing phenotype (Perlmutter et al. 2019). Transgenic expression of *wmk* induces embryonic death and several cytological defects, including DNA damage at sites of dosage compensation activity. Although the mode of action of WMK in male killing is unknown, *wmk* is a very strong candidate for *Wolbachia*'s male-killing factor in *D. melanogaster*.

Wolbachia-induced male killing has also been frequently observed in lepidopteran insects. As the female W chromosome in some lepidopteran species possesses feminizing activity, maternal inheritance of *Wolbachia* is linked with maternal transmission of not only mitochondria but also the W chromosome. The most famous case is a male-killing *Wolbachia* of the nymphalid butterfly *H. bolina* on the Samoan islands Upolu and Savaii (Charlat et al. 2007). In 2001, high prevalence of *Wolbachia w*Bol1, which kills male embryos (early male killing), and a 99% female population were recorded in *H. bolina*. However, in 2005, about half of the individuals were male on the same island, indicating that the sex ratio was drastically altered within 10 generations (Charlat et al. 2007). These *Wolbachia*-infected females were fertile and egg hatch rate was high, showing the absence of male killing. Genotyping of *Wolbachia* revealed that *Wolbachia* strain was not replaced with other non-male-killing strains. Crossing experiments showed that alternation of sex ratios results from the spread of host suppressor genes for male-killer action in this island population (Hornett et al. 2006; Charlat et al. 2007). Such a rapid emergence of host resistance against male killers was also reported in a population of the green lacewing *Mallada desjardinsi* infected with male-killing *Spiroplasma* (Hayashi et al. 2018). These findings strongly suggest that such a rapid emergence and spread of host suppressor genes has occurred frequently after past conflict between cytoplasmic selfish elements and nuclear host genomes.

In *Ostrinia scapulalis* and its congener *Ostrinia furnacalis*, a unique male killing by *Wolbachia* infection was reported (Sugimoto and Ishikawa 2012; Fukui et al. 2015) (Fig. 11.4a, b). Sex ratio distorters in *Ostrinia* species were first discovered in the Japanese population of *O. furnacalis* in 1984 (Miyahara 1984). Subsequent studies revealed that most of the female-biased phenotypes in *O. scapulalis* and *O. furnacalis* were caused by *Wolbachia* infection. *Ostrinia* species have a WZ/ZZ sex chromosome system and all of the progeny of *Wolbachia*-infected mother moths have a W chromosome, showing the existence of male killing (Sugimoto and Ishikawa 2012). Unlike male killing observed in other moths, antibiotic treatment of male-killing strains (all-female strains) results in production of all-male offspring (female killing) (Sugimoto and Ishikawa 2012; Fukui et al. 2015) (Fig. 11.4b), indicating that *Wolbachia* possess a factor that triggers feminization in *Ostrinia*. The *Ostrinia* feminizing factor may be disrupted or inhibited during a long period of infection by this male-killing *Wolbachia*.

The male-type splice variants of *dsx* were not detected in early embryos of *O. scapulalis* and *O. furnacalis* infected with male-killing *Wolbachia*, suggesting that *Wolbachia* establishes sexual manipulation at least from the beginning of the sex determination period (Sugimoto and Ishikawa 2012; Fukui et al. 2015). Our study revealed that RNAi-mediated depletion of *Masc* mRNA in *B. mori* embryos results in male-specific embryonic lethality (Kiuchi et al. 2014). This phenomenon likely mimics the *Wolbachia*-inducing male killing observed in some lepidopteran insects, including *Ostrinia* species (Sugimoto and Ishikawa 2012). We therefore selected *O. furnacalis* embryos as the material for elucidating *Wolbachia*-mediated male killing. RNA-seq revealed that the level of *O. furnacalis Masc* (*OfMasc*) mRNA was reduced and dosage compensation of *O. furnacalis* Z-linked genes failed in *Wolbachia*-infected embryos (Fukui et al. 2015). Moreover, injection of artificially synthesized *OfMasc* cRNA into *Wolbachia*-infected early embryos resulted in the production of male progeny. These results clearly showed that *Wolbachia* disrupts dosage compensation at the early embryonic stage by targeting the host factor Masc,

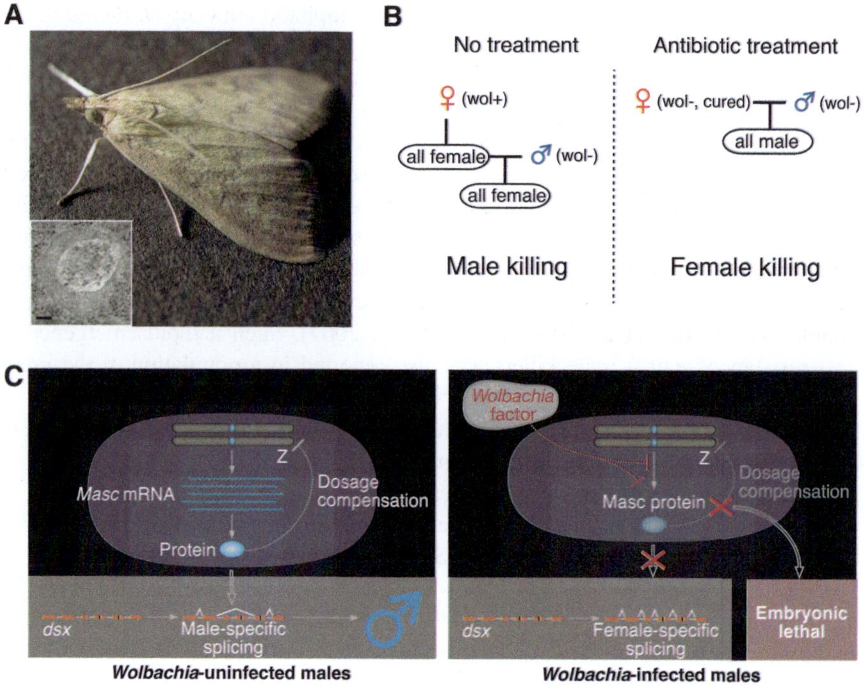

Fig. 11.4 Male killing in *Ostrinia furnacalis* moths. (**a**) *O. furnacalis* female moth infected with male-killing *Wolbachia*. Transmission electron microscopy observation of *Wolbachia* in the *O. furnacalis* ovary is shown in the inset panel. Bar, 100 nm. (**b**) Male killing in a *Wolbachia*-infected matriline and female killing in a cured matriline. (**c**) A proposed model for *Wolbachia*-induced male killing in *O. furnacalis*. (Reprinted or modified from Proc Jpn Acad, Ser B 94 (2018) with permission from The Japan Academy)

which is essential for both masculinization and dosage compensation (Fig. 11.4c). On the other hand, unexpectedly, male larvae hatched from *OfMasc* cRNA-injected *Wolbachia*-infected embryos expressed female-type *dsx* only, indicating that rescued males were not transcriptionally masculinized (Fukui et al. 2018). This is a phenocopy of *Wolbachia*-induced feminization of genetic males. Previous studies have suggested that Masc proteins act as the masculinizing factor in the cytoplasm, whereas they play a role in dosage compensation in the nucleus (Sugano et al. 2016; Katsuma et al. 2019; Hirota et al. 2021). Judging from *Wolbachia*'s cytoplasmic localization of the host cells, we speculate that nuclear Masc can rescue the failure of dosage compensation but cytoplasmic Masc cannot rescue the masculinization process in *Wolbachia*-infected cells.

It was recently reported that the Mediterranean flour moth *Ephestia kuehniella* possesses two *Masc* homologs, either or both of which induce male-type *dsx* splicing (Visser et al. 2021). This suggests that some lepidopteran species likely have several copies of *Masc* homologs. If one copy of *Masc* is essential for masculinization and the other copy plays a role in dosage compensation, feminization without the male-killing phenotype in males can be established when the *Wolbachia* factor interacts with the masculinizing Masc but not with the other. Alternatively, feminization can be established when the *Wolbachia* factor disrupts the interaction between Masc and host cytoplasmic proteins cooperating with Masc for masculinization. The identification of a *Wolbachia* factor targeting Masc will provide evidence on how *Wolbachia* has evolved to hijack the Masc-dependent pathway that controls both lepidopteran insect-specific sex determination and dosage compensation cascades.

11.7 Virus-Induced Male Killing in Insects

Non-bacterial male killers have been discovered in some insects. Late male killing has been reported in the oriental tea tortrix *Homona magnanima* (Morimoto et al. 2001). In this male-killing strain, most male larvae die during the last instar or pupal stage. Pathohistological studies revealed that the degradation of midgut epithelial cells may be the main reason for male-specific death (Hoshino et al. 2008). This male killing is maternally inherited, even when the females are mated with normal males. Antibiotic treatment does not rescue male-specific death, indicating that this male killing is induced by non-bacterial pathogens (Morimoto et al. 2001). Nakanishi et al. (2008) prepared a sucrose density-gradient ultracentrifuge fraction of homogenized male-killing strain larvae and injected the purified fraction into fifth-instar normal strain larvae. The survival rate of the injected male larvae was significantly lower than that of control male larvae, demonstrating involvement of virus-like pathogens in this male killing. They isolated RNA from the male-killing strain and then reverse-transcribed, and identified two novel cDNA sequences (MK1068 and MK1241) by a modified random amplified polymorphic DNA method. These two fragments were amplified from cDNA of the male-killing strain but not from the normal strain, verifying that these RNAs are specific to the male-killing strain. However, the homology and motif search did not match any known sequences or motifs.

Recently, Fujita et al. (2021) performed RNA sequencing of a cDNA library prepared from nuclease-treated female moth homogenates of a late-killing strain. They successfully obtained 27 contigs, which included two previously identified unannotated sequences (MK1068 and MK1241) as well as three different sequences of RNA-dependent RNA polymerases (RdRp). A homology search revealed that all of the three RdRp sequences contained conserved motifs for RdRp and were related to those of *Partitiviridae* viruses (Fujita et al. 2021). These three RdRps (MK-11, MK-25, MKsp30) belonged to novel unclassified *Partitiviridae* species; thus, they were named Osugoroshi virus (OGV) 1, 2, and 3, respectively. Phylogenetic analyses also revealed that OGVs belong to a group of invertebrate partitiviruses, which include viral-like sequences from Coleoptera or Araneae (Fujita et al. 2021). The authors also performed injection of purified viral particles into fourth-instar larvae and traced viral RNAs in the next generation, in which a female-biased sex ratio was observed. The results strongly suggested that OGV is responsible for late male killing in *H. magnanima*, although essential RNA sets of OGV for male killing are still unknown.

Non-bacterial male killing has also been reported in a *Drosophila* species. Kageyama et al. (2017b) discovered a single female *D. biauraria* whose offspring were all female. Unlike the late male killing observed in OGV-infected *H. magnanima* (Fujita et al. 2021), male killing in *D. biauraria* is established at the embryonic stages (Kageyama et al. 2017b). This embryonic male killing is maternally transmitted and is not altered by antibiotic treatment. Moreover, this phenotype can be reproduced by microinjection of filtered homogenate prepared from the all-female strain into a normal strain. This clearly indicates that this male killing is possibly caused by viruses, but not by bacteria or other non-filtrated agents.

Are there any advantages to virus-induced male killing? The roles of viral sexual manipulation in insects have not yet been explored. Similar to bacterial male killing, virus-induced male killing may have some significant effects on the evolution of host insects. A field survey revealed that OGVs are widely distributed and often detected in tea fields in Japan (Fujita et al. 2021). It is speculated that OGV infection may reduce inbreeding and increase the genetic diversity of host populations (Fujita et al. 2021). On the other hand, male-specific embryonic death in *D. biauraria* may trigger female progenies to cannibalize their brothers as nutrients (Kageyama et al. 2017b). Further studies will identify the genes responsible for virus-induced male killing and explore the mechanisms and benefits of this sexual manipulation in insects.

11.8 Conclusion and Perspective

In this chapter, we first described the diversity of insect sex determination systems, which are tightly coupled with dosage compensation systems. Since a failure of dosage compensation during development is likely lethal in insects, the factors involved in these pathways are considered ideal targets for symbiont-induced sexual manipulation. Although few host factors have been identified as the targets for sexually manipulating symbionts, *Spiroplasma* Spaid was shown to induce male killing in *D. melanogaster* by targeting DCC (Harumoto and Lemaitre 2018). We also discovered that the target for *Wolbachia*-induced male killing in *O. furnacalis* is Masc (Fukui et al. 2015), which is essential for both masculinization and dosage compensation in lepidopteran insects (Kiuchi et al. 2014). These results support the hypothesis that sexual manipulation of symbionts either directly or indirectly targets the machineries required for sex determination and/or dosage compensation. Intriguingly, recent studies have revealed that non-bacterial male killers including invertebrate partitiviruses have been found in both lepidopteran and dipteran insects (Kageyama et al. 2017b; Fujita et al. 2021). It is of great interest whether the male-killing viruses also target similar host proteins in different insect orders and what factors they utilize to target host proteins in their male-killing strategies.

What is the origin of symbiont effectors for sexual and reproductive manipulation? All of the CI and male-killing inducer genes identified to date reside within the prophage or plasmid sequences; *Wolbachia* CI-inducing genes (*cifA-cifB*) and a male-killing candidate gene (*wmk*) are located within the eukaryotic association module of prophage WO, whereas *S. poulsonii Spaid* is located on a plasmid (Harumoto and Lemaitre 2018). These facts suggest that symbionts' sexual and reproductive manipulation strategies have been initiated and established by taking advantage of captured or invaded foreign sequences. Further studies will identify more effectors from diverse symbionts, which will provide insights into how the symbionts' sexual manipulation has evolved. Moreover, at present, we have little knowledge about the modes of action of symbiont effectors in sexual manipulation: What host factors do they interact with? Where do they exist and operate within the host? Recent studies on *Wolbachia* revealed that symbiont-induced phenotypes depend on the combination between the host species and symbiont effector homologs (Beckmann et al. 2017; Ote and Yamamoto 2018). Exploring the interacting modes between host and symbiont proteins, including three-dimensional structural insights on their complexes, will enable us to predict symbiont-induced host phenotypes without in vivo infection experiments.

A sometimes-asked question provided after a talk about male killing by *Wolbachia* is "What is the advantage in host insect by *Wolbachia* infection?" Possible answers are that symbiont infection may decrease the chance of inbreeding and increase host genetic diversity, and that in some cases, female progenies may eat their brothers (dead embryos) as nutrients. There may be additional advantages in host physiology and ecology, such as increases in body size and improvement of quality and/or quantity of sex pheromones in females. As a result, host benefits may also have positive effects on symbionts. Further studies are needed to understand how symbiont infection affects host physiological and ecological conditions. On the other hand, however, it is known that in two insect species, male killing can be rapidly suppressed by evolving host suppressors against symbionts (Charlat et al. 2007; Hayashi et al. 2018). In these cases, infected populations showed an extremely female-biased sex ratio, indicating that the host insect was going extinct. Therefore, if the strength of male killing in the host is very high and males are almost extinct, neither hosts nor symbionts have any advantages and the hosts should modify or develop their own factors to suppress the effects of male-killing factors.

A big obstacle to insect symbiont research remains. At present, genome editing and mutant isolation processes for most symbionts, including *Wolbachia* and *Spiroplasma*, are not available. There is only one system to assess symbiont's gene functions in which transgenic insects (e.g., *D. melanogaster*) expressing symbiont-derived genes are generated and examined as to whether they exhibit same phenotypes as symbiont-infected insects. This system is a powerful tool if the symbiont uses a factor that is composed of a single protein and is sufficient for inducing the phenotypes. On the other hand, if this factor is a complex of several proteins, and unlike *Wolbachia* CI factors (CifA–CifB, CidA–CidB) they are not derived from neighbor genes, it is very difficult to estimate whether the focused genes are the real effectors. Very recently, forward genetics of *Wolbachia* using ethyl methanesulfonate mutagenesis and NGS sequencing have identified the genome locus that is involved in *Wolbachia* proliferation and host fitness (Duarte et al. 2021). An alternative approach to investigate the gene functions of symbionts is to generate in vitro materials, i.e., cultured cells, from infected insects. If we establish these, we can also generate cells without symbionts by treatment with antibiotics or other drugs. Genome editing techniques combined with these in vivo and in vitro methods will allow us a chance to establish a new system for assessing symbiont gene functions more precisely in the near future.

Acknowledgments We thank Takahiro Fukui and Fumiko Ishizuna for TEM sample preparation and data acquisition. This work was supported by Grants-in-Aid for Scientific Research on Innovative Areas "Spectrum of the Sex: a continuity of phenotypes between female and male" (17H06431) to S.K.

References

Beckmann JF, Ronau JA, Hochstrasser M (2017) A *Wolbachia* deubiquitylating enzyme induces cytoplasmic incompatibility. Nat Microbiol 2:17007

Berec L, Maxin D, Bernhauerova V (2016) Male-killing bacteria as agents of insect pest control. J Appl Ecol 53:1270–1279

Beye M, Hasselmann M, Fondrk MK et al (2003) The gene *csd* is the primary signal for sexual development in the honeybee and encodes an SR-type protein. Cell 114:419–429

Bordenstein SR, Bordenstein SR (2016) Eukaryotic association module in phage WO genomes from *Wolbachia*. Nat Commun 7:13155

Charlat S, Hornett EA, Fullard JH et al (2007) Extraordinary flux in sex ratio. Science 317:214

Cheng B, Kuppanda N, Aldrich JC et al (2016) Male-killing *Spiroplasma* alters behavior of the dosage compensation complex during *Drosophila melanogaster* embryogenesis. Curr Biol 26: 1339–1345

Clough E, Jimenez E, Kim YA et al (2014) Sex- and tissue-specific functions of *Drosophila* doublesex transcription factor target genes. Dev Cell 31:761–773

Criscione F, Qi Y, Tu Z (2016) GUY1 confers complete female lethality and is a strong candidate for a male-determining factor in *Anopheles stephensi*. Elife 5:e19281

Dannowski J, Flor M, Telschow A et al (2009) The effect of sibmating on the infection dynamics of male-killing bacteria. Evolution 63:2525–2534

Duarte EH, Carvalho A, López-Madrigal S et al (2021) Forward genetics in *Wolbachia*: regulation of *Wolbachia* proliferation by the amplification and deletion of an addictive genomic island. PLoS Genet 17:e1009612

Ellison C, Bachtrog D (2019) Contingency in the convergent evolution of a regulatory network: dosage compensation in *Drosophila*. PLoS Biol 17:e3000094

Elnagdy S, Majerus MEN, Handley LJL (2011) The value of an egg: resource reallocation in ladybirds (Coleoptera: Coccinellidae) infected with male-killing bacteria. J Evol Biol 24:2164–2172

Fujita R, Inoue MN, Takamatsu T et al (2021) Late male-killing viruses in *Homona magnanima* identified as Osugoroshi viruses, novel members of *Partitiviridae*. Front Microbiol 11:620623

Fukui T, Kawamoto M, Shoji K et al (2015) The endosymbiotic bacterium *Wolbachia* selectively kills male hosts by targeting the masculinizing gene. PLoS Pathog 11:e1005048

Fukui T, Kiuchi T, Shoji K et al (2018) In vivo masculinizing function of the *Ostrinia furnacalis Masculinizer* gene. Biochem Biophys Res Commun 503:1768–1772

Hall AB, Basu S, Jiang X et al (2015) A male-determining factor in the mosquito *Aedes aegypti*. Science 348:1268–1270
Harumoto T, Anbutsu H, Fukatsu T (2014) Male-killing *Spiroplasma* induces sex-specific cell death via host apoptotic pathway. PLoS Pathog 10:e1003956
Harumoto T, Anbutsu H, Lemaitre B et al (2016) Male-killing symbiont damages host's dosage-compensated sex chromosome to induce embryonic apoptosis. Nat Commun 7:12781
Harumoto T, Lemaitre B (2018) Male-killing toxin in a bacterial symbiont of *Drosophila*. Nature 557:252–255
Harvey-Samuel T, Norman VC, Carter R et al (2020) Identification and characterization of a *Masculinizer* homologue in the diamondback moth, *Plutella xylostella*. Insect Mol Biol 29: 231–240
Hasimoto H (1933) The role of the W-chromosome in the sex determination of *Bombyx mori*. Jpn J Genet 8:245–247
Hasselmann M, Beye M (2004) Signatures of selection among sex-determining alleles of the honey bee. Proc Natl Acad Sci U S A 101:4888–4893
Hasselmann M, Gempe T, Schiøtt M et al (2008) Evidence for the evolutionary nascence of a novel sex determination pathway in honeybees. Nature 454:519–522
Hayashi M, Nomura M, Kageyama D (2018) Rapid comeback of males: evolution of male-killer suppression in a green lacewing population. Proc Biol Sci 285:20180369
Hilgenboecker K, Hammerstein P, Schlattmann P et al (2008) How many species are infected with *Wolbachia*?--A statistical analysis of current data. FEMS Microbiol Lett 281:215–220
Hiroki M, Kato Y, Kamito T et al (2002) Feminization of genetic males by a symbiotic bacterium in a butterfly, *Eurema hecabe* (Lepidoptera: Pieridae). Naturwissenschaften 89:167–170
Hiroki M, Tagami Y, Miura K et al (2004) Multiple infection with *Wolbachia* inducing different reproductive manipulations in the butterfly *Eurema hecabe*. Proc Biol Sci 271:1751–1755
Hirota K, Matsuda-Imai N, Kiuchi T et al (2021) Characterization of nuclear localization signal in *Ostrinia furnacalis* Masculinizer protein. Arch Insect Biochem Physiol 106:e21768
Hopkins BR, Kopp A (2021) Evolution of sexual development and sexual dimorphism in insects. Curr Opin Genet Dev 69:129–139
Hornett EA, Charlat S, Duplouy AM et al (2006) Evolution of male-killer suppression in a natural population. PLoS Biol 4:e283
Hoshino M, Nakanishi K, Nakai M et al (2008) Gross morphology and histopathology of male-killing strain larvae in the oriental tea tortrix *Homona magnanima* (Lepidoptera: Tortricidae). Appl Entomol Zool 43:119–125
Hurst GDD, Graf von der Schulenburg JH, Majerus TM et al (1999) Invasion of one insect species, *Adalia bipunctata*, by two different male-killing bacteria. Insect Mol Biol 8:133–139
Hurst GDD, Frost CL (2015) Reproductive parasitism: maternally inherited symbionts in a biparental world. Cold Spring Harb Perspect Biol 7:a017699
Jiggins FM, Hurst GD, Jiggins CD et al (2000) The butterfly *Danaus chrysippus* is infected by a male-killing *Spiroplasma* bacterium. Parasitology 120:439–446
Kageyama D, Ohno M, Sasaki T et al (2017a) Feminizing *Wolbachia* endosymbiont disrupts maternal sex chromosome inheritance in a butterfly species. Evol Lett 1:232–244
Kageyama D, Yoshimura K, Sugimoto TN et al (2017b) Maternally transmitted non-bacterial male killer in *Drosophila biauraria*. Biol Lett 13:20170476
Katsuma S, Shoji K, Sugano Y et al (2019) Masc-induced dosage compensation in silkworm cultured cells. FEBS Open Bio 9:1573–1579
Kawaoka S, Minami K, Katsuma S et al (2008) Developmentally synchronized expression of two *Bombyx mori* Piwi subfamily genes, *SIWI* and *BmAGO3* in germ-line cells. Biochem Biophys Res Commun 367:755–760
Kern P, Cook JM, Kageyama D et al (2015) Double trouble: combined action of meiotic drive and *Wolbachia* feminization in *Eurema* butterflies. Biol Lett 11:20150095
Kiuchi T, Koga H, Kawamoto M et al (2014) A single female-specific piRNA is the primary determiner of sex in the silkworm. Nature 509:633–636

Kiuchi T, Sugano Y, Shimada T et al (2019) Two CCCH-type zinc finger domains in the Masc protein are dispensable for masculinization and dosage compensation in *Bombyx mori*. Insect Biochem Mol Biol 104:30–38

Koop JL, Zeh DW, Bonilla MM et al (2009) Reproductive compensation favours male-killing *Wolbachia* in a live bearing host. Proc Biol Sci 276:4021–4028

Krzywinska E, Dennison NJ, Lycett GJ et al (2016) A maleness gene in the malaria mosquito *Anopheles gambiae*. Science 353:67–69

Lechner S, Ferretti L, Schöning C et al (2013) Nucleotide variability at its limit? Insights into the number and evolutionary dynamics of the sex-determining specificities of the honey bee *Apis mellifera*. Mol Biol Evol 31:272–287

Lee J, Kiuchi T, Kawamoto M et al (2015) Identification and functional analysis of a *Masculinizer* orthologue in *Trilocha varians* (Lepidoptera: Bombycidae). Insect Mol Biol 24:561–569

LePage DP, Metcalf JA, Bordenstein SR et al (2017) Prophage WO genes recapitulate and enhance *Wolbachia*-induced cytoplasmic incompatibility. Nature 543:243–247

Liu P, Jin B, Li X et al (2020) *Nix* is a male-determining factor in the Asian tiger mosquito *Aedes albopictus*. Insect Biochem Mol Biol 118:103311

Lucchesi JC (1978) Gene dosage compensation and the evolution of sex chromosomes. Science 202:711–716

Majerus TM, Graf von der Schulenburg JH, Majerus ME et al (1999) Molecular identification of a male-killing agent in the ladybird *Harmonia axyridis* (Pallas) (Coleoptera: Coccinellidae). Insect Mol Biol 8:551–555

Meccariello A, Salvemini M, Primo P et al (2019) Maleness-on-the-Y (MoY) orchestrates male sex determination in major agricultural fruit fly pests. Science 365:1457–1460

Miyahara Y (1984) Abnormal sex ratio in *Ostrinia furnacalis* Guenée (Lepidoptera: Pyralidae). Jap J Appl Ent Zool 28:131–136

Morimoto S, Nakai M, Ono A et al (2001) Late male-killing phenomenon found in a Japanese population of the oriental tea tortrix, *Homona magnanima* (Lepidoptera: Tortricidae). Heredity 87:435–440

Nakanishi K, Hoshino M, Nakai M et al (2008) Novel RNA sequences associated with late male killing in *Homona magnanima*. Proc Biol Sci 275:1249–1254

Narita S, Kageyama D, Nomura M et al (2007) Unexpected mechanism of symbiont-induced reversal of insect sex: feminizing *Wolbachia* continuously acts on the butterfly *Eurema hecabe* during larval development. Appl Environ Microbiol 73:4332–4341

Oishi K (1971) Spirochaete-mediated abnormal sex-ratio (SR) condition in *Drosophila*: a second virus associated with spirochaetes and its use in the study of the SR condition. Genet Res 18:45–56

Ote M, Ueyama M, Yamamoto D (2016) *Wolbachia* protein TomO targets *nanos* mRNA and restores germ stem cells in *Drosophila Sex-lethal* mutants. Curr Biol 26:2223–2232

Ote M, Yamamoto D (2018) Enhancing Nanos expression via the bacterial TomO protein is a conserved strategy used by the symbiont *Wolbachia* to fuel germ stem cell maintenance in infected *Drosophila* females. Arch Insect Biochem Physiol 98:e21471

Perlmutter JI, Bordenstein SR, Unckless RL et al (2019) The phage gene *wmk* is a candidate for male killing by a bacterial endosymbiont. PLoS Pathog 15:e1007936

Penalva LO, Sanchez L (2003) RNA binding protein Sex-lethal (Sxl) and control of *Drosophila* sex determination and dosage compensation. Microbiol Mol Biol Rev 67:343–359

Qi Y, Wu Y, Saunders R et al (2019) *Guy1*, a Y-linked embryonic signal, regulates dosage compensation in *Anopheles stephensi* by increasing X gene expression. eLife 8:e43570

Regassa LB, Gasparich GE (2006) Spiroplasmas: evolutionary relationships and biodiversity. Front Biosci 11:2983–3002

Reynolds KT, Hoffmann AA (2002) Male age, host effects and the weak expression or non-expression of cytoplasmic incompatibility in *Drosophila* strains infected by maternally transmitted *Wolbachia*. Genet Res 80:79–87

Sakai H, Sumitani M, Chikami Y et al (2016) Transgenic expression of the piRNA-resistant *Masculinizer* gene induces female-specific lethality and partial female-to-male sex reversal in the silkworm, *Bombyx mori*. PLoS Genet 12:e1006203

Sasaki T, Kubo T, Ishikawa H (2002) Interspecific transfer of *Wolbachia* between two lepidopteran insects expressing cytoplasmic incompatibility: a *Wolbachia* variant naturally infecting *Cadra cautella* causes male killing in *Ephestia kuehniella*. Genetics 162:1313–1319

Sharma A, Heinze SD, Wu Y et al (2017) Male sex in houseflies is determined by *Mdmd*, a paralog of the generic splice factor gene *CWC22*. Science 356:642–645

Starr DJ, Cline TW (2002) A host parasite interaction rescues *Drosophila* oogenesis defects. Nature 418:76–79

Stouthamer R, Luck RF, Hamilton WD (1990) Antibiotics cause parthenogenetic *Trichogramma* (Hymenoptera/Trichogrammatidae) to revert to sex. Proc Natl Acad Sci U S A 87:2424–2427

Stouthamer R, Breeuwert JA, Luck RF et al (1993) Molecular identification of microorganisms associated with parthenogenesis. Nature 361:66–68

Sugano Y, Kokusho R, Ueda M et al (2016) Identification of a bipartite nuclear localization signal in the silkworm Masc protein. FEBS Lett 590:2256–2261

Sugimoto TN, Ishikawa Y (2012) A male-killing *Wolbachia* carries a feminizing factor and is associated with degradation of the sex-determining system of its host. Biol Lett 8:412–415

Suzuki MG, Funaguma S, Kanda T et al (2003) Analysis of the biological functions of a *doublesex* homologue in *Bombyx mori*. Dev Genes Evol 213:345–354

Suzuki MG, Funaguma S, Kanda T et al (2005) Role of the male BmDSX protein in the sexual differentiation of *Bombyx mori*. Evol Dev 7:58–68

Tabata J, Hattori Y, Sakamoto H et al (2011) Male killing and incomplete inheritance of a novel *Spiroplasma* in the moth *Ostrinia zaguliaevi*. Microb Ecol 61:254–263

Tanaka Y (1916) Genetic studies in the silkworm. J Coll Agric Sapporo 6:1–33

Tinsley MC, Majerus MEN (2006) A new male-killing parasitism: *Spiroplasma* bacteria infect the ladybird beetle *Anisosticta novemdecimpunctata* (Coleoptera: Coccinellidae). Parasitology 132: 757–765

Veneti Z, Bentley JK, Koana T et al (2005) A functional dosage compensation complex required for male killing in *Drosophila*. Science 307:1461–1463

Visser S, Voleníková A, Nguyen P et al (2021) A conserved role of the duplicated *Masculinizer* gene in sex determination of the Mediterranean flour moth, *Ephestia kuehniella*. PLoS Genet 17: e1009420

Wang YH, Chen XE, Yang Y et al (2019) The *Masc* gene product controls masculinization in the black cutworm, *Agrotis ipsilon*. Insect Sci 26:1037–1044

Watanabe K, Yukuhiro F, Matsuura Y et al (2014) Intrasperm vertical symbiont transmission. Proc Natl Acad Sci U S A 111:7433–7437

Watanabe M, Kageyama D, Miura K (2013) Transfer of a parthenogenesis-inducing *Wolbachia* endosymbiont derived from *Trichogramma dendrolimi* into *Trichogramma evanescens*. J Invertebr Pathol 112:83–87

Werren JH, Baldo L, Clark ME (2008) *Wolbachia*: master manipulators of invertebrate biology. Nat Rev Microbiol 6:741–751

Whiting AR (1967) The biology of the parasitic Wasp *Mormoniella vitripennis* [=*Nasonia brevicornis*] (Walker). Q Rev Biol 42:333–406

Williamson DL, Poulson DF (1979) In: Whitcomb RF, Tully JG (eds) In *the mycoplasmas, volume III: plant and insect mycoplasmas*, pp 175–208

Xu J, Chen S, Zeng B et al (2017a) *Bombyx mori* P-element somatic inhibitor (BmPSI) is a key auxiliary factor for silkworm male sex determination. PLoS Genet 13:e1006576

Xu J, Zhan S, Chen S et al (2017b) Sexually dimorphic traits in the silkworm, *Bombyx mori*, are regulated by *doublesex*. Insect Biochem Mol Biol 80:42–51

Zou Y, Geuverink E, Beukeboom LW et al (2020) A chimeric gene paternally instructs female sex determination in the haplodiploid wasp *Nasonia*. Science 370:1115–1118

Chapter 12
The Mechanism for Establishing the Binary Sex with Environmental Signals in the Crustacean *Daphnia magna*

Yasuhiko Kato and Hajime Watanabe

Abstract The cladoceran crustacean *Daphnia magna* produces females parthenogenetically under healthy conditions. In response to environmental stresses such as starvation and higher population density, it produces males that are genetically identical to their sisters. Environment-dependent male production in *D. magna* can be interpreted as the shift of the position of the sex on the spectrum from the female side to the male side by the environmental cues. In this review, we introduce how the environment-dependent male determination is mediated via the endocrine system and the evolutionary conserved sex-determining gene *Doublesex1* (*Dsx1*) in *D. magna*. We then describe how the binary sex can be achieved by regulation of *Dsx1* at transcriptional, epigenetic, and post-transcriptional levels to prevent intersex phenotype.

Keywords Doublesex1 · Environmental sex determination · *Daphnia magna* · Transcriptional regulation · Post-transcriptional regulation · Epigenetic regulation

12.1 Introduction

The healthy population of the cladoceran crustacean *Daphnia magna* is composed of females by parthenogenesis (Fig. 12.1). The female juvenile matures into an adult after several molting. It develops two ovaries aside from the gut and produces eggs without mating (Fig. 12.1, solid arrow). In contrast, under the stressed environment, such as shortened photoperiod, a lack of food, and/or their increased population density, the adult senses those environmental changes and parthenogenetically produces male offspring that are genetically identical with their sisters. Interestingly, the ovulated eggs are already destined to be males, meaning that the commitment of the sex of the offspring to males occurs in the ovaries (Fig. 12.1, dotted arrow). Under the stressed condition, females also switch their reproductive mode from

Y. Kato (✉) · H. Watanabe
Department of Biotechnology, Graduate School of Engineering, Osaka University, Suita, Osaka, Japan
e-mail: kato_yasuhiko@bio.eng.osaka-u.ac.jp; watanabe@bio.eng.osaka-u.ac.jp

M. Tanaka, M. Tachibana (eds.), *Spectrum of Sex*,
https://doi.org/10.1007/978-981-19-5359-0_12

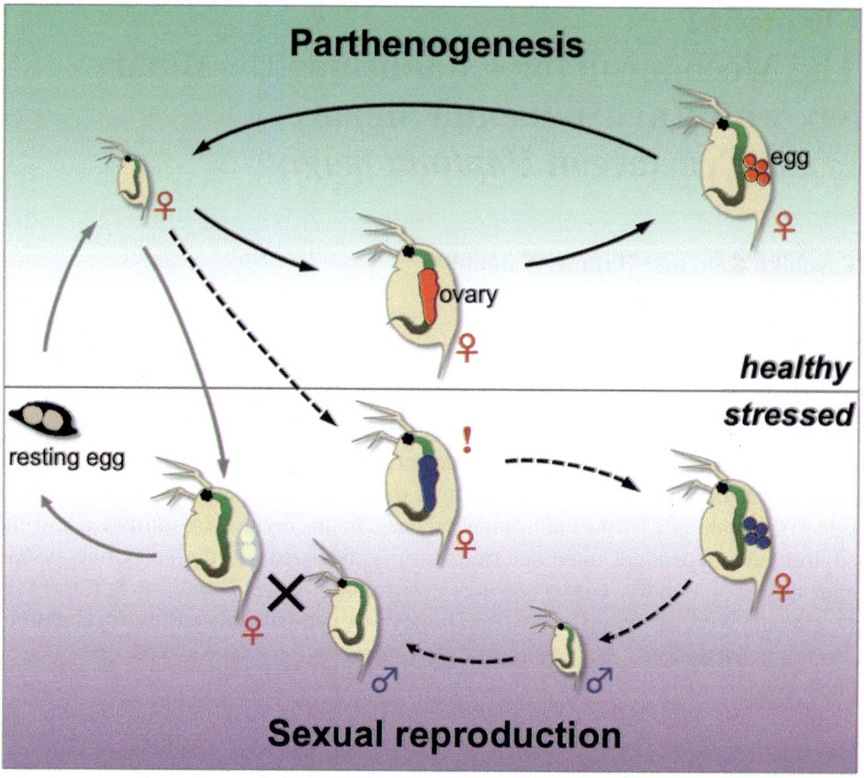

Fig. 12.1 Life cycle of *Daphnia*. This planktonic crustacean engages in cyclical parthenogenesis where it can switch the reproductive mode from parthenogenesis to sexual reproduction under stressed conditions. The ovaries harboring the oocytes that are destined to be developed as females or males are indicated with red or blue color, respectively. With the same color arrangement, the sex of the eggs is shown. In *Daphnia*, the same genotype can produce three different reproductive forms, parthenogenetic females, sexual males, and sexual females. The black arrows are used to show the cycle of female production by parthenogenesis. The dotted arrows indicate the process of male production by parthenogenesis under the stressed condition. The sexual males are necessary for the production of the resting eggs by sexual reproduction. The gray arrows show plastic change of female reproductive mode from parthenogenesis to sexual reproduction, which allows the females to mate with the males for making the resting eggs that can hatch when the surrounding environment is recovered. The exclamation mark refers to the sensing of the stressed environment and secretion of sesquiterpenoid necessary to destine the sex of the offspring to the male

parthenogenesis to sexual reproduction and mate with the males, leading to the production of resting eggs that can withstand desiccation and other adverse conditions. From the resting eggs, only females hatch and can increase their population by parthenogenesis in the subsequent growing seasons (Fig. 12.1, gray arrow) (Hebert 1978; Zaffagnini 1987). Intersex is extremely rare in *D. magna* (Mitchell 2001; Olmstead and LeBlanc 2007). From the concept of the spectrum of sex, environment-dependent male production can be interpreted as the shift of the position of the sex on the spectrum from the female side to the male side by the environmental cues.

Although the females and males are genetically identical, they show sex differences in morphology and behavior as they grow. At the first instar of the juvenile, males have more elongated first antennae, which allows us to distinguish their sex easily by microscopic observation. At this time, gonads develop on both sides of the intestinal tract in both males and females, and morphological sex differences can be already seen. As males grow, their swimming speed increases and they catch females for mating by hooks formed on the first thoracic leg, the most cephalic of the five pairs of thoracic legs. Recent progress of *D. magna* genomics (Orsini et al. 2017; Lee et al. 2019) and gene manipulation methods (Kato et al. 2011b; Törner et al. 2014; Nakanishi et al. 2014, 2016; Naitou et al. 2015) have enabled us to investigate the molecular mechanism underlying the sex differences in this organism. Here, we introduce how the environment-dependent male determination is mediated via the endocrine system and the evolutionary conserved sex-determining gene *Doublesex* (*Dsx*) in *D. magna*. We then introduce how the binary sex can be maintained to prevent intersex phenotype.

12.2 Male-Determining Hormone

In insects and crustaceans, ecdysteroids and sesquiterpenoids function as messengers that transmit information of the surrounding environment to cells in the body. For instance, in insects, both hormones control body size in response to nutrition by their coordinated production via the insulin/insulin-like growth factor signaling (IIS)/ target of rapamycin (TOR) signaling pathway (Koyama et al. 2013). Of those hormones, sesquiterpenoids are known to control species-specific responses to environments called phenotypic plasticity (Miura 2019). In termites, juvenile hormone III (JHIII) is the major sesquiterpenoid and regulates caste differentiation (Cornette et al. 2008). In aphids, JHIII is known to control not only wing-morph differentiation (Ishikawa et al. 2013) but also the reproductive mode (Corbitt and Hardie 1985). In crustacean *Daphnia*, methyl farnesoate (MF), a precursor of JHIII, is the candidate of physiologically active sesquiterpenoid (Toyota et al. 2015) and MF exposure leads to the formation of morphology for defense against its predator (Oda et al. 2011).

MF is also responsible for the male determination of *Daphnia*. Exposure of MF to *D. magna* can induce the production of male offspring (Olmstead and Leblanc 2002; Tatarazako et al. 2003). The juvenile hormone agonists Pyriproxyfen and Fenoxycarb stimulate male production more efficiently than MF in *D. magna* (Tatarazako et al. 2003) and this hormone-dependent sex determination is common in the cladoceran crustaceans (Oda et al. 2005). In *D. magna*, sesquiterpenoid-dependent control of sex is implemented in developing oocytes (Olmstead and Leblanc 2002; Kato et al. 2010). This sesquiterpenoid-sensitive period is consistent with the timing of the male determination trigged by the crowding environment in the other cladoceran *Moina* sex determination (Banta and Brown 1929). Taken together, environmental signals would be processed in the mother's brain and be transmitted to the endocrine system that secrets the sesquiterpenoid, which in turn determines the sexual fate of the offspring before embryogenesis begins (Fig. 12.1, exclamation mark). The chemical structure of the sesquiterpenoid in *Daphnia* needs to be elucidated in the future.

The chemically induced males can be used for experiments to study the molecular mechanism underlying the sex difference because they have male-specific traits and fertility to produce the sexual eggs (Kato et al. 2011a; Nong et al. 2017). In the laboratory culture condition, the offspring of *D. magna* is 100% females while exposure of the Fenoxycarb or Pyriproxyfen leads to 100% male offspring. This chemical manipulation of sex is much more efficient and convenient than that by setting stressful conditions such as crowding, which has contributed to recent advances in the molecular biology of *D. magna* sex determination.

12.3 *Dsx1* Shifts the Position of the Sex Spectrum to the Male Side

12.3.1 Dsx1, *the Master Regulator of Male Development*

Dsx codes for the transcriptional factor harboring the DNA binding domain named the DM domain and was originally found in *Drosophila melanogaster* (Burtis and Baker 1989). Dsx regulates the expression of genes linked to the development of sexually dimorphic traits. The roles of the DM-domain transcription factors in sexual development are conserved not only in the other insects and *C. elegans* (Shen and Hodgkin 1988) but also in vertebrates such as fishes (Matsuda et al. 2002), amphibians (Yoshimoto et al. 2008), and mammals (Raymond et al. 2000).

On the *D. magna* genome, two *Dsx* orthologs, *Dsx1* and *Dsx2*, are tandemly located (Kato et al. 2011a). Dsx1 has a sequence homologous to the oligomerization domain that is conserved among insect Dsx proteins and is known to enhance specific DNA binding through its dimerization (Bayrer et al. 2005). In Dsx2, the oligomerization domain is more diverged and has substitutions of the amino acids on the extensive non-polar interface. Both *Dsx1* and *Dsx2* show strong male-biased expression during embryogenesis and the sexual maturation process (Kato et al. 2011a; Nong et al. 2017), suggesting their roles in the male differentiation. RNAi-mediated knockdown of *Dsx1* in male embryos resulted in sex reversal including not only feminization of organs such as the first antennae and the first thoracic legs but also the development of ovaries. Overexpression of *Dsx1* in females led to the development of the male-specific trait, the elongated first antennae. In contrast, in the case of *Dsx2*, neither gene silencing nor overexpression led to clear phenotypic changes of the sexually dimorphic traits. This less contribution of *Dsx2* to sexual development may attribute to the sequence divergence of the oligomerization domain. These expression and functional analyses demonstrate that *Dsx1* is the master regulator of male development (Kato et al. 2011a).

12.3.2 Spatio-Temporal Expression of Dsx1

In insects, *Dsx* is alternatively spliced in a sex-specific manner and the resulting isoforms control sex-specific development (Wexler et al. 2019). In contrast, *D. magna Dsx1* expression is not controlled by sex-specific splicing despite that this organism shares a common ancestor with insects (Schwentner et al. 2017), showing that the *Dsx1* is regulated at a transcriptional level (Kato et al. 2011a). To understand when and where the *Dsx1* expression is active throughout the life span, the transgenic *D. magna* recapitulating the *Dsx1* expression with the mCherry fluorescence was generated by TALEN-mediated knock-in (Fig. 12.2) (Nong et al. 2017). In this *Dsx1* reporter *Daphnia*, one of the two *Dsx1* alleles has the mCherry ORF sequence. This mCherry reporter was inserted at the position of the endogenous *Dsx1* start codon. Because another *Dsx1* allele is intact, this *Dsx1* hemizygous knock-out males have the reproductive ability and typical male-specific traits even though the slight feminization has been observed in the morphology and frequency of fertilization (Nong et al. 2017). This reporter strain was named Line B (Nong et al. 2020).

Using Line B, no clear mCherry signal was detected both in females and in males until the gastrulation stage. Thereafter, dynamic spatio-temporal expression of *Dsx1* was observed until the late embryonic stage (Nong et al. 2017). In male embryos, during stomodeal invagination (11 hpo), mCherry-positive cells are localized into

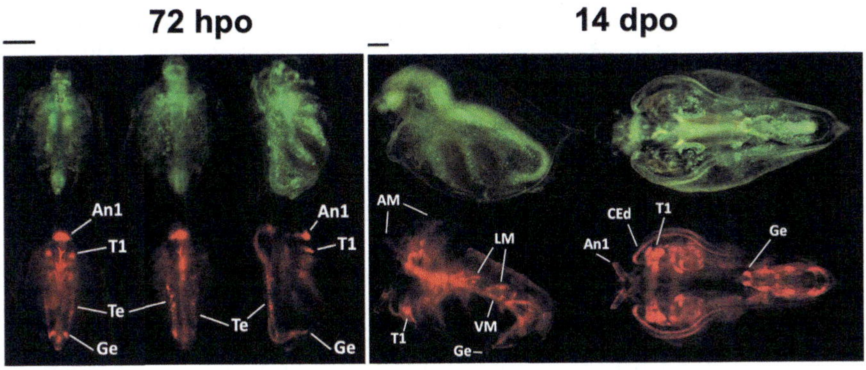

Fig. 12.2 Visualization of spatio-temporal activation of *Dsx1* using the *Dsx1* reporter strain named Line B. 72 hpo and 14 dpo indicate 72 h post-oviposition and 14 days post-oviposition, respectively. Upper panels show pictures of H2B-GFP expression with which cells and tissues are visualized. Lower panels show mCherry fluorescence that recapitulates Dsx1 expression. *A1* first antenna, *T1* first thoracic leg, *Ge* genital (i.e., penis in this case), *Te* testis, *AM* antennal muscle, *LM* lateral muscle, *VM* ventral muscle, *CEd* carapace edge. Scale bars = 200 μm. (The photos were reproduced from Nong et al. (2017))

the cell cumulus and around the blastopore. From 11 hpo to 16 hpo, these cell clusters move on the ventral region and reach the posterior region. The behavior of the mCherry-positive cells resembles that of the primary organizer in spider (Akiyama-Oda and Oda 2003) and chicken (Bénazéraf and Pourquié 2013), suggesting that the primary organizer is sexually plastic and controls sexual development of *D. manga*. When male embryos reach the late embryonic stage, more intense signals were detected. In more detail, at the naupliar segmentation stage (16–18 hpo), in addition to mCherry signals in the posterior region, the intense signals appear in the buds of the first antennae and along the border of the head and body segments. During the thoracic segmentation (18 hpo and later), strong mCherry fluorescence is observed in the first thoracic segment where a hook for capturing females is developed. In juveniles, mCherry signals are detected in organs showing male-specific characteristics such as first antennae, first thoracic legs, testes, and spermiduct-genitalia system (Fig. 12.2, 72 hpo). From the instar 5, consistent with the timing of transformation into mature males, the regions expressing mCherry are expanded into the other male-specific structures including the carapace edge, the tip of the penis, and the skeletal muscle (Fig. 12.2, 14 dpo). These demonstrate the spatio-temporal regulation of *dsx1* transcription for male development.

12.3.3 Dsx1 *Activity-Dependent Formation of the Sex Spectrum*

The Line B harbors a hemizygous *Dsx1* allele and exhibits slight feminization, which suggests that the *Dsx1* mutant can be a model for analyzing the molecular mechanism underlying the sex spectrum. Another *Dsx1* mutant showing the more feminized phenotypes was generated when knock-in of the *mCherry* reporter at the *Dsx1* locus was performed with the TALEN system (Nong et al. 2017, 2020). In this mutant, one *Dsx1* allele was disrupted by integration of the *mCherry,* and the other allele has a 6-bp deletion on the start codon of *Dsx1* ORF, possibly leading to the synthesis of the truncated Dsx1 protein that lacks the N-terminal region but still has intact DM domain and oligomerization domain. This compound heterozygous mutant was designated as Line A (Nong et al. 2020).

In males of Line A, sex-specific traits were more shifted to the female side throughout the life span compared to Line B (Fig. 12.3) (Nong et al. 2020). At the juvenile stages, the first antennae of the Line A male were less elongated than that of the Line B male and at the adult stages, their feminization of the morphology was more profound; larger body size, female-like shapes of the carapace, head region, and genital and anal region (Fig. 12.3). Line A male also developed an ovary-like structure although the eggs were not produced from this feminized gonad at least for several weeks. These phenotypic observations demonstrate that Line A and Line B males are sex intergrades with different degrees of feminization (Nong et al. 2020).

To reveal the gene network underlying the sex spectrum, wild-type females, Line A males, Line B males, and wild-type males were subjected to RNA-seq analysis

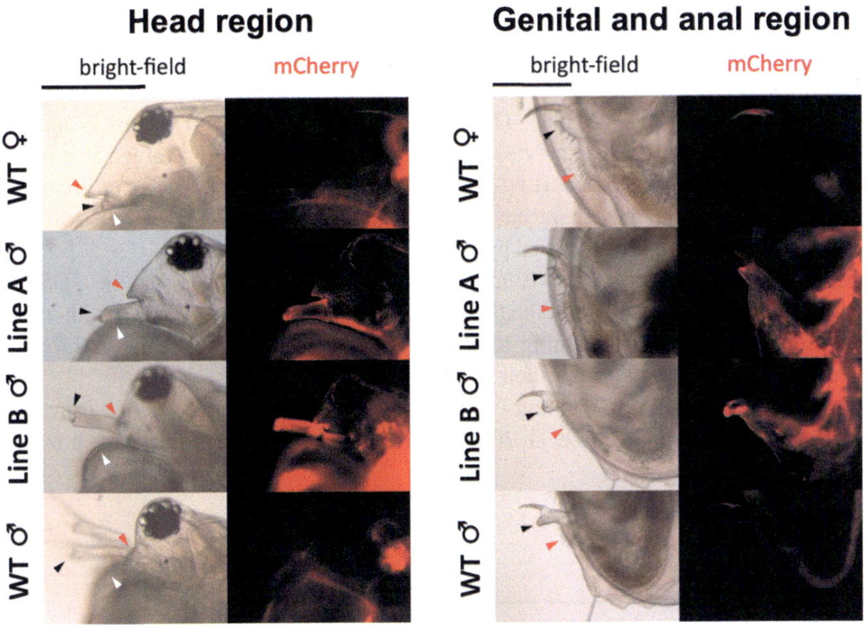

Fig. 12.3 Feminization in the *Dsx1* mutants that have the *mCherry* gene on the *Dsx1* locus. Dsx1 activity is strongest in wild-type males, followed by Line B males, Line A males, and wild-type females. Head region, side view of instar 8 daphniids. Black-arrow heads: first antennae. White arrowheads: carapace curve that reveals copulation hooks in males. Red arrowheads indicate the rostrum. The genital and anal region, side view of instar 8 daphniids. Black arrowheads: genital. Red arrowheads: anus. For each region, the bright-field and mCherry images are shown. Scale bars = 0.5 mm. (The photos were reproduced from Nong et al. (2020))

(Nong et al. 2020). Late embryos at 40 h post-oviposition that exhibit sexually dimorphic *Dsx1* expression were chosen for this transcriptome analysis. There were 1354 differentially expressed (DE) genes between wild-type females and wild-type males. Among those DE genes, 505 and 226 genes showed female- and male-biased expression. Interestingly, 78% and 22% of the female- and male-biased genes exhibited intermediate expression levels in the sex intergrades, Line A and Line B males. The genes linked with transportation and metabolism occupied 22% of the female-biased group. The genes functionally categorized into signaling pathways and embryonic development was more found in the male-biased groups (Nong et al. 2020). These findings indicate that altered Dsx1 expression or activity can change the position of the spectrum of the sex by modification of expression of its target genes.

12.4 Transcriptional Regulation of *Dsx1*

The *Dsx1* gene consists of four exons and occupies about 20 kb of the genome (Fig. 12.4a) (Kato et al. 2011a). The size of the first and second introns is approximately 9 kb and 8 kb, respectively, and is much larger than the average size of the intron (392 bp) in this species (Lee et al. 2019), suggesting the presence of important regulatory elements in these introns. From this locus, the two *Dsx1* mRNA isoforms, *Dsx1-α* and *Dsx1-β*, are produced by the alternative promoter usage. Transcription of *dsx1-α* isoform starts from exon 3 and continues until exon 4 harboring the CDS and 3′ UTR. *Dsx1-β* is encoded by the *β* isoform-specific exons 1 and 2 in addition to the protein-coding exon 4, meaning that both isoforms produce the same amino acid sequence of Dsx1 protein but their regulations are different.

The sesquiterpenoid signaling commits the sex of offspring to males from 4 to 10 h before ovulation (Kato et al. 2010). During this period, *Dsx1* does not show any

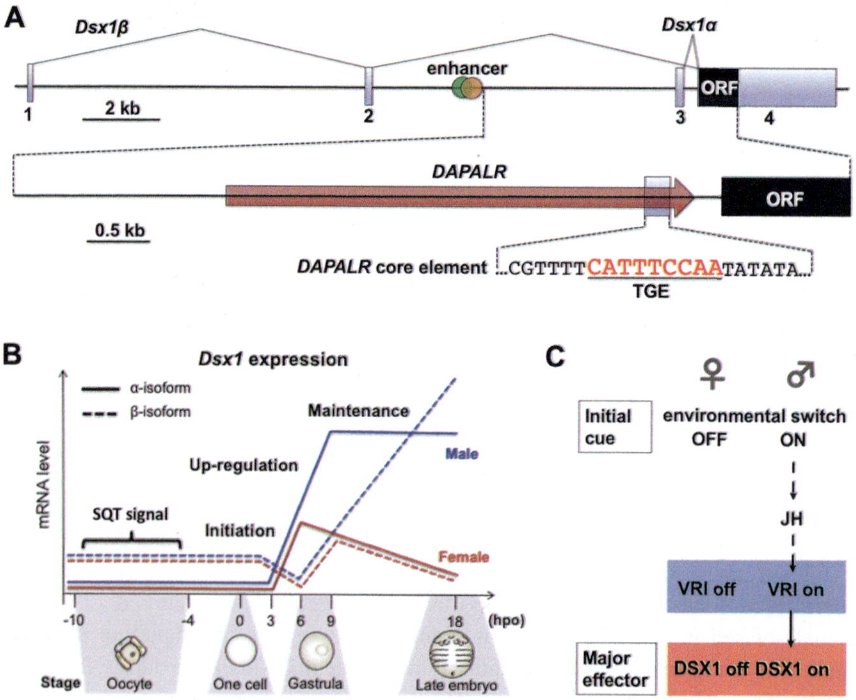

Fig. 12.4 Structure and expression of *Dsx1*. (**a**) The exon-intron structure of Dsx1. Exons of *Dsx1* are indicated with light blue boxes. The red arrow shows the transcribed region of *DAPALR*. *TGE* TRA-2/GLI element. The potential Vri and Dsx1 binding sites are shown in the enhancer with green and orange circles, respectively. (**b**) Temporal expression patterns of Dsx1 isoforms during late oogenesis and early embryogenesis. *hpo* hour post-ovulation, *SQT* sesquiterpenoid. (**c**) A potential sex-determining cascade in *D. magna*. (The figures (**b** and **c**) were reproduced from Mohamad Ishak et al. (2017))

sexually dimorphic expression (Fig. 12.4b). The male-specific expression of *Dsx1* is established via the three processes during embryogenesis: initiation, upregulation, and maintenance (Mohamad Ishak et al. 2017). Non-sex-specific transcription of *Dsx1-α* mRNA occurs before the early gastrulation stage at 6 h post-ovulation (hpo) (initiation). Male-specific activation of *Dsx1-α* expression is implemented during the gastrulation stage from 6 hpo to 9 hpo, and its expression level is maintained in late embryos (18 hpo). *Dsx1-β* isoform is maternally deposited in the ovulated eggs and its male-specific expression begins 3 h later than *Dsx1-α*. These expression patterns of the *Dsx1* isoforms in early embryos explain why male-specific mCherry signals are detected in the *Dsx1* reporter strain Line B after the gastrulation stage. Transcription of *Dsx1-α* isoform is more exclusive in males than that of *Dsx1-β*, suggesting its predominant role in male development (Kato et al. 2011a; Nong et al. 2017).

The male-specific upregulation of the *Dsx1-α* isoform is controlled by the bZIP transcription factor Vrille (Vri) (Mohamad Ishak et al. 2017). The consensus binding element of the mammalian Vrille ortholog, E4BP/NFIL3, is present in intron 2 (Fig. 12.4a) and overlapped with the consensus Dsx binding site. The combinatorial binding of the bZIP protein and Dsx has been reported to control the *Drosophila* yolk protein gene 1, which might suggest conservation of mode of Dsx action between *Daphnia* and *Drosophila*. Importantly, at the early gastrulation stage (6 hpo), just before male-specific *Dsx1* activation occurs, *Vri* transcript is significantly more abundant in male embryos although, in late embryos (18 hpo), it exhibits no sexually dimorphic expression. Knockdown of *Vri* in male embryos reduced *Dsx1* expression whereas its overexpression in female embryos increased the *Dsx1* transcript level. *Dsx1* upregulation was also diminished by the introduction of a mutation into the enhancer harboring the potential Vri and Dsx1 binding sites (Mohamad Ishak et al. 2017). The candidate of the binding site for sesquiterpenoid-dependent transcription factor MET is located in the *Vri* promoter, suggesting the possibility that the sesquiterpenoid-bound MET activates *Vri* expression in early gastrula (6 hpo), and the resulting Vri protein controls upregulation of *Dsx1* in late gastrula (9 hpo) (Fig. 12.4c). Because loss- and gain-of- *Vri* function led to embryonic lethality, it remains unknown whether the Vri is the sole factor for *Dsx1* upregulation. Co-option of Vri as a regulator of *Dsx* expression in the sex-determining pathway has not been reported in other species, supporting the idea for plasticity of the regulatory mechanism of *Dsx* in animals (Herpin and Schartl 2015).

12.5 Epigenetic Regulation of *Dsx1*

Since male and female *Daphnia* are genetically identical, it may be possible that the epigenetic factors are involved in male-specific *dsx1* expression. The result indicating the sex difference of the chromatin structure at the *dsx1* locus has been obtained when the *dsx1* enhancer sequence harboring the potential Vri and Dsx1 binding site in the intron 2 was mutated by CRISPR/Cas in *D. magna* (Fig. 12.4b) (Mohamad Ishak et al. 2017). In this experiment, the two separate gRNAs, gRNA-1 and gRNA-2 that recognize the upstream and downstream sequence of the enhancer respectively, were designed, synthesized, and co-injected with Cas9 protein into the eggs that are destined to develop into females. However, no mutation was introduced both in somatic and in germ cells. In contrast, when the same Cas9 RNP was injected into eggs that develop as males, indel mutations were successfully introduced into the enhancer region, which resulted in the silencing of the *dsx1* gene (Mohamad Ishak et al. 2017). The efficiency of the double-strand break by the CRISPR/Cas in the heterochromatin region has been known to be significantly lower than that in the euchromatin region (Chen et al. 2016; Daer et al. 2016; Jensen et al. 2017). Less DNA accessibility of the CRISPR/Cas was also demonstrated quantitatively by using the genome-wide sequencing approaches (Chung et al. 2020). Thus, the inability of the double-strand break in the enhancer region may suggest heterochromatinization at the *dsx1* locus in the female.

In *D. pulex*, the sex differences of epigenetic modifications including DNA methylation and histone modification have been comprehensively investigated using samples from a mixture of juveniles and adults with different ages (3, 8, and 15 days old) (Kvist et al. 2020). DNA methylation was analyzed by whole-genome bisulfite sequencing. Despite the extremely low methylation level, the differentially methylated CpGs were predominantly found within gene bodies. At the *dsx1* locus, CpG methylation frequency was higher in the female genome in contrast to the overall tendency of higher methylation levels in the male genome (Kvist et al. 2020). For the histone modifications, active and repressive histone marks, histone H3 trimethylation at lysine 4 (H3K4me3) and lysine 27 (H3K27me3), were analyzed by chromatin immunoprecipitation. Among the sex-specific H3K4me3 peaks, 78% and 22% were detected in males and females respectively, suggesting its role for male-specific gene activation. On the other hand, H3K27me3 peaks were predominantly detected in females and the female-specific peaks occupied 84% of the sex-specific peaks. At the *Dsx1* locus on the male genome, the H3K4me3 level was around 300-fold higher but the H3K27me3 level was around 5000-fold lower compared to the female genome (Kvist et al. 2020). These demonstrate that the sex-specific *Dsx1* expression is epigenetically controlled.

12.6 Post-transcriptional Regulation of *Dsx1*

12.6.1 *5′ UTR-Overlapping lncRNA* DAPALR *Activates* Dsx1 *Expression*

In the process of analyzing the role of the *Dsx1* 5′ UTR in the post-transcriptional regulation, a long noncoding RNA that overlaps the 5′ UTR of the *Dsx1-α* isoform in the sense orientation and activates *dsx1* expression has been identified (Kato et al.

2018). In this analysis, the chimeric *DsRed2* reporter mRNA harboring either *Dsx1-α* or *Dsx1-β* 5′ UTR was generated and injected into the eggs that develop into female embryos. Unexpectedly, the reporter RNA with the *Dsx1-α* 5′ UTR led to the development of the male-specific trait, the elongation of the first antennae, although it lacked the *Dsx1* ORF. The RNAs that only coded for the *Dsx1-α* 5′ UTR also induced male differentiation. When those RNAs harboring the *Dsx1-α* 5′ UTR were injected into eggs of the *Dsx1* reporter line (Line B), the mCherry reporter fluorescence was increased, suggesting a potential role of the *Dsx1* 5′ noncoding sequence in the regulation of its own expression (Kato et al. 2018). Interestingly, in the database of the genomic tiling path microarrays from *D. pulex*, there was the long noncoding RNA (lncRNA) that is transcribed from the transcription start site of the *Dsx1-α* and overlaps with the exon 3 encoding its 5′ UTR. The orthologous RNA in *D. magna* was identified, characterized, and named *doublesex1 alpha promoter-associated long RNA* (*DAPALR*). *DAPALR* is 3650 nt of a capped and non-polyadenylated RNA that does not include the *Dsx1* ORF (Fig. 12.4a) (Kato et al. 2018).

Similar to *Dsx1*, *DAPALR* expression increases over 72 h in male embryos until the juvenile stage but its expression level is ten times lower than that of *dsx1* (Kato et al. 2018). The silencing of *Vri*, a gene encoding the transcriptional activator of *dsx1*, resulted in the decrease of the *DAPALR* expression while its overexpression increased the *DAPALR* levels significantly. Disruption of the Vri binding site in the enhancer region reduced *DAPALR* expression, indicating co-regulation of expression of *Dsx1* and *DAPALR* by the Vri protein (Kato et al. 2018). To investigate the role of *DAPALR* in the regulation of male development and *Dsx1* expression, siRNAs specific to *DAPALR* were injected into eggs destined to become males, which resulted in the reduction of *Dsx1* expression and in turn led to the feminization of somatic tissues such as the first antennae and development of ovarian tissues. In addition, more potent and stable stealth siRNAs enabled the injected *Daphnia* to produce eggs that developed normally into females (Kato et al. 2018). Since the RNAi has been known to induce heterochromatin formation at the targeted locus (Martienssen and Moazed 2015), there might be the possibility that DAPALR RNAi induced reduction of *Dsx1* transcript levels indirectly via spreading of heterochromatin region at the *Dsx1* locus. Thus, the effect of overexpression of *DAPALR* on sexual development was also tested. Injection of the *DAPALR* expression plasmid into female embryos of the *Dsx1* reporter strain led to an increase of mCherry fluorescence intensity and masculinization. These demonstrate the role of DAPALR in regulating male-specific *Dsx1* activation (Kato et al. 2018).

12.6.2 DAPALR *Functions as a Decoy of Shep*

The core region of *DAPALR* overlapping with the *Dsx1-α* 5′ UTR (Fig. 12.4a) is sufficient to activate *dsx1* expression (Kato et al. 2018). To identify proteins that bind to the core region, RNAs that code for 205 nt of the overlapping region were synthesized in vitro, conjugated with a Flag-peptide, and used for an immunoprecipitation experiment using a *D. magna* lysate (Perez et al. 2021). Analysis of the co-immunoprecipitated proteins by MS revealed significant binding of the two RNA binding proteins: Alan Shepard (Shep) and CUG binding protein 1 (CUGBP1) to the core element. Shep function was further analyzed because its ortholog Sup-26 in *C. elegans* has been known to regulate the sex-determining gene *Tra-2* at the translation level (Mapes et al. 2010).

Shep mRNA expression did not show any sexual dimorphism in contrast to *Dsx1* and *DAPALR*, suggesting its role in both genders (Perez et al. 2021). To investigate the function of *Shep*, using the *Dsx1* reporter strain Line B, generation of *Shep* mutant has been attempted by introducing indel mutations in the regions that code for the RNA recognition motif (RRM). Although biallelic disruption of *Shep* was likely to be lethal, one monoallelic mutant harboring 15 bp insertion in frame was established. Both males and females of this *Shep* mutant showed significantly higher mCherry fluorescence than the wild-type, suggesting that Shep may suppress *Dsx1* both in males and females. In contrast, *Dsx1* mRNA expression levels were not significantly different between the *Shep* mutant and wild-type in either male or female, suggesting that Shep likely inhibits *Dsx1* translation (Perez et al. 2021).

In *C. elegans*, the Shep ortholog Sup-26 represses translation of *Tra-2* via its binding to the TRA-2/GLI element (TGE) that is located in the *Tra-2* 3′ UTR (Mapes et al. 2010). A highly conserved sequence with TGE was found in the *Dsx1-α* 5′ UTR (Fig. 4a). To investigate this element, either the GFP reporter mRNA with the intact *Dsx1-α* 5′ UTR or the one lacking the potential TGE was injected into female eggs, resulting in higher GFP fluorescence of the embryos injected with the mutated mRNAs than those with the RNAs with the intact TGE. Co-injection of *Shep* mRNA with the intact *Dsx1-α* 5′ UTR:: GFP reporter mRNA led to reduction of the GFP fluorescence. Shep-dependent repression of translation through the potential TGE element was confirmed in vitro by co-incubation of the Luciferase (Luc) reporter mRNA harboring the intact *Dsx1-α* 5′ UTR with the Shep mRNA in the rabbit reticulocyte lysate (Fig. 12.5a). Interaction between the TGE-like element and Shep was further proved to be direct by the pull-down assay. These demonstrate that Shep suppresses the translation of the reporter mRNA through the TGE-like motif (Perez et al. 2021).

DAPALR has the TGE-like motif in the region overlapping with the *Dsx1-α* 5′ UTR (Fig. 12.4a). To examine how *DAPALR* controls *Dsx1* expression, *DAPALR* and *Shep* mRNA were co-incubated with the Luc reporter mRNA harboring the intact *Dsx1-α* 5′ UTR in the reticulocyte lysate. In this reaction, *DAPALR* canceled translational repression by Shep via the TGE element. Importantly, the core element of *DAPALR* harboring the TGE motif had the same ability as the full-length of *DAPALR* (Fig. 12.5a). These rescue efficiencies of *DAPALR* or its core element were dose-dependent (Fig. 12.5b). These results show the function of *DAPALR* as a decoy of Shep (Perez et al. 2021).

For activation and repression of *Dsx1*, the chromatin of this locus is possibly open and closed in males and females, respectively. However, expression noise of *Dsx1* in females might occur due to the nature of stochasticity in gene expression (Kærn et al.

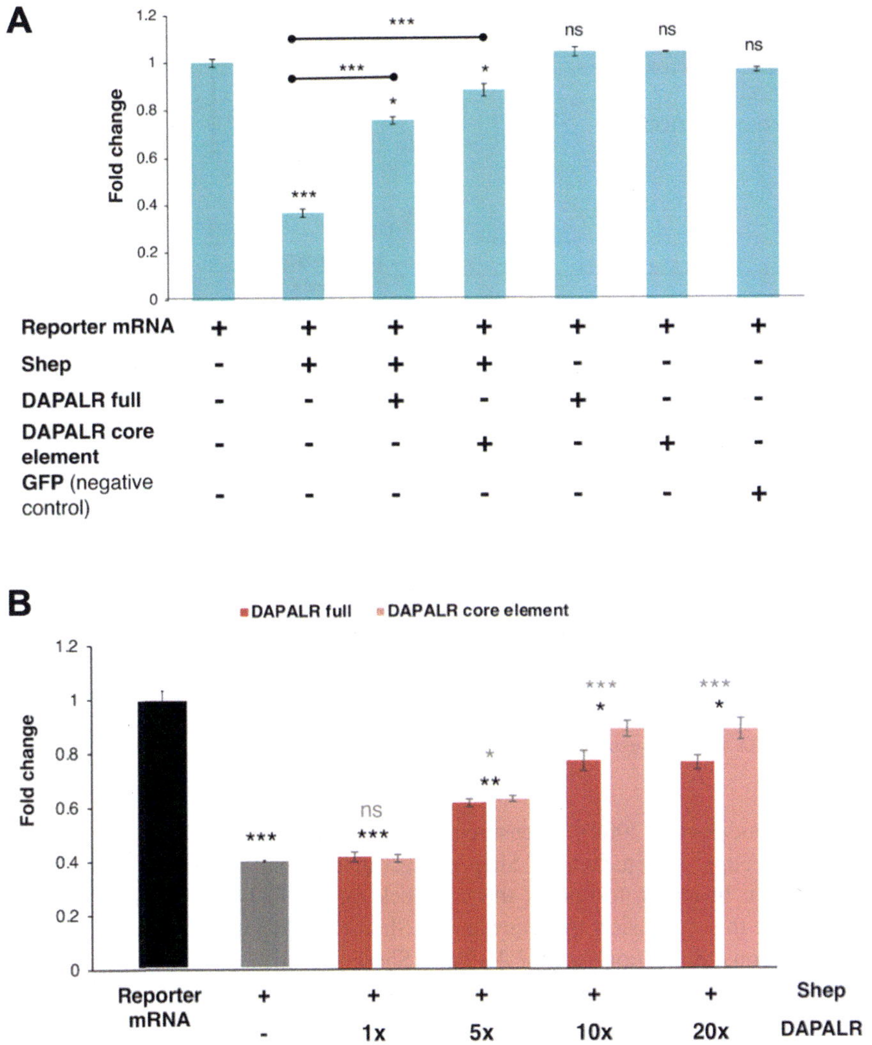

Fig. 12.5 Post-transcriptional regulation of *Dsx1* by *DAPALR* and Shep. (**a**) Relative luciferase activity after in vitro translation assay of *Dsx1* 5′ UTR-*Luc* reporter mRNA with intact TGE upon addition of *Shep* mRNA, *DAPALR* full RNA, *DAPALR* core element, and *GFP* mRNA (negative control). Samples were compared against the expression of the *Dsx1* 5′ UTR-*Luc* reporter mRNA without the addition of any other mRNAs. The endpoints of the line above the bars show which samples were additionally compared statistically. (**b**) Relative luciferase activity after in vitro translation assay of *Dsx1* 5′ UTR-*Luc* reporter mRNA with *Shep* mRNA and different concentrations of full region of *DAPALR* and its core element. Error bars indicate the standard error of the mean, $n = 3$. Black asterisks show significant statistics compared with the expression of the Dsx1 5′ UTR-Luc reporter mRNA. Gray asterisks show significant statistics compared with the Reporter mRNA with Shep. Error bars indicate the standard error of the mean, $n = 3$. $*p < 0.05$, $**p < 0.01$, $***p < 0.001$, ns: not significant (Student's *t*-test). (The figures were reproduced from Perez et al. (2021))

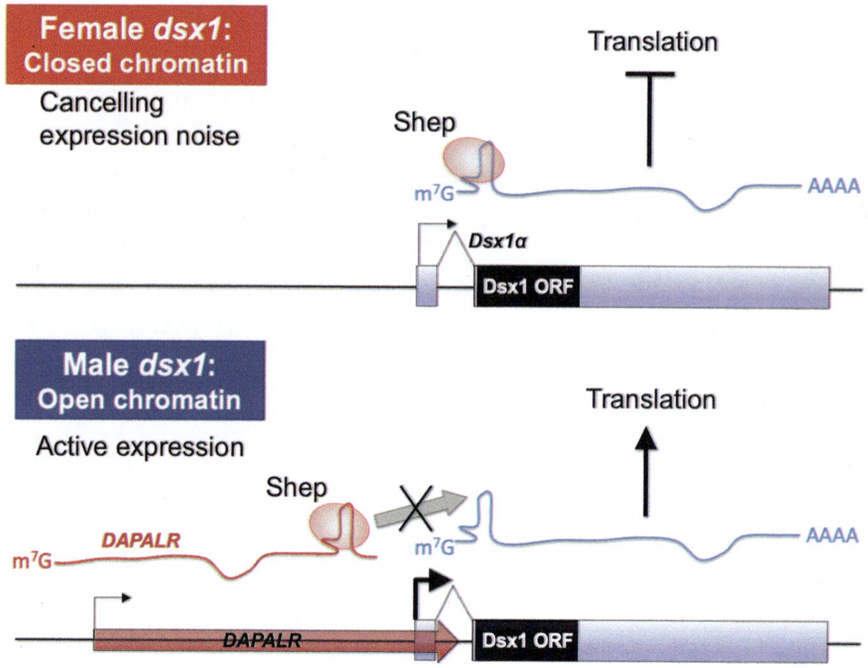

Fig. 12.6 The potential mechanism for establishing the sex-specific *Dsx1* expression. Transcriptional, epigenetic, and post-transcriptional regulations are integrated to achieve male-specific *Dsx1* expression for avoiding sexual ambiguity

2005; Ozbudak et al. 2002). To ensure the inactive state of *Dsx1*, there would be the noise-canceling system where Shep represses unintended expression of *Dsx1* (Fig. 12.6). In males, to unlock the safe lock on *Dsx1* expression, *DAPALR* can counteract this repression by sequestering Shep (Fig. 12.6). To unravel this post-transcriptional regulation by *DAPALR* and Shep, the mode of RNA recognition by Shep and stoichiometry of *DAPALR*, Shep, and *Dsx1* mRNA needs to be elucidated.

12.7 Concluding Remarks and Future Perspectives

In *D. magna*, environmental signals shift the position of the sex on the spectrum from the female side to the male side. Mothers sense severe environments and destine the sex of offspring to males via sesquiterpenoid signaling. This signaling occurs at the late stage of developing oocytes and leads to their commitment to male development possibly by changing the *Dsx1* locus from the closed to open chromatin state, which allows Vri for binding to the *Dsx1* enhancer and leads to upregulation of *DAPALR* and *Dsx1* for initiating the male developmental program. In contrast, in females, *Dsx1* is silenced in the whole body and Shep represses translation of the unintended expression of *Dsx1* derived from noise to prevent intersex phenotype.

Collectively, transcriptional, epigenetic, and post-transcriptional regulation of *Dsx1* could contribute to the establishment of the binary sex in *D. magna.* Factors involving each regulation began to be identified in the last few years. Further studies are needed to address how each regulation is mutually linked. To achieve it, an important step is to analyze the gene regulatory mechanism in the cellular contexts with help of cell sorting and single-cell analysis. It would also be important to investigate the connection between the sesquiterpenoid signaling and *Dsx1* activation. These comprehensive analyses of *Dsx1* regulation will contribute to understanding molecular mechanisms underlying both establishment of the binary sex and the formation of sex spectrum in *D. magna.*

References

Akiyama-Oda Y, Oda H (2003) Early patterning of the spider embryo: a cluster of mesenchymal cells at the cumulus produces Dpp signals received by germ disc epithelial cells. Development 130:1735–1747

Banta AM, Brown LA (1929) Control of sex in cladocera. III. Localization of the critical period for control of sex. Proc Natl Acad Sci U S A 15:71–81

Bayrer JR, Zhang W, Weiss MA (2005) Dimerization of doublesex is mediated by a cryptic ubiquitin-associated domain fold—implications for sex-specific gene regulation. J Biol Chem 280:32989–32996

Bénazéraf B, Pourquié O (2013) Formation and segmentation of the vertebrate body axis. Annu Rev Cell Dev Biol 29:1–26

Burtis KC, Baker BS (1989) Drosophila doublesex gene controls somatic sexual differentiation by producing alternatively spliced mRNAs encoding related sex-specific polypeptides. Cell 56: 997–1010

Chen X, Rinsma M, Janssen JM et al (2016) Probing the impact of chromatin conformation on genome editing tools. Nucleic Acids Res 44:6482–6492

Chung CH, Allen AG, Sullivan NT et al (2020) Computational analysis concerning the impact of DNA accessibility on CRISPR-Cas9 cleavage efficiency. Mol Ther 28:19–28

Corbitt TS, Hardie J (1985) Juvenile hormone effects on polymorphism in the pea aphid, A cyrthosiphon pisum. Entomol Exp Appl 38:131–135

Cornette R, Gotoh H, Koshikawa S, Miura T (2008) Juvenile hormone titers and caste differentiation in the damp-wood termite Hodotermopsis sjostedti (Isoptera, Termopsidae). J Insect Physiol 54:922–930

Daer RM, Cutts JP, Brafman DA, Haynes KA (2016) The impact of chromatin dynamics on Cas9-mediated genome editing in human cells. ACS Synth Biol 6:428–438

Hebert PDN (1978) The population biology of Daphnia (Crustacea, Daphnidae). Biol Rev 53:387–426

Herpin A, Schartl M (2015) Plasticity of gene-regulatory networks controlling sex determination: of masters, slaves, usual suspects, newcomers, and usurpators. EMBO Rep 16:1260–1274

Ishikawa A, Gotoh H, Abe T, Miura T (2013) Juvenile hormone titer and wing-morph differentiation in the vetch aphid Megoura crassicauda. J Insect Physiol 59:444–449

Jensen KT, Fløe L, Petersen TS et al (2017) Chromatin accessibility and guide sequence secondary structure affect CRISPR-Cas9 gene editing efficiency. FEBS Lett 591:1892–1901

Kærn M, Elston TC, Blake WJ, Collins JJ (2005) Stochasticity in gene expression: from theories to phenotypes. Nat Rev Genet 66(6):451–464

Kato Y, Kobayashi K, Oda S et al (2010) Sequence divergence and expression of a transformer gene in the branchiopod crustacean, Daphnia magna. Genomics 95:160–165

Kato Y, Kobayashi K, Watanabe H, Iguchi T (2011a) Environmental sex determination in the branchiopod crustacean Daphnia magna: deep conservation of a Doublesex gene in the sex-determining pathway. PLoS Genet 7:e1001345

Kato Y, Shiga Y, Kobayashi K et al (2011b) Development of an RNA interference method in the cladoceran crustacean Daphnia magna. Dev Genes Evol 220:337–345

Kato Y, Perez CAG, Mohamad Ishak NS et al (2018) A 5′ UTR-overlapping lncRNA activates the male-determining gene doublesex1 in the crustacean Daphnia magna. Curr Biol 28:1811–1817.e4

Koyama T, Mendes CC, Mirth CK (2013) Mechanisms regulating nutrition-dependent developmental plasticity through organ-specific effects in insects. Front Physiol 4:263

Kvist J, Athanàsio CG, Pfrender ME et al (2020) A comprehensive epigenomic analysis of phenotypically distinguishable, genetically identical female and male Daphnia pulex. BMC Genomics 21(1):17

Lee BY, Choi BS, Kim MS et al (2019) The genome of the freshwater water flea Daphnia magna: a potential use for freshwater molecular ecotoxicology. Aquat Toxicol 210:69–84

Mapes J, Chen J-T, Yu J-S, Xue D (2010) Somatic sex determination in Caenorhabditis elegans is modulated by SUP-26 repression of tra-2 translation. Proc Natl Acad Sci U S A 107:18022–18027

Martienssen R, Moazed D (2015) RNAi and heterochromatin assembly. Cold Spring Harb Perspect Biol 7:a019323

Matsuda M, Nagahama Y, Shinomiya A et al (2002) DMY is a Y-specific DM-domain gene required for male development in the medaka fish. Nature 417:559–563

Mitchell SE (2001) Intersex and male development in Daphnia magna. Hydrobiologia 442:145–156

Miura T (2019) Juvenile hormone as a physiological regulator mediating phenotypic plasticity in pancrustaceans. Develop Growth Differ 61:85–96

Mohamad Ishak NS, Nong QD, Matsuura T et al (2017) Co-option of the bZIP transcription factor Vrille as the activator of Doublesex1 in environmental sex determination of the crustacean Daphnia magna. PLoS Genet 13:e1006953

Naitou A, Kato Y, Nakanishi T et al (2015) Heterodimeric TALENs induce targeted heritable mutations in the crustacean Daphnia magna. Biol Open 4:364–369

Nakanishi T, Kato Y, Matsuura T, Watanabe H (2014) CRISPR/Cas-mediated targeted mutagenesis in Daphnia magna. PLoS One 9:e98363

Nakanishi T, Kato Y, Matsuura T, Watanabe H (2016) TALEN-mediated knock-in via non-homologous end joining in the crustacean Daphnia magna. Sci Rep 6:36252

Nong QD, Mohamad Ishak NS, Matsuura T et al (2017) Mapping the expression of the sex determining factor Doublesex1 in Daphnia magna using a knock-in reporter. Sci Rep 7:13521

Nong QD, Matsuura T, Kato Y, Watanabe H (2020) Two Doublesex1 mutants revealed a tunable gene network underlying intersexuality in Daphnia magna. PLoS One 15:e0238256

Oda S, Tatarazako N, Watanabe H et al (2005) Production of male neonates in four cladoceran species exposed to a juvenile hormone analog, fenoxycarb. Chemosphere 60:74–78

Oda S, Kato Y, Watanabe H et al (2011) Morphological changes in Daphnia galeata induced by a crustacean terpenoid hormone and its analog. Environ Toxicol Chem 30:232–238

Olmstead AW, Leblanc GA (2002) Juvenoid hormone methyl farnesoate is a sex determinant in the crustacean Daphnia magna. J Exp Zool 293:736–739

Olmstead AW, LeBlanc GA (2007) The environmental-endocrine basis of gynandromorphism (intersex) in a crustacean. Int J Biol Sci 3:77–84

Orsini L, Gilbert D, Podicheti R et al (2017) Daphnia magna transcriptome by RNA-Seq across 12 environmental stressors. Sci Data 4:17000

Ozbudak EM, Thattai M, Kurtser I et al (2002) Regulation of noise in the expression of a single gene. Nat Genet 31:69–73

Perez CAG, Adachi S, Nong QD et al (2021) Sense-overlapping lncRNA as a decoy of translational repressor protein for dimorphic gene expression. PLoS Genet 17:e1009683

Raymond CS, Murphy MW, O'Sullivan MG et al (2000) Dmrt1, a gene related to worm and fly sexual regulators, is required for mammalian testis differentiation. Genes Dev 14:2587–2595

Schwentner M, Combosch DJ, Pakes Nelson J, Giribet G (2017) A phylogenomic solution to the origin of insects by resolving crustacean-hexapod relationships. Curr Biol 27:1818–1824.e5

Shen MM, Hodgkin J (1988) mab-3, a gene required for sex-specific yolk protein expression and a male-specific lineage in C. elegans. Cell 54:1019–1031

Tatarazako N, Oda S, Watanabe H et al (2003) Juvenile hormone agonists affect the occurrence of male Daphnia. Chemosphere 53:827–833

Törner K, Nakanishi T, Matsuura T et al (2014) Optimization of mRNA design for protein expression in the crustacean Daphnia magna. Mol Gen Genomics 289:707–715

Toyota K, Miyakawa H, Hiruta C et al (2015) Methyl farnesoate synthesis is necessary for the environmental sex determination in the water flea Daphnia pulex. J Insect Physiol 80:22–30

Wexler J, Delaney EK, Belles X et al (2019) Hemimetabolous insects elucidate the origin of sexual development via alternative splicing. Elife 8:e47490

Yoshimoto S, Okada E, Umemoto H et al (2008) A W-linked DM-domain gene, DM-W, participates in primary ovary development in Xenopus laevis. Proc Natl Acad Sci U S A 105:2469–2474

Zaffagnini F (1987) Reproduction in Daphnia. Mem Inst Ital Idrobiol 45:245–284

Chapter 13
Starvation Is a New Component of Sex Reversal in Medaka (*Oryzias latipes*): Significance of Metabolism in Sex Regulation

Yuta Sakae and Minoru Tanaka

Abstract Sexually reproducing animals require proper sex regulation for the production of the next generation. Unlike mammals, many teleost fish switch their sex through life. This suggests that these fish species potentially retain a spectrum of sex. The sex regulation system that ensures this spectrum is useful to adjust "fitness" according to the environment, including social structure. The system is categorized by "sex change" and "sex reversal." In this review, we introduce examples of sex changes and sex reversal. As a novel example of sex reversal recently found in medaka (*Oryzias latipes*), the metabolism underlying starvation-induced sex reversal focuses on fatty acid and lipid. Lastly, we argue the benefit of sex reversal and the contribution of starvation studies on medaka to the research of sex regulation.

Keywords Metabolism · Sex reversal · Teleost Fish

Y. Sakae
Department of Integrated Biosciences, Graduate School of Frontier Sciences, The University of Tokyo, Kashiwa, Japan

Division of Translational Genomics, Exploratory Oncology Research and Clinical Trial Center, National Cancer Center, Kashiwa, Japan

Division of Biological Science, Graduate School of Science, Nagoya University, Nagoya, Aichi, Japan

Present Address:
e-mail: ymkysakae@edu.k.u-tokyo.ac.jp

M. Tanaka (✉)
Division of Biological Science, Graduate School of Science, Nagoya University, Nagoya, Aichi, Japan
e-mail: mtanaka@bio.nagoya-u.ac.jp

M. Tanaka, M. Tachibana (eds.), *Spectrum of Sex*,
https://doi.org/10.1007/978-981-19-5359-0_13

13.1 Introduction

Appropriate regulation of sex is an essential developmental event in sexually reproducing animals. The process of sexual regulation involves the steps of sex determination, sex differentiation, and sexual maturation. Sex determination involves specific genes, environmental factors that include temperature and density, or a combination of both (Devlin and Nagahama 2002; Barske and Capel 2008; Gamble and Zarkower 2012; Capel 2017; Li and Gui 2018; Nagahama et al. 2021; Weber and Capel 2021). After sex determination, the gonads begin to differentiate into those of males or females, with distinctive gene expression and structural development (sex differentiation). As sexual maturation progresses, the increase in the sex steroid hormones secreted from the gonadal somatic cells causes masculinization or feminization of the body and exhibits secondary sexual characteristics (sexual maturation). Interestingly, some vertebrates can switch their sex (Liu et al. 2017; Gemmell et al. 2019).

Among vertebrates, teleosts have diverse mating systems and exhibit many sex switches. Edgecimbe et al. argued the possibility that teleosts display such characteristics because they do not have external reproductive organs and experienced whole-genome duplication approximately 320–350 million years ago during their evolution (Glasauer and Neuhauss 2014; Adolfi et al. 2019a; Edgecombe et al. 2021). Sex changes have not yet been reported in cartilaginous fish with external reproductive organs.

Teleost sex switching is classified into two types. The first is "sex change" in which an individual changes sex after sexual maturation at the adult stage (Liu et al. 2017; Gemmell et al. 2019). Sex change is caused by the size of the individual and/or social status in the population. The second is "sex reversal" in which premature larvae (indicating genetic sex determination) develop into sex different from the genetic sex. Sex reversal is triggered by environmental changes, such as high temperature (Yamamoto 1999; Sato et al. 2005).

Here, we first introduce the types of sexual switching found in teleosts. Next, we outline the mechanism of sex reversal due to starvation discovered in the study of medaka. Finally, we discuss a new perspective of sex switching in view of starvation experiments.

13.2 Many Adult Teleosts Undergo Sex Change

There are three types of sex change in teleosts due to differences in direction (Fig. 13.1): female-to-male sex change (protogyny), male-to-female sex change (protandry), and bidirectional sex change. The evolution of sex change in each species is well explained by the size-advantage model and is closely related to the mating system of that species (Ghiselin 1969; Warner 1988).

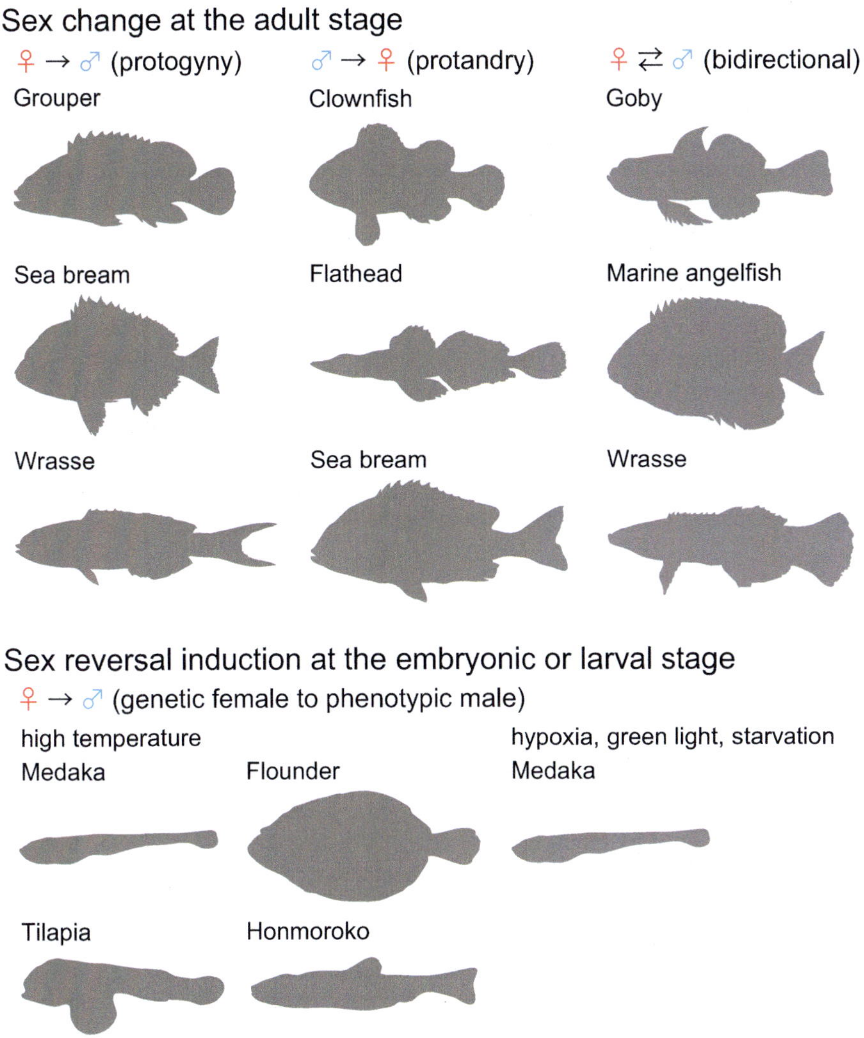

Fig. 13.1 Examples of sex change and sex reversal in teleosts. Two types of sex regulation are present in the teleosts. "Sex change" is observed at the adult stage and is grouped into three types. The first is female-to-male sex change (protogyny), such as grouper (Bhandari et al. 2003), sea bream (Aoyama 1955), and Wrasse (Warner and Swearer 1991; Godwin 2009; Lamm et al. 2015). The second is male-to-female sex change (protandry), such as Clownfish (Fricke and Fricke 1977), Flathead (Sunobe et al. 2016), and sea bream (Wu et al. 2021). The third is bidirectional sex change, such as Goby (Kuwamura and Nakashima 1998), marine angelfish (Kuwamura et al. 2011), and Wrasse (Kuwamura et al. 2011). Another type of sex regulation is called "sex reversal." This is used for the situation where a genetic male or female displays opposite phenotypic sex. It is induced at the larval stage. For instance, medaka, flounder, tilapia, and honmoroko display female-to-male sex reversal by high-temperature treatment at the embryonic or larval stages (Yamamoto 1999; Fujioka 2005; Sato et al. 2005; Rougeot et al. 2008). Female-to-male sex reversal is also observed following hypoxia, exposure to green light, and starvation in medaka (Cheung et al. 2014; Hayasaka et al. 2019; Sakae et al. 2020).

The female-to-male sex change (protogyny) is the most common type of sex change and has been studied in groupers (Bhandari et al. 2003), sea bream (Aoyama 1955), and wrasse (Godwin 2009; Lamm et al. 2015). These creatures display polygamy and form a harem (Liu et al. 2017). In this mating system, males indicate higher "fitness", which is defined as individual reproductive success through life, compared to females as individual sizes increase. Therefore, the largest individual in the population with predominant social status changes sex from female to male. Clownfish are typical species that undergo male-to-female sex changes and have a monogamous mating system (Fricke and Fricke 1977; Godwin and Thomas 1993; Liu et al. 2017). Although females increase fitness depending on body size, males do not change in this case. Therefore, the male-to-female sex change evolves. Bidirectional sex change is a novel type of sex change reported since the 1990s in Goby (Kuwamura and Nakashima 1998), Marine angelfish (Kuwamura et al. 2011), and Wrasse (Kuwamura and Nakashima 1998; Kuwamura et al. 2011). In this type, the largest individual male (male 1) in the population forms a harem, as seen in the protogyny. Male 1 develops into a female again when a new larger male (male 2) enters the mating group (Kuwamura and Nakashima 1998).

13.3 Induction of Sex Reversal at Embryonic or Larval Stage in Teleosts

Several teleosts that include medaka (*Oryzias latipes*), Japanese flounder (*Paralichthys olivaceus*), Nile tilapia (*Oreochromis niloticus*), honmoroko (*Gnathopogon caerulescens*), and sablefish (*Anoplopoma fimbria*) cause female-to-male sex reversal by environmental changes during the process of sex differentiation. Although these species employ a genetic sex determination system, genetic females develop into males when fish are exposed to high temperatures during the embryonic or larval stages (Yamamoto 1999; Fujioka 2005; Sato et al. 2005; Rougeot et al. 2008; Huynh et al. 2019). In addition to high temperatures, medaka also display female-to-male sex reversal by hypoxia and green light conditions (Cheung et al. 2014; Hayasaka et al. 2019). Among them, the molecular mechanism of sex reversal has been demonstrated in high-temperature experiments (Hayashi et al. 2010; Kitano et al. 2012; Uchimura et al. 2019; Castañeda Cortés et al. 2019; Hara et al. 2020; Hattori et al. 2020). The elevated levels of cortisol, a stress steroid hormone, reduce the expression of feminization genes (*aromatase* and *fshr*) and suppress the feminization process (Hayashi et al. 2010; Kitano et al. 2012). On the other hand, cortisol regulates masculinization genes (*dmrt1* and *gsdf*) and the sex differentiation program changes from female to male (Kitano et al. 2012; Adolfi et al. 2019b; Hara et al. 2020). Eventually, sex-reversing individuals mature into functional males that produce sperm. The contribution of cortisol has also been reported in sex changes (Solomon-Lane et al. 2013; Goikoetxea et al. 2017). In fish species that show sex change by social status, cortisol levels differ depending on social status and change according to the alteration of social status (Perry and Grober 2003; Solomon-Lane et al. 2013). In addition, administration of cortisol induces female-to-male sex reversal through an increase in *dmrt1* expression and 11-ketotestosterone (11-KT), and the termination of cortisol administration leads to a re-sex change from male-to-female (Chen et al. 2020). Thus, the mechanism of sex regulation might be conserved in teleosts during the larval and adult stages.

13.4 Contribution of Metabolism and Metabolites to Sex Regulation in Medaka, as Suggested by Studies of Starvation

Sex reversal by environmental factors indicates that the sex of medaka larvae is still plastic during sex differentiation (Hayashi et al. 2010; Cheung et al. 2014; Hayasaka et al. 2019). Interestingly, we found that 20% of genetic female XX medaka larvae cause female-to-male sex reversal during 5 days of starvation just after hatching, which is in the middle of sex differentiation (Sakae et al. 2020). Metabolome analysis was performed to explore the molecular mechanisms underlying this sex reversal from a metabolic perspective. Principle component analysis of metabolome data implied the involvement of the pantothenic acid pathway in sex reversal. Combining metabolome and RT-qPCR analysis indicated suppression of the pantothenic acid pathway by reducing the rate-determining enzyme gene (*Pank*) expression in starved XX medaka. Supporting this, pharmacological inhibition of Pank during the same term as starvation under feeding conditions resulted in female-to-male sex reversal (15%). The pantothenic acid pathway is a synthetic pathway for coenzyme A (CoA), which is required for acetyl-CoA production (Jackowski and Rock 1981; Leonardi et al. 2005).

Acetyl-CoA is used in many metabolic processes, such as the tricarboxylic acid cycle, fatty acid synthesis, and cholesterol synthesis. Fatty acid synthesis is the first step in the synthesis of different types of lipids, including triglycerides and phospholipids. Fatty acid synthase (FAS) is a multifunctional enzyme with seven domains. It is a rate-limiting enzyme in the fatty acid synthesis pathway. Pantetheine, consisting of pantothenic acid and cysteamine, is essential for fatty acid synthesis and is required for fatty acids to bind the acyl carrier protein domain of FAS (Liu et al. 2010). Fatty acids and lipids modulate gene expression via transcription factors that include peroxisome proliferator-activated receptor (PPAR), liver X receptor (LXR), hepatocyte nuclear factor 4 (HNF4), nuclear factor-kappa B (NF-κB), and sterol regulatory-element binding protein (SREBP) (Jump 2004; Tao et al. 2020). Knockout analyses of these genes revealed the essential role of lipids in mouse early embryo development (Yamashita 1999; Chirala et al. 2003; Tatsumi et al. 2018; Ye et al. 2021). In *Caenorhabditis elegans*, fatty acids (myristic acid) are involved in germline sex determination (Tang and Han 2017).

We evaluated the amount of lipids, including cholesterol esters, triacylglycerols (TG), cholesterols, cardiolipins, phosphatidylethanolamines, phosphatidylserine, phosphatidylcholines, and sphingolipids, extracted from the whole body of medaka larvae 5 days post-hatching by high-performance thin-layer chromatography. The finding that starved XX larvae displayed a marked reduction in TG (Sakae et al. 2020) strongly suggests that fatty acid synthesis contributes to sex differentiation. Therefore, medaka larvae just after hatching were treated with FAS inhibitor (C75) for 5 days to investigate the effect of fatty acid synthesis on sex differentiation. Sex typing at the adult stage of treated XX medaka indicated female-to-male sex reversal (13%). These pharmacological inhibitions (Pank inhibitor and C75) revealed the involvement of metabolism (pantothenic acid pathway and/or fatty acid synthesis) in sex differentiation.

Next, we aimed to elucidate whether both metabolic pathways were locally activated in the gonads. We examined the expression pattern of *Pank* (Pank gene: pantothenic acid pathway) and *Fasn* (FAS gene: fatty acid synthesis) in the body by whole-mount in situ hybridization. *Pank* expression was mainly observed in the liver and was scarcely detected in the gonads, whereas *Fasn* expression was observed in both the liver and gonads. Therefore, the pantothenic acid pathway may be operative mainly in the liver, while fatty acid synthesis is locally activated in the liver and also in the gonads.

Expression analysis of the sex differentiation-related genes (*aromatase*, *foxl2*, *gsdf*, and *dmrt1*) has provided insight into the link between starvation, metabolism, and sex differentiation. Starvation was shown to repress the expression of feminization genes (*aromatase* and *foxl2*) and induce the expression of the masculinization gene *dmrt1*. An increase in *dmrt1* expression was also observed during treatment with Pank inhibitor and C75. These results collectively suggest that starvation and inhibition of metabolism(s) (pantothenic acid pathway and/or fatty acid synthesis) caused female-to-male sex reversal through the elevation of the masculinization gene (*dmrt1*) in XX medaka larvae. Consistent with this, starvation treatment for the *dmrt1* mutant did not result in female-to-male sex reversal (Sakae et al. 2020).

Based on the results of starvation and pharmacological treatment experiments (Fig. 13.2), we next discuss the location of synthesis of the factors (CoA, fatty acid/lipid, and factor X) regulating *dmrt1* expression. *Pank* encodes an essential enzyme for the production of CoA, which is mainly observed in the liver at the mRNA level. A human study suggested that acetyl-CoA synthesized in the liver is secreted and transported into the gonad via blood (Speziale et al. 2018). Therefore, CoA in the gonad may be supplied as acetyl-CoA from the liver through blood circulation, although the possibility that CoA is synthesized locally in the gonad cannot be excluded. CoA is an essential metabolite in many metabolic pathways including fatty acid synthesis that is, as described above, markedly affected under starvation. Since *fatty acid synthase* (*Fasn*) is expressed in the liver, gonadal somatic cells, and germ cells, these cells/tissues are likely capable of synthesizing fatty acids/lipids. Interestingly, transcription factor X (such as PPAR, LXR, HNF4, NFkB, and SREBP) receives fatty acid/lipid as a ligand (Jump 2004; Tao et al. 2020). The expression of the masculinization gene *dmrt1* in gonadal somatic cells may be

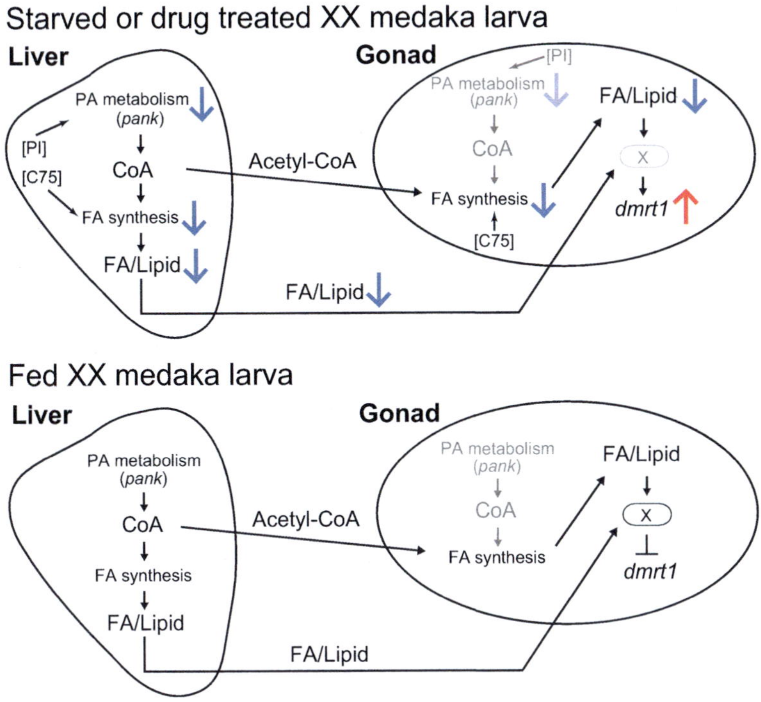

Fig. 13.2 Model and new perspective of sex regulation based on data from starvation experiments involving medaka. The synthesis of CoA via pantothenic acid metabolism mainly occurs in the liver, although the possibility of local CoA synthesis in gonadal cells cannot be excluded. CoA synthesized in the liver may be secreted and transported into the gonad via acetyl-CoA in the blood. Fatty acids/lipids are synthesized in the liver, gonadal somatic cells, and germ cells. Fatty acid/lipids synthesized in the liver may also be secreted and transported into the gonad via blood. Finally, the expression of the masculinization gene *dmrt1* in gonadal somatic cells is suppressed by transcription factor X according to the binding of fatty acid/lipid derived from liver or gonadal somatic cells or germ cells. Under starvation or drug treatment conditions, CoA synthesis is reduced by starvation or a Pank inhibitor, leading to a decline in CoA and fatty acid/lipid. Fatty acid synthesis is inhibited by C75. The transcription factor X cannot bind to the *dmrt1* promoter region to suppress *dmrt1* expression due to the low level of fatty acid/lipid. Finally, *dmrt1* is expressed in gonadal somatic cells, and female-to-male sex reversal occurs. *PI* Pank inhibitor, *C75* fatty acid synthesis inhibitor, *FA* fatty acid. The white box denotes transcription factor X

suppressed by the transcription factor X. Interestingly, it has been reported that PPARα contributes to sex reversal at high temperatures in medaka (Hara et al. 2020). In this context, CoA and fatty acid/lipid act are biased to female sex differentiation by repressing the expression of *dmrt1* in wild-type females, implying the importance of sex-biased metabolism with multicellular interconnection in sex regulation (Fig. 13.2). Our results suggest that sex-biased metabolism and multicellular interconnections are components of the phenotypic spectrum of sex.

13.5 New Perspective from the Study of Starvation-Induced Sex Reversal

What are the benefits of female-to-male sex reversal under starvation conditions for medaka? Apart from forced sex change by artificial exposure to a sex steroid hormone, it has not been reported that genetically modified medaka display sex change (the XX mutant medaka seemingly displaying sex change to male should be categorized as sex reversal because it exhibits *foxl2* expression in a larval stage; Zhou et al. 2016). Sex changes in adults require a long time and a large energy cost. The lifespan of many sex-changing fishes is for several years or several decades, while that of wild medaka is approximately 1 year (Shima and Mitani 2004). In addition, the duration of the opportunity for mating is limited to approximately 3–4 months during spring and summer in wild medaka. The sex change in adults could not have been allowed through the evolutionary process in medaka. Sex reversal is likely to increase fitness more in an environment where undernutrition is expected.

In addition to lifespan, body size is another factor that affects fitness and restricts the ability to change sex. The size-advantage model explains sex change well, such as for Clownfish and Wrasse, where visual information or aggressive behavior induces sex change based on body size through the neuroendocrine system (hypothalamic–pituitary–gonadal [HPG] axis and hypothalamic–pituitary–adrenal axis) and sex steroid hormones (Perry and Grober 2003; Liu et al. 2017).

What is the mechanism underlying body size? Our results suggest that fatty acids/lipids are a component of the mechanism. In addition to functioning as a ligand for transcription factors, fatty acids and lipids are associated with the HPG axis and epigenome (Pietrocola et al. 2015; Keating and El-Osta 2015; Su et al. 2016; Manfredi-Lozano et al. 2018; Wang et al. 2020). Leptin may be a candidate factor that connects body size and sex change. Leptin is a sensor hormone of lipids that is secreted from adipocytes and is involved in the onset of puberty and the reproductive cycle through gonadotropin-releasing hormone (GnRH) regulation (Sanchez-Garrido and Tena-Sempere 2013; Childs et al. 2021). Leptin signals, depending on the amount of lipid in black porgy, may be linked to sex change through GnRH modulation (Wu et al. 2021).

Another aspect of fatty acid/lipid on the mechanism of sex change is their involvement in epigenome alternations, such as euchromatinization by histone acetylation, through the production of acetyl-CoA, which is an acetyl group donor (Pietrocola et al. 2015; Keating and El-Osta 2015; Su et al. 2016). Although the involvement of histone acetylation in sex change has not been well studied, DNA methylation of the aromatase promoter region has been suggested in sex change in black porgy, European sea bass, and bluehead Wrasse (Navarro-Martín et al. 2011; Wu et al. 2016; Todd et al. 2019). It is possible that female-specific epigenetic status is mediated by lipid metabolism and histone acetylation, which may consequently result in changes in steroid production described in the size-advanced model.

To date, comprehensive genetic analyses have revealed the conflicts between sex-related genes (e.g., mutual antagonism between the female gene *foxl2* and the male gene *dmrt1*) (Schlessinger et al. 2010; Jiménez et al. 2021). Starvation experiments predicted that fatty acid/lipid can repress *dmrt1* expression in the gonadal somatic cells of fed genetic female medaka larvae (Fig. 13.2, Sakae et al. 2020; Sakae and Tanaka 2021). Our discovery emphasizes the importance of fatty acid/lipid as critical intermediates, located on the crossing pathway to sex change by body size through the HPG axis, epigenome alterations, and *dmrt1* expression, as well as in the conflict between males and females.

13.6 Conclusion

The sex determination genes of teleosts are more diverse than those of mammals (Pan et al. 2016). Three sex determination genes (*DMY*/*dmrt1bY*, *gsdf*, and *sox3*) have been reported in medaka-related species in Southeast Asia, including Japan, suggesting that the sex determination genes have transitioned during evolution (Matsuda and Sakaizumi 2016). The diversity of sex determination genes is likely related to the presence of a mechanism that ensures the two sexes and would be the mechanism that causes sex reversal during the larval stage, such as medaka. Sex change and sex reversal take part in the reproductive strategy that has evolved to maximize fitness in response to changes in the environment, including nutritional conditions. Uncovering the mechanism of sex change and sex reversal in various fish species impacts not only basic studies on sex regulation but also the way in which we understand sex.

Acknowledgments We would like to express our gratitude to our lab members for continuous support and apologize to those whose work is not cited.

Conflict of Interest Statement The authors have no conflicts of interest to declare.

Funding Sources This work was supported by a Grand-in-Aid for Scientific Research on Innovative Areas (17H06430, MT).

Author Contributions Y.S. wrote the manuscript and drew all figures. M.T. supervised the content.

References

Adolfi MC, Nakajima RT, Nóbrega RH, Schartl M (2019a) Intersex, hermaphroditism, and gonadal plasticity in vertebrates: evolution of the Müllerian duct and Amh/Amhr2 signaling. Annu Rev Anim Biosci 7:149–172. https://doi.org/10.1146/annurev-animal-020518-114955

Adolfi MC, Fischer P, Herpin A, Regensburger M, Kikuchi M, Tanaka M, Schartl M (2019b) Increase of cortisol levels after temperature stress activates *dmrt1a* causing female-to-male sex reversal and reduced germ cell number in medaka. Mol Reprod Dev 86(10):1405–1417. https://doi.org/10.1002/mrd.23177

Aoyama T (1955) On the hermaphroditism in the yellow sea bream, Taius tumifrons. Jpn J Ichthyol 4:119–129. [In Japanese with English summary]. https://doi.org/10.11369/jji1950.4.119

Barske LA, Capel B (2008) Blurring the edges in vertebrate sex determination. Curr Opin Genet Dev 18(6):499–505. https://doi.org/10.1016/j.gde.2008.11.004

Bhandari RK, Komuro H, Nakamura S, Higa M, Nakamura M (2003) Gonadal restructuring and correlative steroid hormone profiles during natural sex change in protogynous honeycomb grouper (*Epinephelus merra*). Zool Sci 20(11):1399–1404. https://doi.org/10.2108/zsj.20.1399

Capel B (2017) Vertebrate sex determination: evolutionary plasticity of a fundamental switch. Nat Rev Genet 18(11):675–689. https://doi.org/10.1038/nrg.2017.60

Castañeda Cortés DC, Arias Padilla LF, Langlois VS, Somoza GM, Fernandino JI (2019) The central nervous system acts as a transducer of stress-induced masculinization through corticotropin-releasing hormone B. Development 146(8):dev172866. https://doi.org/10.1242/dev.172866

Chen J, Peng C, Yu Z, Xiao L, Yu Q, Li S, Zhang H, Lin H, Zhang Y (2020) The administration of cortisol induces female-to-male sex change in the protogynous orange-spotted grouper, Epinephelus coioides. Front Endocrinol 11:12. https://doi.org/10.3389/fendo.2020.00012

Cheung CHY, Chiu JMY, Wu RSS (2014) Hypoxia turns genotypic female medaka fish into phenotypic males. Ecotoxicology 23:1260–1269. https://doi.org/10.1007/s10646-014-1269-8

Childs GV, Odle AK, MacNicol MC, MacNicol AM (2021) The importance of leptin to reproduction. Endocrinology 162(2):1–18. https://doi.org/10.1210/endocr/bqaa204

Chirala SS, Chang H, Matzuk M, Abu-Elheiga L, Mao J, Mahon K, Finegold M, Wakil SJ (2003) Fatty acid synthesis is essential in embryonic development: fatty acid synthase null mutants and most of the heterozygotes die in *utero*. Proc Natl Acad Sci U S A 100(11):6358–6363. https://doi.org/10.1073/pnas.0931394100

Devlin RH, Nagahama Y (2002) Sex determination and sex differentiation in fish: an overview of genetic, physiological, and environmental influences. Aquaculture 208(3–4):191–364. https://doi.org/10.1016/S0044-8486(02)00057-1

Edgecombe J, Urban L, Todd EV, Gemmell NJ (2021) Might gene duplication and neofunctionalization contribute to the sexual lability observed in fish? Sex Dev 15(1): 122–133. https://doi.org/10.1159/000515425

Fricke H, Fricke S (1977) Monogamy and sex change by aggressive dominance in coral reef fish. Nature 266:830–832. https://doi.org/10.1038/266830a0

Fujioka Y (2005) Thermolabile sex determination in honmoroko. J Fish Biol 59:851–861. https://doi.org/10.1111/j.1095-8649.2001.tb00156.x

Gamble T, Zarkower D (2012) Sex determination. Curr Biol 22(8):257–262. https://doi.org/10.1016/j.cub.2012.02.054

Gemmell NJ, Todd EV, Goikoetxea A, Ortega-Recalde O, Hore TA (2019) Natural sex change in fish. In: Current topics in developmental biology, vol 134, 1st edn. Elsevier Inc. https://doi.org/10.1016/bs.ctdb.2018.12.014

Ghiselin MT (1969) The evolution of hermaphroditism among animals. Q Rev Biol 44(2):189–208. https://doi.org/10.1086/406066

Glasauer SMK, Neuhauss SCF (2014) Whole-genome duplication in teleost fishes and its evolutionary consequences. Mol Gen Genomics 289(6):1045–1060. https://doi.org/10.1007/s00438-014-0889-2

Godwin J (2009) Social determination of sex in reef fishes. Semin Cell Dev Biol 20(3):264–270. https://doi.org/10.1016/j.semcdb.2008.12.003

Godwin JR, Thomas P (1993) Sex change and steroid profiles in the protandrous anemonefish *Amphiprion melanopus* (Pomacentridae, Teleostei). Gen Comp Endocrinol 91(2):144–157. https://doi.org/10.1006/gcen.1993.1114

Goikoetxea A, Todd EV, Gemmell NJ (2017) Stress and sex: does cortisol mediate sex change in fish? Reproduction 154(6):R149–R160. https://doi.org/10.1530/REP-17-0408
Hara S, Furukawa F, Mukai K, Yazawa T, Kitano T (2020) Peroxisome proliferator-activated receptor alpha is involved in the temperature-induced sex differentiation of a vertebrate. Sci Rep 10(1):1–11. https://doi.org/10.1038/s41598-020-68594-y
Hattori RS, Castañeda-Cortés DC, Arias Padilla LF, Strobl-Mazzulla PH, Fernandino JI (2020) Activation of stress response axis as a key process in environment-induced sex plasticity in fish. Cell Mol Life Sci 77(21):4223–4236. https://doi.org/10.1007/s00018-020-03532-9
Hayasaka O, Takeuchi Y, Shiozaki K, Anraku K, Kotani T (2019) Green light irradiation during sex differentiation induces female-to-male sex reversal in the medaka *Oryzias latipes*. Sci Rep 9(1): 1–8. https://doi.org/10.1038/s41598-019-38908-w
Hayashi Y, Kobira H, Yamaguchi T, Shiraishi ERI, Yazawa T, Hirai T (2010) High temperature causes masculinization of genetically female medaka by elevation of cortisol. Mol Reprod Dev 77:679–686. https://doi.org/10.1002/mrd.21203
Huynh TB, Fairgrieve WT, Hayman ES, Lee JSF, Luckenbach JA (2019) Inhibition of ovarian development and instances of sex reversal in genotypic female sablefish (Anoplopoma fimbria) exposed to elevated water temperature. Gen Comp Endocrinol 279:88–98. https://doi.org/10.1016/j.ygcen.2018.12.013
Jackowski S, Rock CO (1981) Regulation of coenzyme A biosynthesis. J Bacteriol 148(3): 926–932. https://doi.org/10.1016/0098-2997(94)90011-6
Jiménez R, Burgos M, Barrionuevo FJ (2021) Sex maintenance in mammals. Genes 12(7):1–14. https://doi.org/10.3390/genes12070999
Jump DB (2004) Fatty acid regulation of gene transcription. Crit Rev Clin Lab Sci 41(1):41–78. https://doi.org/10.1080/10408360490278341
Keating ST, El-Osta A (2015) Epigenetics and metabolism. Circ Res 116(4):715–736. https://doi.org/10.1161/CIRCRESAHA.116.303936
Kitano T, Hayashi Y, Shiraishi E, Kamei Y (2012) Estrogen rescues masculinization of genetically female medaka by exposure to cortisol or high temperature. Mol Reprod Dev 79:719–726. https://doi.org/10.1002/mrd.22080
Kuwamura T, Nakashima Y (1998) New aspects of sex change among reef fishes: recent studies in Japan. Environ Biol Fish 21:125–135. https://doi.org/10.1023/a:1007389519702
Kuwamura T, Suzuki S, Kadota T (2011) Reversed sex change by widowed males in polygynous and protogynous fishes: female removal experiments in the field. Naturwissenschaften 98(12): 1041–1048. https://doi.org/10.1007/s00114-011-0860-6
Lamm MS, Liu H, Gemmell NJ, Godwin JR (2015) The need for speed: neuroendocrine regulation of socially-controlled sex change. Integr Comp Biol 55(2):307–322. https://doi.org/10.1093/icb/icv041
Leonardi R, Zhang YM, Rock CO, Jackowski S (2005) Coenzyme A: back in action. Prog Lipid Res 44(2):125–153. https://doi.org/10.1016/j.plipres.2005.04.001
Li XY, Gui JF (2018) Diverse and variable sex determination mechanisms in vertebrates. Sci China Life Sci 61(12):1503–1514. https://doi.org/10.1007/s11427-018-9415-7
Liu H, Liu JY, Wu X, Zhang JT (2010) Biochemistry, molecular biology, and pharmacology of fatty acid synthase, an emerging therapeutic target and diagnosis/prognosis marker. Int J Biochem Mol Biol 1(1):69–89
Liu H, Todd EV, Lokman PM, Lamm MS, Godwin JR, Gemmell NJ (2017) Sexual plasticity: a fishy tale. Mol Reprod Dev 84(2):171–194. https://doi.org/10.1002/mrd.22691
Manfredi-Lozano M, Roa J, Tena-Sempere M (2018) Connecting metabolism and gonadal function: novel central neuropeptide pathways involved in the metabolic control of puberty and fertility. Front Neuroendocrinol 48:37–49. https://doi.org/10.1016/j.yfrne.2017.07.008
Matsuda M, Sakaizumi M (2016) General and comparative endocrinology evolution of the sex-determining gene in the teleostean genus Oryzias. Gen Comp Endocrinol 239:80–88. https://doi.org/10.1016/j.ygcen.2015.10.004

Nagahama Y, Chakraborty T, Paul-Prasanth B, Ohta K, Nakamura M (2021) Sex determination, gonadal sex differentiation, and plasticity in vertebrate species. Physiol Rev 101(3):1237–1308. https://doi.org/10.1152/physrev.00044.2019

Navarro-Martín L, Viñas J, Ribas L, Díaz N, Gutiérrez A, Croce LD, Piferrer F (2011) DNA methylation of the gonadal aromatase (cyp19a) promoter is involved in temperature-dependent sex ratio shifts in the European sea bass. PLoS Genet 7(12):e1002447. https://doi.org/10.1371/journal.pgen.1002447

Pan Q, Anderson J, Bertho S, Herpin A, Wilson C, Postlethwait JH et al (2016) Vertebrate sex-determining genes play musical chairs. C R Biol 339(7–8):258–262. https://doi.org/10.1016/j.crvi.2016.05.010

Perry AN, Grober MS (2003) A model for social control of sex change: interactions of behavior, neuropeptides, glucocorticoids, and sex steroids. Horm Behav 43(1):31–38. https://doi.org/10.1016/S0018-506X(02)00036-3

Pietrocola F, Galluzzi L, Pedro JMBS, Madeo F, Kroemer G (2015) Acetyl coenzyme A: a central metabolite and second messenger. Cell Metab 21(6):805–821. https://doi.org/10.1016/j.cmet.2015.05.014

Rougeot C, Prignon C, Kengne CVN, Mélard C (2008) Effect of high temperature during embryogenesis on the sex differentiation process in the Nile tilapia, *Oreochromis niloticus*. Aquaculture 276:205–208. https://doi.org/10.1016/j.aquaculture.2008.02.001

Sakae Y, Tanaka M (2021) Metabolism and sex differentiation in animals from a starvation perspective. Sex Dev 15(1):168–178. https://doi.org/10.1159/000515281

Sakae Y, Oikawa A, Sugiura Y, Mita M, Nakamura S, Nishimura T et al (2020) Starvation causes female-to-male sex reversal through lipid metabolism in the teleost fish, medaka (*Oryzias latipes*). Biol Open 9(4):1–12. https://doi.org/10.1242/bio.050054

Sanchez-Garrido MA, Tena-Sempere M (2013) Metabolic control of puberty: roles of leptin and kisspeptins. Horm Behav 64(2):187–194. https://doi.org/10.1016/j.yhbeh.2013.01.014

Sato T, Endo T, Yamahira K, Hamaguchi S, Sakaizumi M (2005) Induction of female-to-male sex reversal by high temperature treatment in Medaka, Oryzias latipes. Zool Sci 22:985–988. https://doi.org/10.2108/zsj.22.985

Schlessinger D, Garcia-Ortiz JE, Forabosco A, Uda M, Crisponi L, Pelosi E (2010) Determination and stability of gonadal sex. J Androl 31(1):16–25. https://doi.org/10.2164/jandrol.109.008201

Shima A, Mitani H (2004) Medaka as a research organism: past, present and future. Mech Dev 121(7–8):599–604. https://doi.org/10.1016/j.mod.2004.03.011

Solomon-Lane TK, Crespi EJ, Grober MS (2013) Stress and serial adult metamorphosis: multiple roles for the stress axis in socially regulated sex change. Front Neurosci 7:210. https://doi.org/10.3389/fnins.2013.00210

Speziale B, Montesano C, Leonibus MLD, Bonelli F, Fezzardi P, Beconi MG, Monteagudo E, Elbaum D, Orsatti L (2018) Determination of acetyl coenzyme A in human whole blood by ultra-performance liquid chromatography-mass spectrometry. J Chromatogr B 1083:57–62. https://doi.org/10.1016/j.jchromb.2018.02.039

Su X, Wellen KE, Rabinowitz JD (2016) Metabolic control of methylation and acetylation. Curr Opin Chem Biol 30:52–60. https://doi.org/10.1016/j.cbpa.2015.10.030

Sunobe T, Sakaida S, Kuwamura T (2016) Random mating and protandrous sex change of the platycephalid fish Thysanophrys celebica (Platycephalidae). J Ethol 34(1):15–21. https://doi.org/10.1007/s10164-015-0439-3

Tang H, Han M (2017) Fatty acids regulate germline sex determination through ACS-4-dependent myristoylation. Cell 169(3):457–469.e13. https://doi.org/10.1016/j.cell.2017.03.049

Tao LJ, Seo DE, Jackson B, Ivanova NB, Santori FR (2020) Nuclear hormone receptors and their ligands: metabolites in control of transcription. Cell 9(12):2606. https://doi.org/10.3390/cells9122606

Tatsumi T, Takayama K, Ishii S, Yamamoto A, Hara T, Minami N, Miyasaka N, Kubota T, Matsuura A, Itakura E, Tsukamoto S (2018) Forced lipophagy reveals that lipid droplets are required for early embryonic development in mouse. Development 145(4):dev161893. https://doi.org/10.1242/DEV.161893

Todd EV, Ortega-Recalde O, Liu H, Lamm MS, Rutherford KM, Cross H, Black MA, Kardailsky O, Marshall Graves JA, Hore TA, Godwin JR, Gemmel NJ (2019) Stress, novel sex genes, and epigenetic reprogramming orchestrate socially controlled sex change. Sci Adv 5(7):eaaw7006. https://doi.org/10.1126/sciadv.aaw7006

Uchimura T, Hara S, Yazawa T, Kamei Y, Kitano T (2019) Involvement of heat shock proteins on the transcriptional regulation of corticotropin-releasing hormone in medaka. Front Endocrinol 10:1–9. https://doi.org/10.3389/fendo.2019.00529

Wang L, Tran A, Lee J, Belsham DD (2020) Palmitate differentially regulates Spexin, and its receptors Galr2 and Galr3, in GnRH neurons through mechanisms involving PKC, MAPKs, and TLR4. Mol Cell Endocrinol 518:110991. https://doi.org/10.1016/j.mce.2020.110991

Warner PR (1988) Sex change and the size-advantage model. Trends Ecol Evol 3(6):133–136. https://doi.org/10.1016/0169-5347(88)90176-0

Warner RR, Swearer SE (1991) Social control of sex change in the bluehead wrasse, *Thalassoma bifasciatum* (Pisces: Labridae). Biol Bull 181(2):199–204. https://doi.org/10.2307/1542090

Weber C, Capel B (2021) Sex determination without sex chromosomes. Philos Trans R Soc B Biol Sci 376(1832):20200109. https://doi.org/10.1098/rstb.2020.0109

Wu GC, Li HW, Huang CH, Lin HJ, Lin CJ, Chang CF (2016) The testis is a primary factor that contributes to epigenetic modifications in the ovaries of the protandrous black porgy, Acanthopagrus schlegelii. Biol Reprod 94(6):1–13. https://doi.org/10.1095/biolreprod.115.137463

Wu GC, Dufour S, Chang CF (2021) Molecular and cellular regulation on sex change in hermaphroditic fish, with a special focus on protandrous black porgy, Acanthopagrus schlegelii. Mol Cell Endocrinol 520:111069. https://doi.org/10.1016/j.mce.2020.111069

Yamamoto E (1999) Studies on sex-manipulation and production of cloned populations in hirame, *Paralichthys olivaceus* (Temminck et Schlegel). Aquaculture 173(1–4):235–246. https://doi.org/10.1016/S0044-8486(98)00448-7

Yamashita T (1999) A vital role for glycosphingolipid synthesis during development and differentiation. Proc Natl Acad Sci U S A 96(16):9142–9147. https://doi.org/10.1073/pnas.96.16.9142

Ye Q, Zeng X, Cai S, Qiao S, Zeng X (2021) Mechanisms of lipid metabolism in uterine receptivity and embryo development. Trends Endocrinol Metab 32(12):1015–1030. https://doi.org/10.1016/j.tem.2021.09.002

Zhou L, Charkraborty T, Zhou Q, Mohapatra S, Nagahama Y, Zhang Y (2016) Rspo1-activated signalling molecules are sufficient to induce ovarian differentiation in XY medaka (*Oryzias latipes*). Sci Rep 6:1–12. https://doi.org/10.1038/srep19543

9789811953583